Silicon-Germanium Heterojunction Bipolar Transistors for mm-Wave Systems: Technology, Modeling and Circuit Applications

RIVER PUBLISHERS SERIES IN ELECTRONIC MATERIALS AND DEVICES

Indexing: All books published in this series are submitted to the Web of Science Book Citation Index (BkCI), to CrossRef and to Google Scholar.

The "River Publishers Series in Electronic Materials and Devices" is a series of comprehensive academic and professional books which focus on the theory and applications of advanced electronic materials and devices. The series focuses on topics ranging from the theory, modeling, devices, performance and reliability of electron and ion integrated circuit devices and interconnects, insulators, metals, organic materials, micro-plasmas, semiconductors, quantum-effect structures, vacuum devices, and emerging materials. Applications of devices in biomedical electronics, computation, communications, displays, MEMS, imaging, micro-actuators, nanoelectronics, optoelectronics, photovoltaics, power ICs and micro-sensors are also covered.

Books published in the series include research monographs, edited volumes, handbooks and textbooks. The books provide professionals, researchers, educators, and advanced students in the field with an invaluable insight into the latest research and developments.

Topics covered in the series include, but are by no means restricted to the following:

- Integrated circuit devices
- Interconnects
- Insulators
- Organic materials
- Semiconductors
- Quantum-effect structures
- Vacuum devices
- Biomedical electronics
- Displays and imaging
- MEMS
- Sensors and actuators
- Nanoelectronics
- Optoelectronics
- Photovoltaics
- Power ICs

For a list of other books in this series, visit www.riverpublishers.com

Silicon-Germanium Heterojunction Bipolar Transistors for mm-Wave Systems: Technology, Modeling and Circuit Applications

Editors

Niccolò Rinaldi

University of Naples
Italy

Michael Schröter

Technische Universität Dresden
Germany

Published 2018 by River Publishers
River Publishers
Alsbjergvej 10, 9260 Gistrup, Denmark
www.riverpublishers.com

Distributed exclusively by Routledge
4 Park Square, Milton Park, Abingdon, Oxon OX14 4RN
605 Third Avenue, New York, NY 10017, USA

Silicon-Germanium Heterojunction Bipolar Transistors for mm-Wave Systems: Technology, Modeling and Circuit Applications / by Niccolò Rinaldi, Michael Schröter.

Printed on acid-free paper.

Routledge is an imprint of the Taylor & Francis Group, an informa business

ISBN 978-87-93519-61-9 (print)

Contents

Preface

The demand for high-speed circuits and systems has steadily increased over time. Led initially by automotive safety (radar), silicon-germanium (SiGe) heterojunctiom bipolar transistors (HBTs) and their combination with complementary metal-oxide-semiconductor (CMOS) transistors into highly integrated BiCMOS process technologies have become a very attractive solution for a plethora of present and prospective applications operating at frequencies from 30 GHz up to several 100 GHz (so-called mm-waves). Such applications range from communications (presently driven by 5G) to imaging in public transportation (security), medicine and biology, just to name a few. Requiring device performance beyond CMOS, above applications were traditionally placed into the realm of III–V technologies. With the goal of providing highly integrated and cost effective mm-wave electronic systems employing high-speed SiGe HBT front-end circuitry integrated with high density CMOS digital processing capability, the two European Commission funded joint research projects DOTFIVE (2007–2010) and DOTSEVEN (2013–2016) were instrumental in making SiGe HBTs and BiCMOS technology competitive and attractive for the above mentioned applications.

DOTSEVEN was a highly successful project, which not only lifted SiGe HBT performance to an unprecedented level but also provided the theoretical understanding of this new HBT technology as well as the demonstration of its capabilities. Key to the project's success was the excellent cooperation within the consortium consisting of partners from the different and complementary areas of process engineering, device modeling and circuit design. This book summarizes the important results of DOTSEVEN in more detail than the many associated publications and thus addresses not only expert readers familiar with the technology but also students and others who like to learn more about SiGe HBT technology or need a concise overview in one of the associated research areas. In each chapter, the research described is motivated and put in perspective by an introduction that provides sufficient background and additional literature for supporting the understanding.

Not covered though are fundamentals of device physics, process technology and circuit design. Here, the reader is referred to the standard text books in each area.

This book is organized as follows. The Introduction provides the motivation for pursuing HBTs, in particular those based on SiGe, and for setting-up the DOTSEVEN project. Its organization and a brief description of the various work packages are presented, which mirror the relevance of the respective topics for the project. Here, also the most important results are briefly summarized in a single place.

Chapter 1 "SiGe HBT Technology" first introduces the relevant metrics used to evaluate transistor performance. Then, those HBT device and process architectures are explored that have been pursued within DOTSEVEN. It is shown why the conventional double-polysilicon self-aligned selective epitaxial growth approach limits the simultaneous increase of transit and power gain cut-off frequency. The solutions overcoming these issues and leading to the new vertical and lateral architecture developed within DOTSEVEN is described and documented by experimental results.

Chapter 2 "Device Simulation" addresses different types of numerical simulation and visualizes the results through many examples. First, the simulation of isothermal (in terms of the device temperature) carrier transport is discussed. In particular, deterministic solutions for the Boltzmann transport equation and the trade-offs necessary for applying computationally more efficient simulation approaches for device optimization are described as they were pursued within the project. The second part covers electro-thermal simulation based on coupling the impact of phonon scattering and corresponding self-heating with carrier transport simulation. The resulting simulated temperature increase is used to develop methods for accurately determining the thermal resistance from the simulated DC characteristics. Third, hot-carrier effects in advanced SiGe HBTs are investigated employing microscopic simulation.

Chapter 3 "SiGe HBT Compact Modeling" starts with an overview on the standard HBT compact model HICUM Level 2. Afterwards, the most recent extensions related to DOTSEVEN, which were mostly targeted towards the intrinsic device operation are described. This is followed by a detailed discussion of the corresponding parameter extraction methods, including bias, geometry and temperature dependence. Geometry scaling and modeling of the intra-device substrate coupling concludes this chapter.

Chapter 4 "(Sub)mm-wave Calibration" addresses the electrical high-frequency on-wafer characterization of the fabricated HBTs, which becomes inaccurate beyond 220 GHz when using traditional methods. The conventional calibration and deembedding techniques are reviewed, followed by a discussion of the signal propagation modes in transmission lines on lossy substrates as they are encountered on a silicon wafer. Direct on-wafer calibration up to the device-under-test ports is then introduced, using a special transmission line configuration. As a demonstration, a comparison of measurements using this technique with HICUM is shown for frequencies up to 325 GHz.

Chapter 5 “Reliability” reports on the medium- and long-term degradation of the fabricated advanced HBTs due to hot carrier stress. The measured impact of hot carriers generated during device operation on DC and low-frequency noise characteristics is shown and compared to the results of a correspondingly extended HICUM version. Furthermore, self-heating is investigated, which is becoming increasingly important in downscaled high-performance devices and plays an important role in device degradation. A method for determining the thermal resistance is described and corroborated on thermal simulation data. Finally, different analytical models for describing the thermal resistance as a function of geometry are compared.

Chapter 6 “Millimeter-wave Circuits and Applications” is divided into three parts. First, simple basic building blocks termed benchmark circuits are considered which serve mostly for verifying the compact models in a circuit environment, but can also be used for benchmarking the technology performance. For each of the selected examples a sensitivity study reveals the most important transistor parameters. The second part reports on larger circuit building blocks as they would be used in a system. Examples are given that demonstrate competitive high-frequency circuit operation down to 0.5 V supply voltage and thus very low power consumption. The third part describes the three mm-wave demonstrator systems and their building blocks that were realized with DOTSEVEN technology, namely a 240 GHz Transceiver for ultra-high data rate wireless communication, a 210–270 GHz circularly polarized radar, and a 0.5 THz computer tomography system. The latter is the first-ever purely silicon based tomography system operating at such frequency.

Chapter 7 “Future of SiGe HBT Technology and Its Applications” puts the DOTSEVEN results in perspective by comparing, based on the standard performance metrics, the different technology options for mm-wave and future sub-mm-wave applications. A more intriguing perspective is provided when looking at the transistor operation within circuits, where the load comprised of parasitics from connections to other devices and the latter themselves paints a different picture of transistor speed as given by standard metrics. In view of the ITRS/IRDS roadmap for SiGe HBTs, which grew out of the modeling effort in DOTFIVE and DOTSEVEN, it is shown that HBTs in general have a bright future for realizing high-frequency systems. Their near future is discussed for different (sub-)mm-wave application areas based on DOTSEVEN results as a reference.

We hope that this book is a useful reference for a wide range of readers interested in mm- and sub-mm-wave technology, devices, and applications.

Naples and Dresden/San Diego
February 2018

Niccolo Rinaldi
Michael Schröter

Acknowledgements

This work was supported by the European Commission under the contract No. 316755-DOTSEVEN. Fruitful collaboration with all partners of the DOTSEVEN consortium is gratefully acknowledged. In particular, the authors gratefully acknowledge the numerous contributions of the HBT development teams of IHP and Infineon with special thanks to: R. Barth, A. Fox, J. Korn, T. Lenke, S. Marschmeyer, A. Scheit, D. Schmidt, and D. Wolansky of IHP, and J. Böck, S. Boguth, K. Knapp, R. Lachner, W. Liebl, D. Manger, T. F. Meister, and A. Pribil of Infineon. We thank W. Skorupa and T. Schumann of Helmholtz-Zentrum Dresden-Rossendorf and S. Häberlein of FHR Anlagenbau GmbH for their support with millisecond-annealing capabilities. The authors of chapter 3 like to thank P. Sakalas for his measurement support. The authors of chapter 4 would also like to thank the European Metrology Programme for Innovation and Research (EMPIR) Project 14IND02 "Microwave measurements for planar circuits and components", for partially supporting the reported work. The EMPIR program is co-financed by the participating countries and from the European Unions Horizon 2020 research and innovation program.

We, the editors, greatly appreciate the effort of our (co-)authors to contribute the various chapters and to spend a certainly significant portion of their busy schedule to finally bring this book to completion. We would also like to thank Mark deJong for enabling the publication of this book and Junko Nagajima who tirelessly worked through several iterations of corrections for assembling the diverse contributions into a homogeneous final version.

List of Contributors

Anindya Mukherjee, *Chair for Electron Devices and Integrated Circuits, Technische Universität Dresden, Helmholtzstr, 18, Barkhausenbau, 01062 Dresden, Germany*

Andreas Pawlak, *Chair for Electron Devices and Integrated Circuits, Technische Universität Dresden, Helmholtzstr, 18, Barkhausenbau, 01062 Dresden, Germany*

Alessandro Magnani, *Department of Electrical Engineering and Information Technology, University Federico II, via Claudio 80125, Naples, Italy*

Bertrand Ardouin, *XMOD Technologies, 74 rue G. Bonnac, 3300 Bordeaux, France*

Bernd Heinemann, *IHP, Im Technologiepark 25, 15236 Frankfurt (Oder), Germany*

Christoph Jungemann, *RWTH Aachen University, 52056 Aachen, Germany*

Cristell Maneux, *Laboratory of Integration of Material to System (IMS), University of Bordeaux, Bordeaux, France*

Gerald Wedel, *Chair for Electron Devices and Integrated Circuits, Technische Universität Dresden, Helmholtzstr, 18, Barkhausenbau, 01062 Dresden, Germany*

Gerhard G. Fischer, *IHP, Frankfurt (Oder), Germany*

Holger Rücker, *IHP, Im Technologiepark 25, 15236 Frankfurt (Oder), Germany*

Janusz Grzyb, *Institute for High-Frequency and Communication Technology, University of Wuppertal, Wuppertal, Germany*

Klaus Aufinger, *Infineon Technologies AG, Neubiberg, Germany*

Luca Galatro, *1. Electronic Research Laboratory, Delft University of Technology, Mekelweg 4, 2628CD, Delft, The Netherlands*
2. Vertigo Technologies B.V., Mekelweg 4, 2628CD, Delft, The Netherlands

Michael Schröter, *1. Chair for Electron Devices and Integrated Circuits, Technische Universität Dresden, Helmholtzstr, 18, Barkhausenbau, 01062 Dresden, Germany*
2. Department of Electrical and Computer Engineering, University of California at San Diego, 9500 Gilman Drive, La Jolla, CA 92093-0407, USA

Marco Spirito, *Electronic Research Laboratory, Delft University of Technology, Mekelweg 4, 2628CD, Delft, The Netherlands*

Niccolò Rinaldi, *Department of Electrical Engineering and Information Technology, University Federico II, via Claudio 80125, Naples, Italy*

Philipp Hillger, *Institute for High-Frequency and Communication Technology, University of Wuppertal, Wuppertal, Germany*

Ritesh Jain, *Institute for High-Frequency and Communication Technology, University of Wuppertal, Wuppertal, Germany*

Salvatore Russo, *Department of Electrical Engineering and Information Technology, University Federico II, via Claudio 80125, Naples, Italy*

Ullrich Pfeiffer, *Chair for the Institute for High-Frequency and Communication Technology (IHCT), Universiry of Wuppertal, Wuppertal, Germany*

Vincenzo d'Alessandro, *Department of Electrical Engineering and Information Technology, University Federico II, via Claudio 80125, Naples, Italy*

Wenfeng Liang, *Chair for Electron Devices and Integrated Circuits, Technische Universität Dresden, Helmholtzstr, 18, Barkhausenbau, 01062 Dresden, Germany*

List of Figures

List of Tables

List of Abbreviations

ATSF	Aging Time Scale Factor
BBA	Broadband Amplifier
BC	Base–collector
BEOL	Back-end-of-line
BiCMOS	Bipolar Complementary Metal Oxide Semiconductor
BJT	Bipolar Junction Transistor
BTB	Band-to-band
BTE	Boltzmann transport equation
CB	Common base
CE	Common emitter
CL-ICPW	Capacitively loaded inverted coplanar waveguide
CM	Compact model
CMC	Compact model coalition
CML	Current mode logic
CMOS	Complementary Metal-Oxid-Semiconductor
CMP	Chemical mechanical polishing
CPW	Coplanar wave guide
DC	Direct current
DPSA	Double-poly-Si self-aligned
DT	Deep trench
DUT	Device under test
DUV	Deep ultraviolet
EB	Emitter–base
EBL	Epitaxial base link
EM	electro-magnetic
FEM	Finite-element method
fmax	Maximum oscillation frequency
FoM	Figure of Merit
fps	Frames per second
fT	Transit frequency
Gbps	Gigabit per second

GCPW	grounded coplanar wave guide
GICCR	General integral charge-control relation
G-R	Generation-Recombination
HBT	Heterojunction Bipolar Transistor
HBT	Heterojunction bipolar transistor
HC	Hot carrier
HD	Hydrodynamic
HF	High-frequency
HICUM	High Current Model
IFX	Infineon Technologies AG
IHP	Innovations for High Performance microelectronics
II	Impact ionization
IIP3	Third-order input intercept point
IMD	Inter metal dielectric
InP	Indium phosphide
ITRS	International technology roadmap for semiconductors
LDD	Low-doped drain
LNA	Low-noise amplifier
LRM	Line reflect match
LSQ	Least squares
MM	Mixed-mode
mm-wave	Millimeter wave
MOSFET	Metal-Oxid-Semiconductor Field Effect Transistor
NSEG	Non-selective epitaxial growth
P_DC	Power dissipation
PA	Power amplifier
PDK	Process design kit
PPW	Parallel plate waveguide
RF	Radio frequency
RTP	Rapid thermal processing
RTS	Random telegraph signal
SCR	Space charge region
SEG	Selective epitaxial growth
SEM	Scanning electron microscope
SH	Self-heating
Si	Silicon
SIC	Selectively implanted collector
SiGe	Silicon germanium
SOI	Silicon on insulator

SOLR	Short open load reciprocal
s-parameters	Scattering parameters
SPICE	Simulation Program with Integrated Circuit Emphasis
SRH	Shockley-Read-Hall
ST	Shallow trench
STI	Shallow trench isolation
STM	STMicroelectronics
TCAD	Technology computer-aided design
TE	Transverse electric
TEM	Transmission electron microscopy
TM	Transverse magnetic
TRL	Thru reflect line
VCO	Voltage controlled oscillator
VNA	Vector network analyzer
W-band	90 to 110 GHz frequency range
WP	Work package

Introduction

M. Schröter[1,2]

[1]Chair for Electron Devices and Integrated Circuits, Technische Universität Dresden, Germany
[2]Department of Electrical and Computer Engineering, University of California at San Diego, USA

The semiconductor industry is the fundamental building block of the new economy. There is no area of modern life untouched by the progress of micro- and nanoelectronics. The electronic chip is becoming an ever-increasing portion of system solutions, starting initially from less than 5% in the 1970 microcomputer era, to more than 60% of the final cost of a mobile telephone, 50% of the price of a personal computer (representing nearly 100% of the functionalities), and 30% of the price of a monitor. In addition to their value in terms of cost, semiconductor components are the enablers of new applications, such as ABS, location detection (e.g., GPS positioning), smart cards, autonomous vehicles, and high-rate data communications (e.g., machine to machine), where electronics take over a lot of the essential functionalities and offer safety and security. Thanks to advances in nanoelectronics with their more than proportional increase of operational capabilities in relation to cost, new market opportunities are emerging worldwide, characterized by the needs of products and services offering, e.g., mobility, connectivity, and security.

Interest in utilizing the (sub-)mm-wave frequency spectrum for commercial and research applications has been steadily increasing. Such applications, which constitute a diverse but sizeable future market, span a large variety of areas such as health, material science, mass transit, industrial automation, communications, and space exploration. For the deployment of the respective high-performance and partially highly integrated circuits and systems in

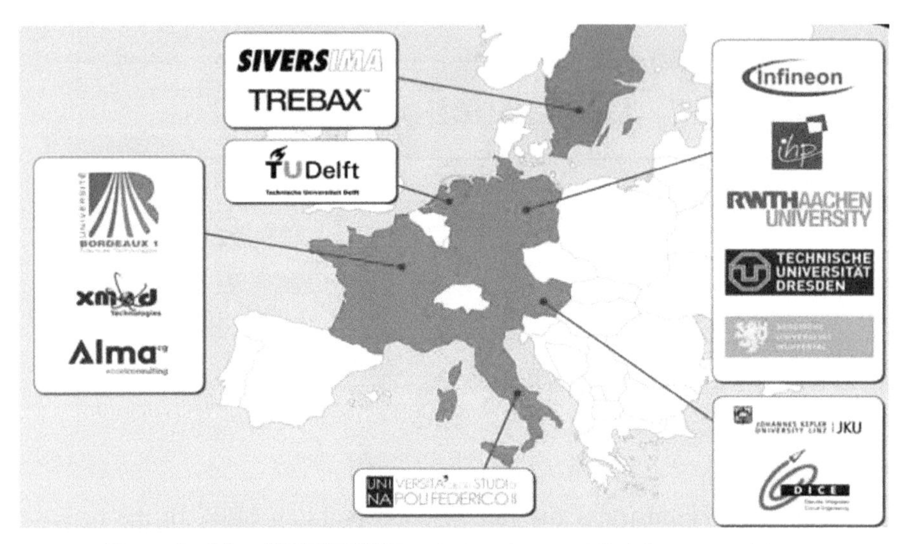

Figure 1 List of DOTSEVEN project partners and their home countries.

commercial markets, silicon-germanium (SiGe) heterojunction bipolar transistor (HBT) BiCMOS technology is well-suited due to the combination of HBT device speed for the high-frequency front-ends with the high integration levels of CMOS for digital signal processing [Che17, Sch17].

Based on a successful research cooperation in the predecessor project DOTFIVE, which produced the first half-THz SiGe HBTs [Hei10], the project DOTSEVEN was launched in late 2012 by the European Commission, targeting a variety of ambitious goals such as pushing HBT performance to 700 GHz (maximum oscillation frequency) and demonstrating working systems at 240 GHz. More and specific goals will be detailed later. For accomplishing the project's goals, a consortium consisting of partners with complementary expertise was assembled from industry, a research institute, and academia (cf. Figure 1). The project was subdivided into four technical work packages (WPs), one dissemination WP, and one administrative WP.

This book provides an overview of the research results of DOTSEVEN. It starts in this chapter with the motivation at the beginning of the project and a summary of its major achievements. The subsequent chapters provide a detailed description of the obtained research results in the various areas of process development, device simulation, compact device modeling, experimental characterization, reliability, (sub-)mm-wave circuit design and systems.

Motivation and Objectives of the DOTSEVEN Project

The DOTSEVEN project proposal was motivated by the increasing interest in utilizing the mm-wave frequency spectrum within the so-called THz gap[1], which ranges from 0.3 to 30 THz (cf. Figure 2), for a wide variety of applications. Examples for these applications at the beginning of the project were >120 GHz industrial sensors including mm-wave scanning and radar, extremely broadband ADCs with 50–100 Gs/s and >25 GHz signal bandwidth at 5–6 bit resolution, 400 Gb/s optical (backbone) transmission, as well as highly linear amplifiers, e.g., for 5G mobile communications. These circuits and systems serve a large variety of markets [Sie02, Sie04, Ton07, Kuk10, Coo11, Tay11, Son11, Kem11, Aji11, Eis11] such as health care and biology (e.g., medical equipment, patient monitoring, tissue and genetic screening), infrastructure and construction (e.g., structural safety), mass transportation (e.g., security screening, automotive radar, in-seat entertainment), industrial automation (e.g., sensors), and communications (e.g., high-bandwidth terrestrial point-to-point wireless, satellites). The deployment of the associated high-performance circuits and systems in commercial and military markets is driven mainly by cost, form-factor, and energy-efficiency.

The rapidly increasing interest in THz-applications was documented by the start of a new IEEE Journal, namely the "Transactions on Terahertz Science and Technology" as well as first business reports (e.g., [Thi11]), according to which applications operating in the mm- and sub-mm-wave range constitute a diverse but quite sizeable future market.

The design and implementation of high-speed circuits such as those mentioned above requires individual transistors to be able to operate, i.e., maintain power gain, at 3–10 times higher frequencies. This puts their characteristic operating frequencies well beyond the previous 500 GHz. Therefore, the main objectives of the DOTSEVEN project were:

- The realization of SiGe:C HBTs operating at a maximum (oscillation) frequency up to 700 GHz (i.e., 0.7 THz) at room temperature.
- The evaluation, understanding, and modeling of the relevant physical effects occurring in such high-speed devices and circuits for supporting process development and circuit design.

[1]The designation "gap" results from the strong drop in signal output power generated from both electronic and optical devices in this frequency range.

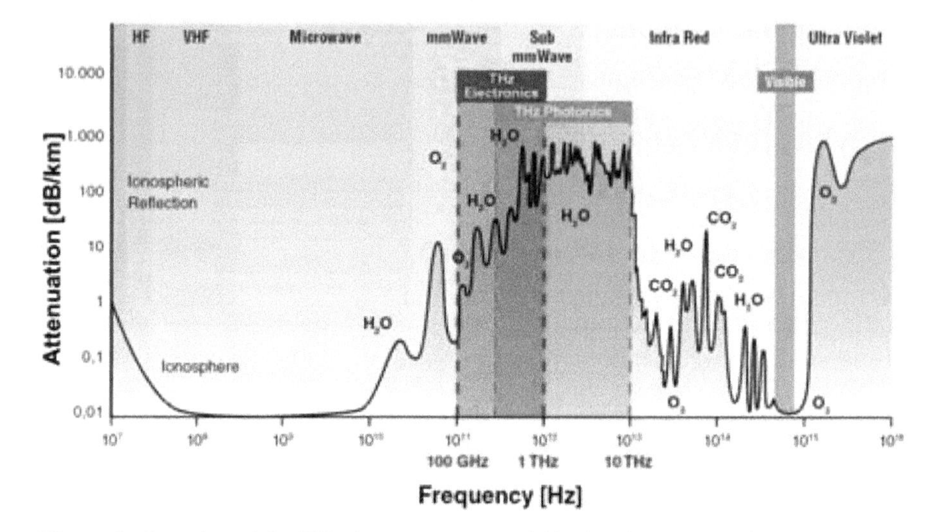

Figure 2 Location of the THz frequency range within the electromagnetic spectrum. The overlap region (THz-gap) between electronic and photonic approaches around 1 THz is indicated.

- The design and demonstration of working integrated mm- and sub-mm-wave circuits using such HBTs for specific applications as specified further below.
- Establishing and maintaining the European leadership in sub-mm-wave SiGe HBT process technology and opening up the mm- and sub-mm-wave market to the broader European and international industry.

In this book, the designations "maximum frequency" or "operating speed" of a transistor are synonymous for maximum oscillation frequency f_{max} and CML gate delay τ_{CML}, which – compared to the often used common-emitter current gain transit frequency f_{T} – are more relevant transistor-related figures of merit (FoMs) for circuit applications and provide more detailed performance information.[2] Nevertheless, the project strived for a balanced device design with a reasonable ratio $f_{\mathrm{max}}/f_{\mathrm{T}}$ not exceeding two in order to make the developed process technologies applicable to a wide range of applications.

[2]Note that f_{T} and f_{max} values are always extrapolated from a single-pole low-pass behavior of the respective gain. This definition enables a reliable comparison of device-related speed performance between different technologies.

Approach toward Achieving the Ambitious Goals

The research and development aspects of the DOTSEVEN project were tackled by four clearly defined and well-connected WPs during a period of 45 months. Below is a brief description of each WP. Details are discussed in the subsequent technical chapters.

Work package 1 comprised advanced and revolutionary process development with the goal to create a "SiGe HBT process technology platform," subdivided into two major directions. First, at the research institute novel ideas were pursued and possibly revolutionary process modules were developed with a focus on device architectures that had a high probability of significantly improving the device performance. Key aspects addressed here were in the areas of collector formation, vertical profile optimization, and emitter/base architecture. Second, at the industrial fabrication partner the most promising research results were further considered regarding their potential integration in an advanced industrial bipolar and later on also in a BiCMOS production process flow. In parallel to the above activities, the additional passive and active devices necessary for fully integrated mm-wave circuits were developed and integrated into the existing process flow. At certain stages of the project, wafers and process design kits (PDKs) were delivered to the other partners in (i) WP3 for elaborate electrical characterization, model parameter extraction, and compact model development, (ii) WP2 for the calibration and verification of numerical simulation tools, and (iii) WP4 for circuit design and characterization as well as building the demonstrators.

Work package 2 was dedicated to "Computational modeling tools" using physics-based predictive simulation tools for solving fundamental equations numerically, which allow the simulation of electrical characteristics of the fabricated structures. The device simulation subset, solving the semiconductor equations, is also known as "Technology Computer Aided Design (TCAD)". This modeling activity aimed at the continuous support of not only the technology development work in WP1 by providing guidelines for optimizing device performance, but also the development of compact models and the generation of virtual electrical characteristics for extracting preliminary model parameters for WP4. Advanced hydro-dynamic (HD) transport models were employed as a compromise between acceptable computation time and accuracy of simulations, while the Boltzmann transport equation (BTE) was utilized for predicting the electrical performance and calibrating the HD transport model. In addition, three-dimensional thermal and electromagnetic simulations, partially based on in-house tools developed previously, were applied for predicting and investigating distributed parasitic effects resulting from self-heating and HF operation.

Work package 3 emphasized on "Device characterization and compact modeling" and linked process technology (WP1) with circuit design (WP4). This WP addressed the following tasks. First, a common set of optimized test structures and new model parameter extraction strategies were developed. The second task was dedicated to compact model development for overcoming the deficiencies of existing models for sub-mm-wave frequencies and applications; also, the large-signal compact model was demonstrated to be valid all the way up to 1 THz [Sch14]. Within the third task, based on existing hardware and in collaboration with WP1 and WP2, realistic compact models representing the target process (i.e., possible target profiles for a 0.7 THz HBT) and its performance were predicted in order to establish a rough guideline for process development. The fourth task was to set up the methodology for accurate transistor measurements as well as suitable de-embedding and calibration methods at the targeted mm- and sub-mm-wave frequencies. The fifth task, again in strong cooperation with WP2 and WP1, was to do preliminary investigations of the reliability of the newly developed HBTs.

Work package 4 comprised "Millimeter-wave circuit applications and demonstrators". The objective of this work package was twofold. First, simple benchmark circuits were designed in cooperation with WP3 to evaluate the developed models under realistic circuit conditions, to allow a comparison with other process technologies, and to give circuit designers a good idea of the process capabilities for practical applications. The simple nature of these circuits provided valuable feedback regarding the impact of specific physical effects and the accuracy of the delivered models. In parallel, automated procedures for the initial design of these mm-wave benchmark circuits were developed to enable their implementation by modeling and process engineers. Second, as the major goal, WP4 demonstrated the viability of next-generation mm-wave and THz applications by exploiting the developed advanced process technology and modeling capabilities. In particular, WP4 aimed at building the demonstrators described later in more detail and hence indispensable design expertise to spur economic growth in emerging mm-wave and THz markets, such as communications, safety, health care environment, and security, by taking into account industrial design objectives and specifications.

In addition to the four technical and scientific work packages, the dissemination and presentation of the project results were organized by the separate WP5. Besides journal and conference publications, project results were presented in tutorials and workshops at various conferences. That material can be found at www.iee.et.tu-dresden.de/iee/eb/res/dot7/dot7.html.

Overview of Results and Their Impact

DOTSEVEN turned out to be highly successful as it even exceeded some of its goals at an extremely competitive cost. Significant improvements against the previous state-of-the-art were accomplished in all areas of SiGe HBT research, which has been documented by the following (non-exhaustive) list of technical achievements:

- Demonstration of SiGe:C HBTs with new room temperature world record performance of 720 GHz f_{max} and 1.34 ps minimum gate delay as well as the best ever balanced combination with an f_{T} of 505 GHz and a BV_{CEO} of 1.86 V, all based on a 130 nm lithography.
- Proof of concept of an industrial 130 nm SiGe BiCMOS process with (f_{max}, f_{T}) = (500, 300) GHz and pre-development of a cost-efficient industrial 130 nm SiGe BiCMOS process with leading edge performance of (f_{max}, f_{T}) = (360, 250) GHz.
- Establishment of a full suite of technology computer-aided design (TCAD) tools for accurately simulating and modeling advanced SiGe:C HBTs as well as for predicting their future performance along with the first SiGe HBT technology roadmap for the ITRS/IRDS.
- Delivery of accurate HICUM/L2 SiGe HBT models for first-time-right mm-wave designs.
- Demonstration of several world record and benchmark circuits at frequencies up to 500 GHz (triple push oscillator) and fundamental circuits up to 240 GHz (fundamental oscillators and amplifiers), including the realization of mm-wave power amplifiers above 200 GHz with output power up to 10 dBm and the successful demonstration of power combining techniques in SiGe-based PAs above 200 GHz for the first time.
- Demonstration of the first all-silicon computer tomograph operating at 490 GHz as well as of a 210–270 GHz 3D imaging system.
- Demonstration of 240 GHz transmitter and receiver for communications.

- Increase of output power and reduction of power dissipation by about a factor two for an industry-based 77 GHz automotive radar transmitter.

Emerging mm-wave and THz markets will benefit from the developed 0.7 THz SiGe HBT transistors in two ways: (1) Their superior HF performance and possible combination with CMOS integration capabilities make them an enabling technology for applications that historically have exhibited low integration levels and yield. (2) Economies of scale can be used to provide a cost-effective platform for highly integrated subsystems and single-chip solutions at mm-wave frequencies and beyond. As such, a 0.7 THz SiGe HBT technology targets the system miniaturization at very high frequencies to enable the "mm-wave System-On-Chip" in the future.

In summary, SiGe:C HBTs have proved their capability to support large bandwidth and high data rates for high-speed/high-frequency systems. Devices with impressive operating frequency now have been demonstrated that only a couple of years ago would have been believed to be reserved for III–V technologies. The higher operating speed of SiGe:C HBTs developed within the proposed project DOTSEVEN can be leveraged for advanced circuits and systems in different ways:

- They can open up new applications at very high frequencies (THz) using harmonics while still providing higher output power than passives used today.
- Their speed can be traded for lower power dissipation.
- They can be used to mitigate the impact of process, voltage, and temperature variations (PVT-variations) at lower frequencies for higher yield and improved reliability (e.g., in case of automotive radar application or high-bandwidth communications requiring highly linear front-ends).
- The resulting BiCMOS technologies enable the fabrication of complex (sub-)mm-wave systems for medium- and high-volume applications.

The results listed above were obtained through a tight cooperation and efficient communication between technology, modeling, and design partners within and across work packages. For instance, the close cooperation between the two technology providers made a fast transfer of IHP's research results into IFAG's production technology pre-development cycle possible. DOTSEVEN has cemented the international leadership of the European mm-wave community. The results mentioned above were achieved with a total EU contribution of just 8.6 M€ and an overall cost of just 12.3 M€, which is extremely small compared to the development of III–V technologies and in particular RF-CMOS, especially when considering the fact that as of today still no RF-CMOS process exists with HF performance that is even close to the one of DOTSEVEN. A more detailed technology comparison is given in the Outlook chapter.

In summary, DOTSEVEN has played a unique key role in the development of a new generation of high-speed silicon technologies and has pushed Europe to the forefront of the world-wide competition. The project benefited significantly from the strong links between research and industry. It enabled to pursue the broad range of activities from device and material research to the system implementation and demonstration simultaneously and with sufficient critical mass, thus opening up the path to a rapid exploitation of the results and economic growth on the one hand and providing the means for educating a highly skilled work force in the mm- and sub-mm-wave application area, on the other hand. Overall, DOTSEVEN demonstrated clearly that seriously advancing high-frequency electronics is possible with relatively low cost (compared to CMOS and III–V technology) by assembling a suitable multidisciplinary consortium, efficient organization, and defining meaningful and realistic goals.

References

[Aji11] Ajito, K., and Ueno, Y. (2011). THz chemical imaging for biological applications. *IEEE Trans. Terahertz Sci. Technol.* 1, 293–300.

[Che17] Chevalier, P., Schröter, M., Bolognesi, C. R., d'Alessandro, V., Alexandrova, M., Böck, J. (2017). Si/SiGe:C and InP/GaAsSb heterojunction bipolar transistors for THz Applications. *Proc. IEEE*, 105, 1035–1050. doi: 10.1109/ JPROC.2017.2669087.

[Coo11] Cooper, K., et al. (2011). THz imaging radar for standoff personnel screening. *IEEE Trans. Terahertz Sci. Technol.* 1, 169–182.

[Eis11] Eisele, H. (2010). State of the art and future of electronic sources at terahertz frequencies. *Electron. Lett.* 46, S8–S11.

[Hei10] Heinemann, B., Barth, R., Bolze, D., Drews, J., Fox, A., Fursenko, O. (2010). SiGe HBT technology with fT/fmax of 300GHz/500GHz and 2.0 ps CML gate delay. *IEDM Tech. Dig.* 2010, 688–691.

[Kem11]Kemp, M. (2011). Explosives detection by Terahertz spectroscopy – A bridge too far? *IEEE Trans. Terahertz Sci. Technol.* 1, 282–292.

[Kuk10] Kukutsu, N., Hirata, A., Yaita, N., Ajito, K., Takahasi, H., Kosugi, T., et al. (2010). Towards practial applications above 100 GHz. *IEEE IMS* 2010, 134–137.

[Sch17] Schröter, M., Rosenbaum, T., Chevalier, P., Heinemann, B., Voinigescu, S., Preisler, E., Boeck, J. (2017). SiGe HBT technology: future trends and TCAD based roadmap. *Proc. IEEE* 105, 1068–1086. doi: 10.1109/JPROC.2015.2500024.

[Sch14] Schröter, M., Pawlak, A. (2014). "Physics-based nonlinear compact modeling of HBTs for mm-wave applications," in *Proceedings of the IEEE MTT-IMS, Short Course (WMF)*, Tampa, FL.

[Sie02] Siegel, P. (2002). Terahertz technology. *IEEE Trans. Microw. Theory Tech.* 50, 910–928.

[Sie04] Siegel, P. (). Terahertz technology in biology and medicine. *IEEE Trans. Microwave Theory Techn.* 52, 2438–2447.

[Son11] Song, H.-J., and Nagatsuma, T. (2011). Present and future of terahertz communications. *IEEE Trans. Terahertz Sci. Technol.* 1, 256–263.

[Tay11] Taylor, Z., et al. (2011). THz medical imaging: in vivo hydration sensing. *IEEE Trans. Terahertz Sci. Technol.* 1, 201–219.

[Thi11] Thintri Inc (2011). *Millimieter Wave Systems Opening up Billion-Dollar Markets in Security Consumer Products Telecommunications.* Available at: http://www.thintri.com

[Ton07] Tonouchi, M. (2007). Cutting-edge terahertz technology. *Nat. Photonics* 1, 97–105.

1

SiGe HBT Technology

H. Rücker and B. Heinemann

IHP, Germany

1.1 Introduction

Advances in silicon–germanium (SiGe) heterojunction bipolar transistor (HBT) technologies resulted in an impressive increase in high-frequency performance during the last decade extending the addressed application frequencies into the mm- and sub-mm-wave bands. Today, SiGe HBTs are widely used for applications like automotive radar, high-speed wireless and optical data links, and high-precision analog circuits. BiCMOS technologies which comprise high-speed SiGe HBTs in a radio-frequency (RF) CMOS technology environment combine the excellent RF performance of SiGe HBTs with the high level of integration and the high computing power of Si CMOS. These technologies became a key enabler for demanding mm-wave systems which integrate radio front-end circuits together with digital control circuits and signal processing on a single chip. Previous development has demonstrated that SiGe HBTs continue to offer significantly higher cutoff frequencies, higher output power, and superior analog characteristics compared to CMOS transistors of the same lithography node. Thus, the integration of SiGe HBTs in a CMOS platform represents a very attractive option to boost the RF performance of a given technology node.

The state of the art of SiGe HBT technology before the start of the DOTSEVEN project in October 2012 was reviewed in [Che11]. Developments performed within the predecessor project DOTFIVE resulted in the first demonstration of SiGe HBTs with maximum oscillation frequencies, f_{MAX}, of 500 GHz together with transit frequencies, f_{T}, of 300 GHz and minimum ring oscillator gate delays of 2.0 ps [Hei10]. This was the starting point of the DOTSEVEN project addressing the challenging target for SiGe HBTs with peak f_{MAX} values of 700 GHz and minimum gate delays of

1.4 ps. Figure 1.1 summarizes published peak f_T and f_{MAX} values of selected high-speed SiGe HBT processes from the last decade. BiCMOS technologies with peak f_{MAX} values between 300 GHz and 400 GHz and peak f_T values of 230–320 GHz are in production or pre-production at several companies now. Recent research results have demonstrated that this performance can be increased much further. The values obtained within the DOTSEVEN project are indicated as red diamonds in Figure 1.1. In 2015, two separate investigations demonstrated new record values for f_{MAX} [Boe15] and f_T [Kor15]. Further technology optimization finally enabled the demonstration of the DOTSEVEN goal including the simultaneous realization of peak f_T and f_{MAX} values of 505 GHz and 720 GHz, respectively [Hei16].

The reminder of this chapter is organized as follows. Major performance factors of SiGe HBTs are reviewed in the section "HBT Performance Factors." Fundamental dependencies of typical high-frequency figures of merit (FoMs) on device parameters are discussed here. The section "HBT Device and Process Architectures Explored in the DOTSEVEN Project" addresses device architectures for high-performance SiGe HBTs and process integration aspects. Favored process options for HBTs with selective epitaxial growth (SEG) and with non-selective epitaxial growth (NSEG) of the SiGe base layer are analyzed in detail. The focus is on the work performed in the DOTSEVEN project concerning the development of the high-performance SiGe BiCMOS technology platform B11HFC at Infineon (see the section "DPSA-SEG Device Architecture"), the investigation of an advanced HBT

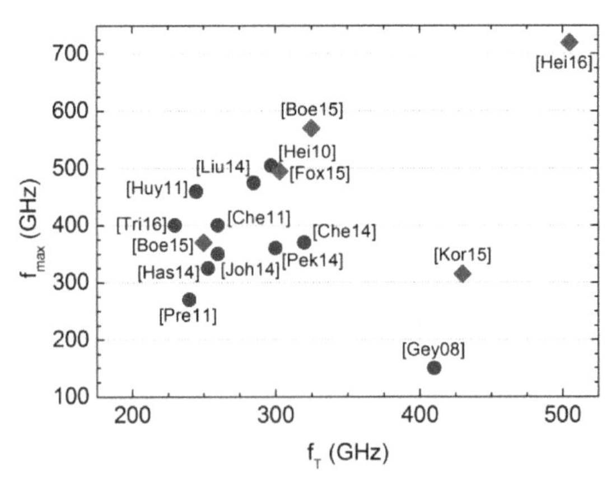

Figure 1.1 Peak f_T and f_{MAX} values of high-speed SiGe HBT technologies. Red diamonds indicate results of the DOTSEVEN project.

process with SEG of the base and epitaxial base link (EBL) regions (see the section "Approaches to Overcome Limitations of the DPSA-SEG Architecture"), and the optimization of a process with NSEG of the base (see the section "Non-selective Epitaxial Growth of the Base"), which was finally utilized by IHP to reach the DOTSEVEN goal. The section "Optimization of the Vertical Doping Profile" addresses the optimization of the vertical doping profile for f_{T} improvement. The final technology optimization for minimum device parasitics and balanced f_{T} and f_{MAX} improvement is discussed in the section "Optimization towards 700 GHz f_{MAX}."

1.2 HBT Performance Factors

Typical FoMs characterizing a process technology in terms of high-frequency performance are the transit frequency f_{T} and the maximum oscillation frequency f_{MAX}. The transit frequency f_{T} is defined as the frequency for which the small-signal current gain $|h_{21}|$ falls to unity, i.e.,

$$\left|h_{21}\left(f\right)\right|_{f=f_{\mathrm{T}}} = 1. \tag{1.1}$$

The frequency f_{MAX} is defined as the maximum frequency for which the transistor can amplify power. In this context, Mason's unilateral power gain U is widely used, and f_{MAX} is defined by:

$$U\left(f\right)|_{f=f_{\mathrm{MAX}}} = 1. \tag{1.2}$$

While the frequency f_{MAX} represents a speed metric for circuits such as amplifiers and oscillators, f_{T} gives a measure of the speed of switching circuits such as dividers. Ring-oscillator gate delays are relevant FoMs for digital high-speed circuits. Here, we use the current mode logic (CML) ring-oscillator gate delay time τ_{CML}. In addition, the base–collector breakdown voltage BV_{CEs} and the open-base emitter–collector breakdown voltage BV_{CEo} are important since they determine the maximum output power that can be provided by a transistor. Further characteristics of relevance for evaluating potential applications include the minimum noise figure, the linearity, and the gain of a transistor.

In the following, we are going to discuss the impact of different device regions and their electronic properties on RF performance. Basic device regions are indicated in the generic cross section of a high-performance SiGe HBT shown in Figure 1.2. For the analysis of the contribution of the individual device region to the delay time of the transistors

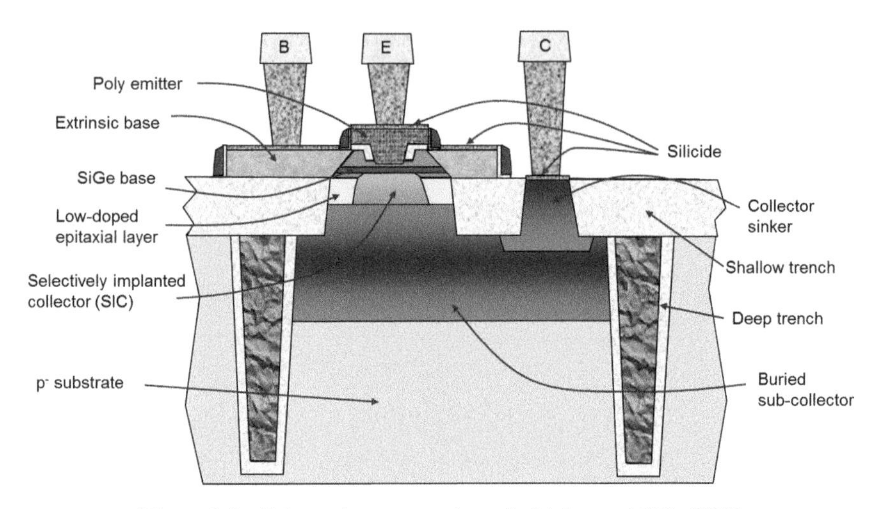

Figure 1.2 Schematic cross section of a high speed SiGe HBT.

response to an RF signal, it is helpful to relate the resistances and capacitances of the device regions to a simplified small-signal compact model.

Figure 1.3 indicates the resistances and capacitances of the individual device regions and relates them to a small signal equivalent circuit for the transistor operation in forward active mode. The model includes the resistances R_E, R_B, and R_C, of the emitter, base, and collector, respectively. The base resistance is divided into an intrinsic contribution R_{Bi} and extrinsic

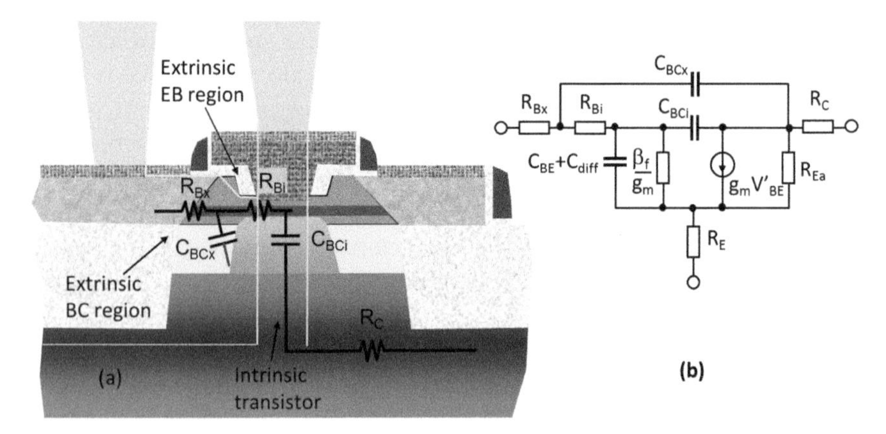

Figure 1.3 Device cross section with parasitic resistances and capacitances associated with different device regions (a) and a corresponding small signal equivalent circuit (b).

contribution R_{Bx} originating from the link region which contacts the intrinsic base. The base–collector capacitance (C_{BC}) is divided into an intrinsic part C_{BCi} and an extrinsic part C_{BCx} related to the base link region. C_{BE} includes the depletion capacitance as well as the oxide capacitance of the base–emitter junction, C_{diff} is the diffusion capacitance related to the storage of minority charges in the forward operation mode, β_{f} is the forward DC current gain, g_{m} is the transconductance, R_{Ea} is the output resistance related to the Early effect, and V'_{BE} is the intrinsic base–emitter voltage.

The frequency-dependent small signal current gain $h_{21}(f)$ of the model depicted in Figure 1.3 is approximately given by:

$$\frac{1}{h_{21}(f)} = \frac{1}{\beta_{\mathrm{f}}} + j2\pi f\left(\frac{C_{\mathrm{diff}} + C_{\mathrm{BE}} + C_{\mathrm{BC}}}{g_{\mathrm{m}}} + (R_{\mathrm{E}} + R_{\mathrm{C}})\,C_{\mathrm{BC}}\right). \quad (1.3)$$

In the limit of large frequencies, h_{21} is inversely proportional to the frequency f. The corresponding unit gain transit frequency f_{T} is given as:

$$\frac{1}{2\pi f_{\mathrm{T}}} = \frac{C_{\mathrm{diff}} + C_{\mathrm{BE}} + C_{\mathrm{BC}}}{g_{\mathrm{m}}} + (R_{\mathrm{E}} + R_{\mathrm{C}})\,C_{\mathrm{BC}}. \quad (1.4)$$

The transconductance g_{m} is proportional to the collector current I_{C} in the bias region of ideal exponential slope according to $g_{\mathrm{m}} = qI_{\mathrm{C}}/k_{\mathrm{B}}T$, where q is the elementary charge, k_{B} is Boltzmann's constant, and T is the junction temperature. The diffusion capacitance C_{diff} accounts for the storage of locally compensated minority carriers during forward transistor operation. The contribution to C_{diff} can be analyzed in a charge-control model [Tau98]. This analysis relies on the fact that any variation of the bias point of the device is related to changes of the carrier densities within the device which are fed by currents into the device contacts. The corresponding forward transit time

$$\tau_{\mathrm{F}} = \frac{C_{\mathrm{diff}}}{g_{\mathrm{m}}} = \tau_{\mathrm{E}} + \tau_{\mathrm{EB}} + \tau_{\mathrm{B}} + \tau_{\mathrm{BC}} \quad (1.5)$$

can be divided into contributions accounting for charge storage in the emitter, base–emitter junction, base, and base–collector junction regions, respectively. According to the charge-control model, these contributions are approximately given by:

$$\tau_{\mathrm{E}} = \frac{C_{\mathrm{E}}}{g_{\mathrm{m}}}, \quad (1.6)$$

$$\tau_{\mathrm{EB}} = \frac{C_{\mathrm{N}}}{g_{\mathrm{m}}}, \quad (1.7)$$

$$\tau_{\mathrm{B}} = \frac{w_{\mathrm{B}}^2}{2D_{\mathrm{nB}}} + \frac{w_{\mathrm{B}}}{v_{\mathrm{sat}}}, \tag{1.8}$$

$$\tau_{\mathrm{BC}} = \frac{w_{\mathrm{BC}}}{2v_{\mathrm{sat}}}. \tag{1.9}$$

Here, w_{B} is the width of the neutral base region, w_{BC} is the depletion width of the base–collector junction, D_{nB} is the electron diffusion coefficient in the base, and v_{sat} is the saturation velocity of electrons. C_{E} and C_{N} denote the parts of the diffusion capacitance related to neutral charge storage in the emitter and base–emitter junction regions, respectively. The compensated charge C_{N} stored in the base–emitter junction can account for a significant contribution to τ_{F} in particular at high current densities [Hue96]. The emitter delay time τ_{E} is of minor importance for typical SiGe HBTs since the amount of holes stored in the emitter is inversely proportional to the current gain. The magnitude of C_{E} is determined by the emitter properties. The maximum oscillation frequency of the equivalent circuit of Figure 1.3 is approximately given by:

$$f_{\mathrm{MAX}} \cong \sqrt{\frac{f_{\mathrm{T}}}{8\pi\left(\left(R_{\mathrm{Bx}} + R_{\mathrm{Bi}}\right)C_{\mathrm{Bi}} + R_{\mathrm{Bx}}C_{\mathrm{Bx}}\right)}}. \tag{1.10}$$

This relation is reduced to:

$$f_{\mathrm{MAX}} \cong \sqrt{\frac{f_{\mathrm{T}}}{8\pi R_{\mathrm{B}}C_{\mathrm{BC}}}}, \tag{1.11}$$

if R_{B} and C_{BC} are not separated into extrinsic and intrinsic contributions.

Based on the Equations (1.4) to (1.10), the following scenario can be envisioned for the enhancement of the cutoff frequencies f_{T} and f_{MAX} by scaling vertical and lateral device dimensions. The transit frequency f_{T} is predominantly determined by the vertical doping profile. Figure 1.4 illustrates qualitatively the directions of profile optimization for f_{T} enhancement.

Reduction of the width w_{B} of the boron-doped base reduces the base transit time τ_{B} according to Equation (1.7). A minimum width of the boron-doped region has to be ensured together with a low base sheet resistance. Today, base layers with typical sheet resistances of about 2 kΩ/sq can be grown epitaxially with widths of less than 5 nm. In addition to the deposition of a thin base, its diffusion during subsequent processes has to be kept as small as possible. A widely applied approach to minimize B diffusion is the additional doping of the SiGe layer with carbon [Lan96, Ost97, Rue99]. Moreover, the thermal budget of post-epi processing has to be kept low.

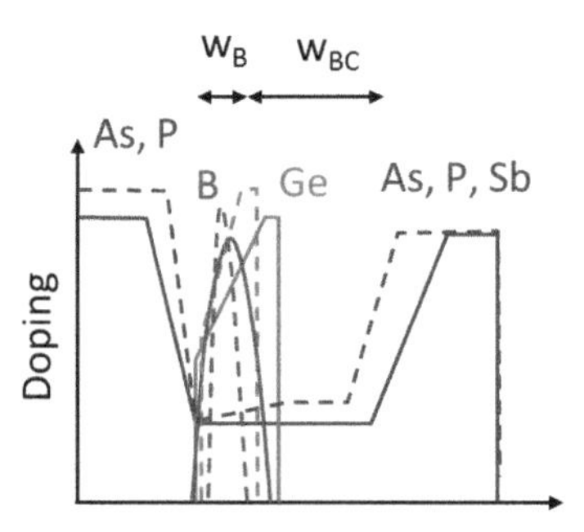

Figure 1.4 Schematic vertical doping profile of a SiGe HBT. The dashed lines indicate a scaled profile for enhanced f_T.

The challenge here is to realize simultaneously high dopant activation in heavily doped device regions and minimum diffusion broadening of the base.

Together with the base width, the width of the Ge profile is also shrunk. This allows one to increase the peak Ge concentration without exceeding the critical thickness of the SiGe layer above which the SiGe layer becomes thermodynamically unstable against the formation of dislocations. For low base transit time τ_B, it is beneficial to realize a steep gradient of the Ge concentration across the non-depleted base width w_B. The grading of the Ge profile as indicated in Figure 1.4 causes a built-in electric field due to the decrease of the band gap with increasing Ge content. This field accelerates minority electrons in the base and reduces τ_B.

Reduction of depletion width w_{BC} of the base–collector junction is a measure to reduce the base–collector transit time τ_{BC}. This reduction of τ_{BC} has to be traded off against an increased base–collector capacitance C_{BCi} and a reduced base–collector breakdown voltage due to reduced w_{BC}.

The neutral charge storage C_N in the base–emitter region can be reduced by decreasing of the base–emitter depletion width w_{EB} in conjunction with an optimized Ge-profile in the base–emitter junction region. However, reduction of w_{EB} results also in a reduction of the base–emitter breakdown voltage BV_{EB0} and in tunnel currents at low base–emitter voltages. These effects have to be traded off against the reduction of C_N.

The doping profiles of the non-depleted emitter and collector regions are optimized for low resistivity due to high concentrations of electrically active dopants. The lowest emitter resistances are typically achieved with mono-crystalline emitters. In addition to the above mentioned measures for f_T enhancement by vertical profile engineering, one also has to minimize contributions to the base–emitter and base–collector capacitances originating from the edges of the device for realizing higher f_T values according to Equation (1.4).

For realizing high f_{MAX} values, it is crucial to minimize the base resistance and the base–collector capacitance together with high f_T values as indicated by Equation (1.9). The required reduction of these device parasitics is typically addressed by scaling lateral device dimensions and by optimizing the base-link regions. Figure 1.5 illustrates relevant lateral device dimensions and major contributions of the base-link region to R_B and C_{BC}.

Contributions to the extrinsic base resistance originate from the extension of the base layer below the base–emitter spacer, from the adjacent mono-crystalline or poly-crystalline p-doped region, the contact resistance between silicide and base poly-Si, the silicide resistance, the contact resistance between silicide and metal contact plug, as well as from the resistance of subsequent metal regions. The extrinsic base–collector capacitance includes capacitances of the mono- and poly-crystalline extrinsic base regions to the selectively implanted collector (SIC), the buried collector layer in the active region, and the buried collector below the base–collector isolation layer. Reduction of these parasitic resistances and capacitances is addressed by reducing the corresponding lateral device dimensions such as the width of the base–emitter spacer d_{Sp}, the width of the emitter poly-Si w_{EP}, and the width of the active collector region w_{Col}. However, depending on the details of the device architecture, there are several tradeoffs between the different parameters. For example, the reduction of w_{Col} can lead not only to a reduction of C_{BCx} but also to an increase of R_{Bx}.

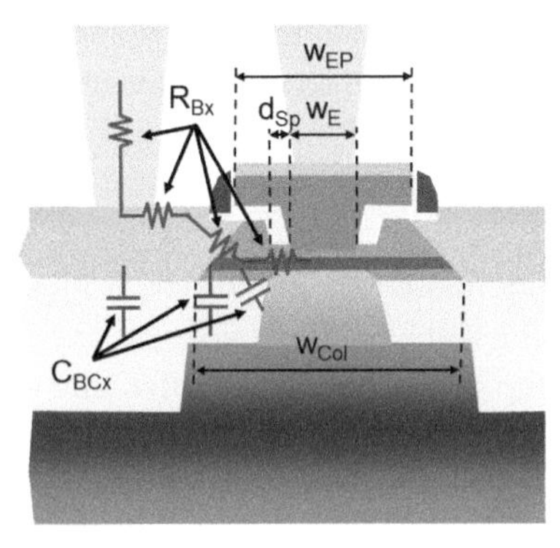

Figure 1.5 Device cross section with lateral device dimensions and major contributions of the base-link region to R_B and C_{BC}.

The intrinsic contribution to the base resistance R_{Bi} can be reduced for a given base sheet resistance R_{sbi} by narrowing the emitter window width w_{E}. In a typical scaling scenario, lateral scaling of w_{E} is accompanied by vertical scaling of the doping profile and increased current densities at peak f_{T}. Under these conditions, it is a major challenge to maintain low emitter and collector resistances as well as thermal resistances when the emitter width is scaled down.

Scaling of the HBT device dimensions under the boundary conditions of minimum base, emitter, and collector resistances imposes complex requirements on device architecture and fabrication process. These challenges have stimulated various innovations of the HBT fabrication process which addressed the reduction of the individual device parasitics by structural improvements as well as by improved material properties such as reduced specific and contact resistances. Approaches explored in the DOTSEVEN project will be reviewed in the following sections.

1.3 HBT Device and Process Architectures Explored in the DOTSEVEN Project

Innovations of the device architecture and of the fabrication processes have been major factors for the improvement of the RF performance of SiGe HBTs during the last decades. Fundamental requirements on the device architecture for high-speed HBTs are minimum access resistances to the intrinsic emitter, base, and collector regions together with low contributions of the extrinsic device regions to the base–collector and base–emitter capacitances. The development of device and process architectures which facilitate the simultaneous realization of low R_{B} and low C_{BC} has been a major challenge in this context. The realization of devices with low thermal resistances is a further requirement in order to limit self-heating.

The above-mentioned device targets have to be realized in fabrication processes which are manufacturable in high volumes with high yield. A further fundamental requirement on the HBT fabrication process is the compatibility with the addressed CMOS technology platform. The integration of SiGe HBTs and other RF-enabling passive or active devices into a BiCMOS technology platform has to be realized without degrading HBT or CMOS device characteristics or yield. The large potential of advanced CMOS processes for geometry scaling opens new options also for the HBT fabrication. However, new challenges arise for the integration of SiGe HBTs in continuously shrinking CMOS nodes from tight constraints on the thermal budget and on device topology.

As regards the SiGe HBT device concepts, all current production-related high-speed transistors take advantage of the so-called double-poly-Si (DP) architecture. This configuration provides access from the contact region to the intrinsic base and emitter region by poly-Si layers which are dielectrically isolated against the surrounding transistor regions. It is a powerful means to keep extrinsic parasitics, such as R_{Bx}, C_{BCx}, R_{E}, and C_{BE}, small. It is therefore evident that the basic structure of modern SiGe HBTs is becoming more similar. Nevertheless, we are faced with quite different approaches for device manufacturing resulting in different consequences of their potential electrical performance.

A key differentiator for SiGe HBT fabrication is the way in which the SiGe base is formed. Existing SiGe HBT technologies use either selective or non-selective epitaxial growth of the base. Both approaches have been used for the development of high-performance SiGe HBT processes and found their way into industrial mass production. In the DOTFIVE project, technological solutions were developed promising further speed enhancements for HBT concepts with selective as well as with non-selective base epitaxy. Due to their specific implications on the process complexity and the self-alignment of the transistor regions, various technologies were investigated also in the DOTSEVEN project using the different base-epitaxy methods. Opportunities and challenges of the two approaches will be discussed in detail in the following subsections. This applies also to process options regarding the lateral collector isolation by deep trenches or by the standard shallow trenches of the CMOS process, the formation of the highly conductive sub-collector, and the formation of the base–emitter structure. The choice of the substrate, i.e., bulk or silicon-on-insulator (SOI), is another criterion to differentiate SiGe HBT technologies. Driven by the continuous development of SOI-based CMOS technologies several publications have been devoted to the issue of a suitable technology and device concept for high-speed SiGe HBTs on SOI wafers [Was00, Rue04, Ave05, Thi13]. Here, we will not address this architectural aspect because it was outside the focus of the DOTSEVEN project.

1.3.1 Selective Epitaxial Growth of the Base

The classical DP-self-aligned (SA) SiGe HBT technology with SEG of the base represents the most attractive process architecture from the point of view of manufacturing and degree of self-alignment. As described in the next section, the DOTFIVE project partners Infineon and STMicroelectronics as well as FreeScale (now NXP) worked intensively on this concept in the last decade to push its performance. However, substantial improvements beyond the current level will hardly be feasible with this approach as discussed in the section "Approaches to Overcome Limitations of the DPSA-SEG Architecture." In the DOTFIVE project, alternative SEG process flows were developed to overcome limitations of the conventional DPSA-SEG technology. It will be reported below in detail on joint activities of Infineon and IHP in the DOTSEVEN project to test one of these approaches within Infineon's 130 nm BiCMOS platform.

1.3.1.1 DPSA-SEG device architecture

The conventional double-poly-Si self-aligned SiGe HBT technology with SEG enjoys large popularity and has been applied in production for long time by several companies [Boe04, Ave09, Joh07]. This process takes advantage from the fact that only one lithographic step is needed to completely construct the internal transistor. In principle, no further mask step is necessary to form the SIC region or the isolation between emitter and base. Usually, this process starts with the deposition of a layer stack comprising a bottom oxide, a p^+ poly-Si layer, an upper oxide, and a capping nitride. The emitter window is opened by dry etching which stops at the bottom oxide (Figure 1.6(a)). Nitride spacer formation will prevent from pulling back the upper oxide during the subsequent oxide wet etching. By this step, the intrinsic collector region is exposed and an overhang of the p^+ poly-Si is created (Figure 1.6(b)). At this point the SIC can be formed which provides a low-ohmic connection to the highly doped sub-collector. In addition, the nitride inside spacers protect against Si seeding of the p^+ poly-Si during the following base epitaxy. With this step, the link between the intrinsic base and the extrinsic part (p^+ poly-Si = base poly) is formed (Figure 1.6(c)). Whether the nitride layers are removed at a later stage or not is handled differently [Che11].

The remaining steps are very common also for other HBT processes such as technologies with non-selective base epi. Inside base–emitter spacers are formed and the in situ doped emitter layer is grown with a non-selective epitaxial step. Therefore, at least partly, a mono-crystalline emitter region can be found adjacent to the substrate surface already after deposition. The HBT flow is continued by patterning the emitter and base poly-Si layers. Finally, a short-term high-temperature treatment is needed in order to out-diffuse dopants from the highly doped emitter layer into the base cap layer before the process flow is completed by salicidation and backend fabrication (Figure 1.6(d)). The final annealing step contributes not only to improved

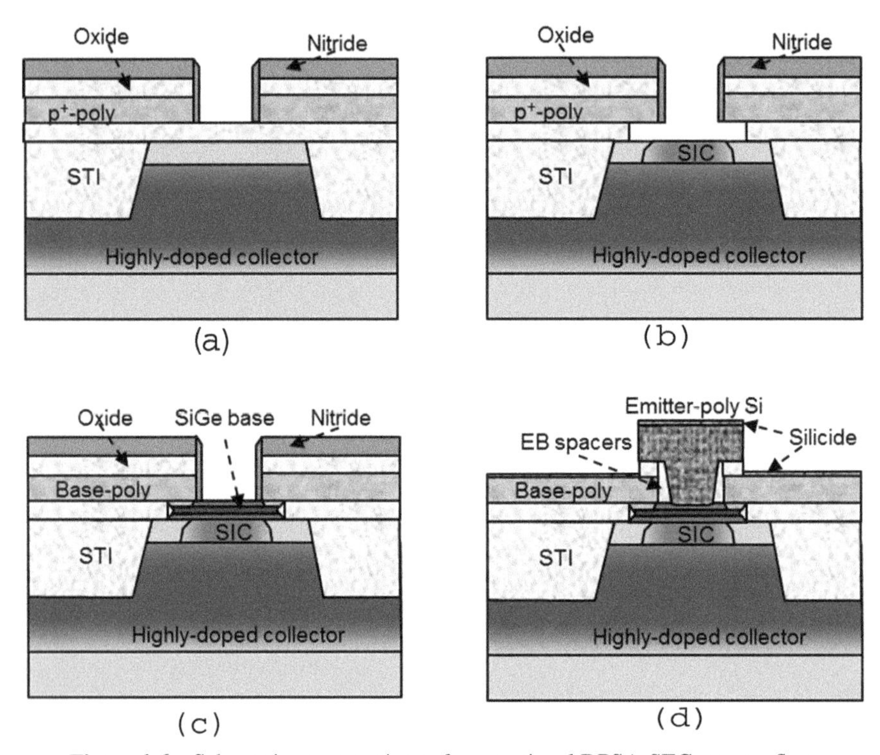

Figure 1.6 Schematic cross sections of conventional DPSA-SEG process flow.

base-current idealities but it also determines the emitter resistance. In the case of a bipolar-only technology, e.g., Infineon's B7HF200 [Boe04], the thermal budget is largely governed by the needs of the HBT. In BiCMOS processes, as a rule, the minimum requirements of the source–drain anneal determine the crucial thermal budget of the HBT assuming that a 'source/drain-after-HBT' integration scheme is realized.

The basic structure of the aforementioned DPSA-SEG process flow was already employed for the first demonstrations [Sat92, Mei95, Pru95]. Essentially, it has been maintained to date. In Figure 1.7, a TEM cross section of Infineon's latest SiGe HBT transistor generation is shown.

In [Che11], certain differences between the developments of different companies are pointed out. For example, Infineon utilized temporary auxiliary spacers to adjust the area of the SIC whereas the SIC was performed right after the emitter window opening at STMicroelectronics. There are also deviations in the annealing regime of the base poly. In the Infineon

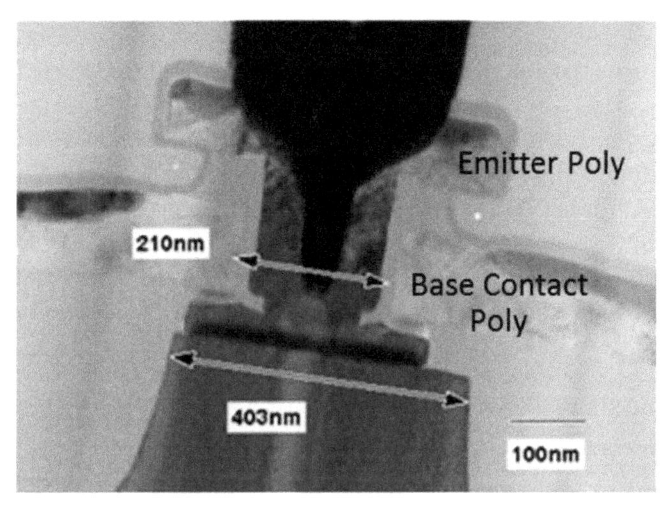

Figure 1.7 TEM cross section of a DPSA-SEG HBT of Infineon's B11HFC technology [Boe15].

flow, an extra thermal treatment after base epitaxy is introduced for outdiffusing B from the base poly toward the intrinsic base layer. Consequently, the base resistance can be reduced but the base tends to broaden causing lower f_T values. Nevertheless, variations of this annealing showed room for optimizations to increase f_{MAX} with tolerable constrains for f_T [Can12].

The collector construction used by Hitachi [Has14], Infineon [Boe15], and STMicroelectronics [Che14] includes all elements which are typical for a high-speed Si bipolar transistor: an epitaxially buried, highly doped subcollector isolated laterally by deep trenches and a low-ohmic connection to the collector contact realized by a so-called collector sinker. In order to save fabrication efforts and to reduce complexity of the BiCMOS process, NXP (former FreeScale) developed a Sub-Isolation Buried Layer (SIBL) collector structure using only shallow trench isolation (STI) [Joh07]. A common feature of all these technologies is the shallow-trench isolation (STI) between the internal transistor region and the collector contact which enables simultaneously a low capacitive base link.

In the DOTFIVE project comprehensive efforts were made by Infineon and STMicroelectronics to improve the high-frequency behavior of the conventional DPSA-SEG technology as reported in [Che11]. Clear progress was achieved by changing the vertical profile, thermal treatments, as well as the lateral transistor dimensions. At that time, Infineon was able to increase its initial f_T/f_{MAX} values of 190 GHz/250 GHz (B7HF200, [Boe04]) finally to 230 GHz/350 GHz while STMicroelectronics increased these FoMs from 230 GHz/280 GHz (BiMOS9MW, [Ave09]) to 260 GHz/400 GHz. Later, both companies could demonstrate this performance level in a BiCMOS environment too. The f_{MAX} results obtained in the DOTFIVE project and in the recent past for the DPSA-SEG concept [Has14, Che14, Tri16] indicate that it is difficult to reach values beyond 400 GHz. The limited possibilities to decrease R_{Bx} have been proved as the key bottleneck for advancements of the overall RF performance. Alternative concepts which were conceived to overcome this issue will be presented in the next subsection.

1.3.1.2 Approaches to overcome limitations of the DPSA-SEG architecture

If the geometry of a conventional DPSA-SEG SiGe HBT (see Figure 1.6(d)) is shrunk according to scaling scenarios for next technology nodes, as it was carried out in TCAD studies [Sch11b, Sch17], there is no serious indication for an imminent end of performance progress compared to other technology approaches. In practice, however, we are confronted with the fact that the attempts to increase f_{MAX} did not go beyond values of 400 GHz while alternative concepts indeed surpassed this level. Unfortunately, the main reason for this deficiency is closely connected with the key advantage of the conventional DPSA-SEG process, namely the straightforward manufacturing of the link between the intrinsic transistor region and the base poly-Si layer.

The vertical gap between the substrate surface and the base poly is bridged during the selective growth of the base layer (Figure 1.8(a)). Obviously, the base layer and the p+ base poly are separated after base epitaxy

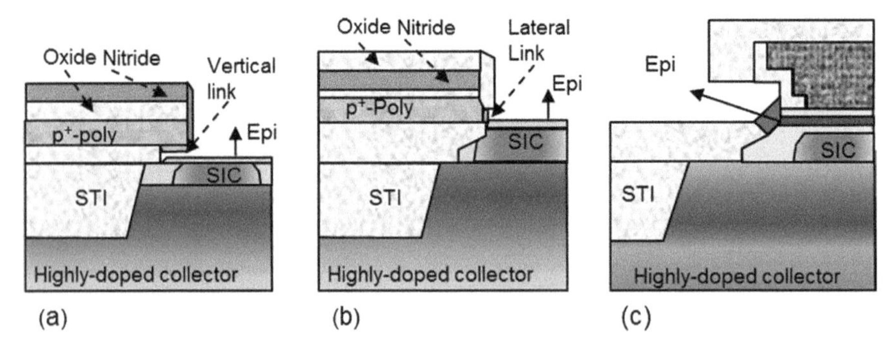

Figure 1.8 Different base link configurations of selective epitaxial growth (SEG) HBTs.

by a higher ohmic region which has to be eliminated by B in-diffusion from the base poly. Additionally, it would be beneficial if dopants could be introduced in the un-doped Si cap layer beneath the emitter–base spacers. Several attempts have been made to solve this problem. For example, boronsilicate-glass EB-spacers and a dedicated anneal at 800°C for 10 min were utilized by NEC [Sat92] to overcome this issue for one of the first DPSA-SEG technology developments delivering f_T/f_{MAX} values of about 50 GHz. STMicroelectronics tested soak annealing for a few seconds at 1,010°C to 1,040°C between base and emitter deposition for an advanced DPSA-SEG version [Can12]. An improvement of the f_T/f_{MAX}/CML gate delay values from 300 GHz/ 370 GHz/2.33 ps to 320 GHz/390 GHz/2.2 ps was shown for a split with additional annealing at 1,010°C in combination with a reduced B dose of the base layer which was applied to compensate for stronger B broadening compared to the reference process.

Other concepts tried to include the region which surrounds the base layer for a lateral connection (Figure 1.8(b)) in addition to the standard configuration with a vertical link to the base poly. Such an approach could mitigate the effect of the decreasing contact area between the intrinsic base layer and the base poly-Si with ongoing lateral scaling. For this purpose, a second poly-Si layer was inserted in the layer stack enclosing the emitter window in [Was03]. An extra selective epi step was implemented in [Fox08] for positioning of the base poly in lateral direction related to the SiGe base. A simple and meanwhile widely used measure to increase the contact area between the base poly and the base layer is, to some extent, the introduction of 45° rotated substrates. In this way, the emitter windows are aligned along the <100> crystal orientation and the formation of unfavorable facets during epitaxial growth of the base is minimized. However, the progress of above mentioned approaches on RF performance was limited apart from the increased process complexity or the disadvantage of a higher thermal budget. In particular, these attempts did not achieve the progress that might be theoretically possible for highly conductive mono-crystalline base-link region. In this respect, the HBT module with EBL (Figure 1.8(c)) represents an unconventional SEG approach which brought substantial progress for decreasing the specific R_{Bx} contribution without dampening the prospects for high f_T values.

A detailed description of the EBL fabrication process can be found in [Fox11]. The major steps are described below. Figure 1.9(a) shows a cross section after emitter formation. The emitter window was etched in a layer

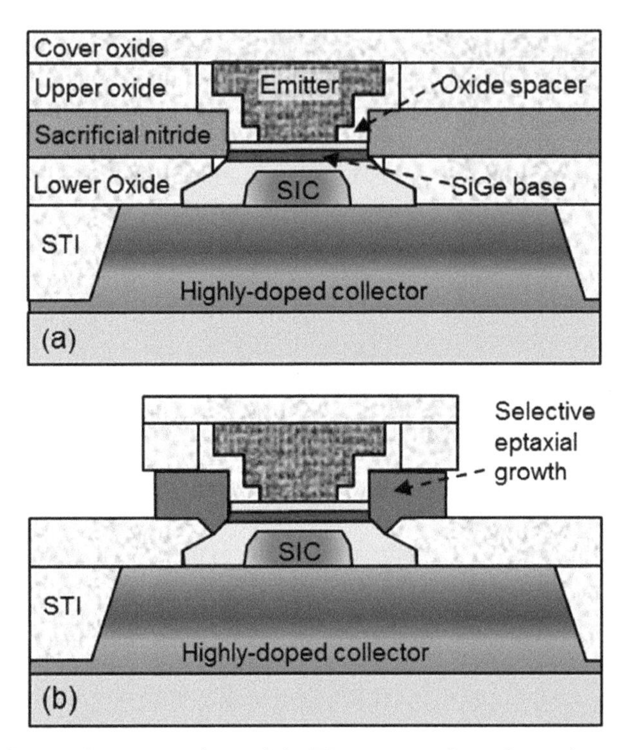

Figure 1.9 Schematic cross sections of the EBL process flow after emitter structuring (a) and after selective growth of the EBL (b).

stack consisting of a lower oxide, a sacrificial nitride, an upper oxide, and a top nitride layer that is already removed at the state of Figure 1.9(a). At this point, there are two specific features which differ from a conventional DPSA-SEG flow. First, a sacrificial nitride layer is deposited instead of the base poly-Si. Second, an overhang of the emitter poly-Si over the emitter window was created in a self-aligned manner by pulling back the upper oxide layer before EB spacer formation. Emitter poly-Si structuring is completed by chemical–mechanical polishing (CMP) similar to [Fox08]. The key idea to form the EBL is illustrated in Figure 1.9(b). After emitter CMP, the devices are covered by an oxide layer, and then the cover oxide and the upper oxide are patterned by a masked dry etching step before the sacrificial nitride is removed by wet etching. The resulting cavities are filled by SEG of B-doped Si as illustrated in Figure 1.9(b) followed by non-selective growth of B-doped Si.

In the first publication on this technology concept [Fox11] f_T/f_{MAX}/CML gate delay values of 300 GHz/480 GHz/1.9 ps were presented. However, a detailed comparison of the EBL HBT performance against standard DPSA-SEG results was not in the focus of this first demonstration. This assessment has been addressed in the DOTSEVEN project. The EBL HBT was compared directly with the conventional DPSA-SEG approach based on identical collector designs, transistor layouts, and measurement conditions. For this purpose, the EBL HBT module was implemented in Infineon's 0.13 μm BiCMOS environment which includes the standard collector concept of an epitaxially buried sub-collector and deep-trench (DT) isolation combined with a mm-wave Cu back-end-of-line (BEOL). In contrast, the original EBL process comprised IHP's DT-free collector approach [Hei02] using STI-isolated, highly doped collector regions as well as an Al-based BEOL.

For this exercise, EBL HBTs and conventional DPSA-SEG HBTs were compared in two ways. First, IHP manufactured its novel device on Infineon wafers in a bipolar-only flow. The joint fabrication started at Infineon by forming the buried sub-collector, the deep trench and STI, and the MOS gates (Table 1.1). Then, the wafer processing was continued at IHP with the EBL module. The CMOS fabrication steps after the HBT module, which could deteriorate the HBT performance, were skipped in these experiments. All process steps for the bipolar devices including the final activation annealing and salicidation were done in these runs at IHP. Compared to [Fox11], the emitter–base spacer process was slightly modified to assist the formation of reduced emitter widths. To eliminate the risk of poly-Si residues, an extra mask was introduced to remove the emitter poly-Si outside of the transistor regions before emitter planarization. Otherwise, we preserved the original EBL flow including the thermal treatment and doping of the SIC, SiGe base, and emitter.

In a second cycle, the full BiCMOS flow was applied. The HBT fabrication was finished at IHP with removing the CMOS protection layer. The further processing corresponded to Infineon's 0.13-μm BiCMOS process including low-doped-drain implantation and annealing, CMOS gate spacer deposition, source/drain implantation and annealing, and salicidation. Table 1.1 shows which process modules were carried out by Infineon and which by IHP for the bipolar-only flow and for the full BiCMOS process, respectively.

Table 1.1 Process modules done by Infineon and IHP for the bipolar-only runs (left) and for the full BiCMOS process (right)

Bipolar-only		BiCMOS	
Infineon	IHP	Infineon	IHP
Buried layer, Epi		Buried layer, Epi	
Isolation		Isolation	
Wells		Wells	
Gate		Gate	
	CMOS protection		CMOS protection
	SiGe HBT		SiGe HBT
		LDD	
		MOS spacers	
		S/D	
	Final anneal	Final anneal	
	Salicide	Salicide	
	IMD	IMD	
Contact		Contact	
BEOL		BEOL	

To assist the BiCMOS integration, several adjustments of the EBL HBT module compared to the bipolar-only runs were made:

- The number of SIC implants for the high-speed HBTs and subsequently the total dose was reduced.
- The base profile thickness was slightly increased to make the profile more robust against the additional thermal budget caused by the CMOS integration.
- The emitter doping was reduced and the Si-cap thickness of the SiGe base deposition was adjusted to compensate for potentially enhanced emitter diffusion due to CMOS integration.
- The effective emitter width was slightly reduced using optimized processes for emitter window lithography and etching to support lateral scaling.
- A new laterally scaled emitter–base spacer process was introduced to reduce the base link resistance.

Figure 1.10 shows the resulting cross sections of the emitter–base complex in the bipolar-only process and the full BiCMOS runs. The effective emitter width amounts to 130 nm for the bipolar-only process and 120 nm in the BiCMOS flow.

In the following, electrical properties of the joint Infineon/IHP HBT fabrications, i.e., bipolar-only and BiCMOS EBL, are evaluated in comparison

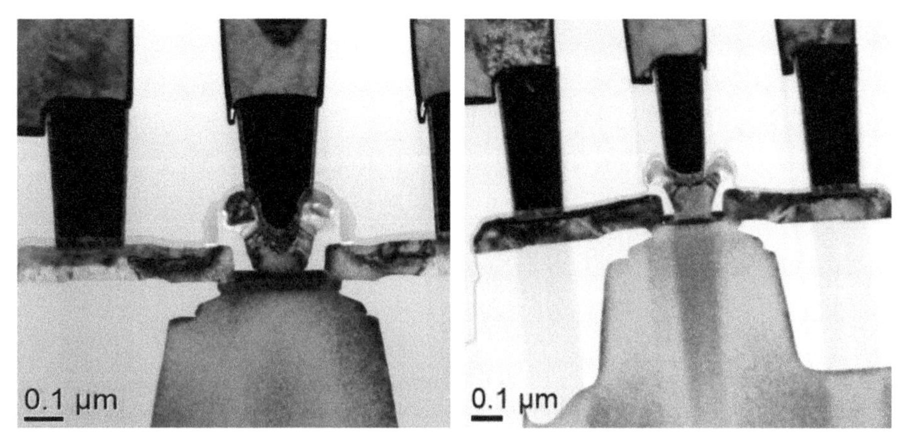

Figure 1.10 TEM cross section of IHP's EBL-HBT module on Infineon's 130 nm platform for the bipolar-only run (left) and full BiCMOS flow (right).

to those of Infineon's DOTFIVE DPSA-SEG results [Che11] and of the IHP reference [Fox11]. Static and dynamic parameters are summarized in Table 1.2.

Concerning the emitter–base (BV_{EB0}) and collector–base (BV_{CB0}) breakdown voltages, it should be noted that the collector–base and base–emitter profiles of the reference EBL HBT [Fox11], and consequently also those of the joint bipolar-only runs, are more aggressively scaled than those of the DPSA reference transistor of Infineon [Che11]. Due to the modifications listed above, the corresponding profiles of the BiCMOS version are relaxed resulting in similar values of BV_{EB0} but also of the current density at peak f_{T} compared to the Infineon reference.

Now, we turn to the evaluation of the high-frequency behavior. For the determination of f_{T} and f_{MAX}, OPEN and SHORT de-embedded *s*-parameters up to 50 GHz were used. f_{T} and f_{MAX} are extrapolated from the small-signal current gain $|h_{21}|$and the unilateral gain U, respectively, with –20 dB decay per frequency decade. Regarding the transistor layout, the focus will be on the double-base contact (BEBC) design because it has been proved superior in terms of high-frequency performance. For the Infineon reference transistor only single-base contact (BEC) data are available [Che11]. Therefore, a BEC configuration of the joint bipolar-only preparation was included in Table 1.2 to facilitate the comparison with the conventional DPSA data. IHP's reference device consists of an 8-emitter BEC configuration with comparatively short emitter lengths optimized for the used unconventional collector construction.

Table 1.2 HBT parameters of EBL HBTs fabricated in joint Infineon/IHP flows in comparison to results of EBL [Fox11] and DPSA [Che11] reference flows

	Unit	Measuring Conditions	Infineon/IHP [Lie16]	Infineon/IHP [Fox15]		IHP [Fox11]	Infineon [Che11]
Layout			BEBC	BEBC	BEC	BEC	BEC
Technology			BiCMOS	Bipolar-only			
No. transistor			3	3		8	3
$w_E \times L_E$	μm^2		0.12 × 2.69	0.13 × 2.69		0.155 × 1.0	0.13 × 2.70
f_T	GHz	V_{CB} = 0.5 V, T = 298 K	240	300	305	320	240
f_{MAX}	GHz		500	500	465	445	380
j_C(peak f_T)	$mA/\mu m^2$		11	17		16	10
Gate delay	ps	ΔV = 200 mV	1.94	1.83	1.86	1.9	2.4
Peak β			650	1,000		450	1,300
BV_{CE0}	V	I_B reversal, V_{BE} = 0.7 V	1.7	1.5		1.75	1.5
BV_{CB0}	V	j_C = 3 $\mu A/\mu m^2$	4.9	4.8		4.1	5.5
BV_{EB0}	V	j_E = 3 $\mu A/\mu m^2$	2.2	1.5		1.35	2.3
$(R_B + R_E) \times L_E$	$\Omega \times \mu m$	y_{11} circle fit @ peak f_T	n.a.	46	51	52	86
$R_C \times L_E$	$\Omega \times \mu m$	b forced to 1	n.a.	55		40	n.a.
R_{TH}	K/W	[Rus09]	n.a.	2,100		1,500	n.a.
C_{CB}/L_E	fF/μm	*s*-parameter	1.38	1.45	1.4	2.2	1.3
C_{BE}/L_E	fF/μm	*s*-parameter	1.9	2.1	1.9	2.4	2.1
C_{CS}/L_E	fF/μm	Array	n.a.	0.9		0.6	0.9
R_{SBi}	kΩ	Tetrode	2.3	3.0		2.6	2.6

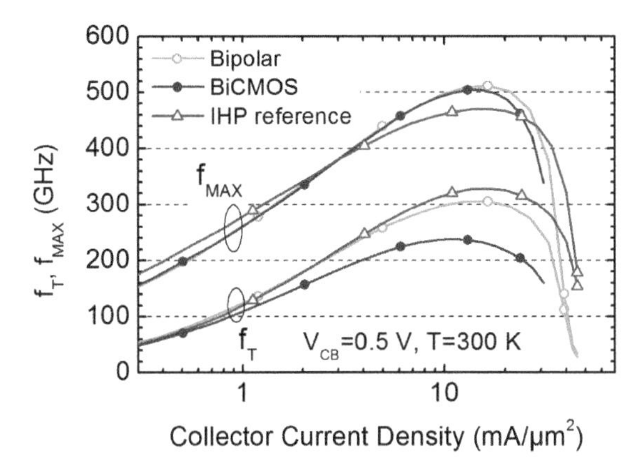

Figure 1.11 Transit frequency f_T and maximum oscillation frequency f_{MAX} vs. the collector current density j_C for Infineon/IHP EBL fabrication in a bipolar-only process and in a BiCMOS run vs. the IHP EBL reference.

Figure 1.11 shows f_T and f_{MAX} as a function of the collector-current density for transistors of the two Infineon/IHP EBL versions and of IHP's reference preparation. Looking at the Infineon/IHP bipolar-only results, the high-frequency parameters (peak f_T/f_{MAX} values of 300 GHz/500 GHz at V_{CB} = 0.5 V for the BEBC device, and 305 GHz/465 GHz for the BEC transistor) represent a substantial progress compared to those of the BEC Infineon device (240 GHz/380 GHz) or to results of other DPSA processes [Che14, Tri16]. In general, these figures fit well to the data of IHP's reference transistor, although the BEBC layout shows even a 55 GHz higher peak f_{MAX} whereas peak f_T is 20 GHz lower compared to the IHP reference (Table 1.2). In the case of f_{MAX}, these deviations are explained by the lower C_{CB} of Infineon's collector isolation scheme while in the case of f_T, the lower R_C and R_{TH} of the IHP transistor design have to be considered. Note that the IHP reference was re-measured at Infineon under company typical measuring conditions. The somewhat lower f_{MAX} (445 GHz [Fox15]; 480 GHz [Fox11]) is primarily attributed to the changed extrapolation frequency (20 GHz [Fox15]; 38 GHz [Fox11]).

One important factor for the enhanced RF performance of the HBTs from the joint bipolar-only process vs. that of the Infineon reference is the increase of f_T by about 25% due to an advanced vertical doping profile. Nevertheless, the main effect of the EBL process on f_{MAX} is the marked reduction of R_B relative to the Infineon reference value of a conventional DPSA-SEG process. Already for the BEC configuration of the EBL HBT, a decrease of the length-specific input resistance (R_B + R_E) by 40% relative to the Infineon reference is observed. Similar relations are true also of the BEBC BiCMOS device. It should be stressed at this point again how important identical RF transistor layouts, measurement tools, and extraction procedures are for reliable evaluations. For example, (R_B + R_E) values given in Table 1.2 are extracted from y_{11} circle fit. This leads to 17% lower values compared to the procedure applied in [Fox 11] based on a circular fit of s_{11}.

Considering the high-frequency behavior of the HBTs, promising results were demonstrated with respect to a reduced base link resistance. However, the f_T of 240 GHz realized for the EBL module within Infineon's 130 nm BiCMOS platform is significantly below the ambitious targets for next SiGe HBT generation. Certainly, the effect of a higher thermal budget of the post-HBT BiCMOS steps at Infineon compared to those of the original IHP flow has to be considered. In addition, more aggressive EB and BC doping profiles have to be applied for further f_T enhancement.

It remains the question whether there are architecture- or flow-related reasons which make it more difficult or impossible to approach the performance values presented in the section "Optimization towards 700 GHz f_{MAX}" for the NSEG HBT also with the EBL concept. However, the finally achieved enhanced high-frequency parameters of the NSEG HBT were paid partially with increased process complexity. The search for an HBT architecture and corresponding process flow which combine best performance and reliable, cost-effective processing is in this context of continuing interest.

1.3.2 Non-selective Epitaxial Growth of the Base

Non-selective epitaxial growth of the SiGe base layer is widely used in SiGe HBT fabrication. Examples are production processes of IBM/ Globalfoundries [Orn03, Pek14] and TowerJazz [Pre11] as well as several technology generations developed by IHP [Kno04, Hei07, Rue10] and by NXP and imec [Don07, Huy11]. These processes have in common a layer stack consisting of a Si buffer layer, a SiGe layer containing the boron-doped base, and a Si cap layer deposited across the whole wafer. This layer stack grows mono-crystalline in active HBT areas where the Si surface is exposed while it grows poly-crystalline in all other areas covered with oxide or nitride. This so-called differential growth mode is in contrast to the SEG where the deposition occurs only in the exposed Si regions.

An implication of the non-selective growth mode is that the poly-crystalline layer which is grown on the isolation layers adjacent to the active HBT can be used to form the extrinsic base regions. Typically this approach is combined with an additional ion implantation into the extrinsic base regions to enhance their conductivity. This approach was applied, e.g., in the 0.25-μm BiCMOS process SG25H1 of IHP which provides peak f_T/f_{MAX} values of 180 GHz/220 GHz [Hei07]. In such an approach, the thickness of the extrinsic base is defined by the layer stack grown to form the intrinsic base. This limits the achievable sheet resistance of the extrinsic base and in particular the conductivity of the extrinsic base region below the poly-emitter overhang. That is why several approaches have been developed to enhance the conductivity of the extrinsic base region by deposition of additional Si layers. It turned out that those elevated extrinsic base regions were necessary for extending f_{MAX} of NSEG HBTs beyond 300 GHz. An NSEG process with elevated extrinsic base regions self-aligned to the emitter window is used, e.g., in IBM's 90 nm SiGe BiCMOS technology [Pek14] exhibiting peak f_T/f_{MAX} values of 300 GHz/360 GHz. In a variant of the process, the RF performance could be further improved to f_T/f_{MAX} of 285 GHz/475 GHz with the help of millisecond annealing [Liu14].

An alternative NSEG HBT process with elevated extrinsic base regions is used in IHP's 130 nm BiCMOS technology [Rue10, Rue12]. This HBT concept was the starting point for the performance optimization toward 500 GHz f_{MAX} performed in the DOTFIVE project. It turned out to be a promising concept for even further performance improvement in the DOTSEVEN project. In the following, we review the main features of the corresponding HBT process flow. The elevated extrinsic base regions are formed by an additional epitaxial step after emitter structuring as first published in [Rue03]. The implementation described below corresponds to the technology SG13G2 of IHP offering HBTs with f_T/f_{MAX}/gate-delay values of 300 GHz/450 GHz/2.0 ps.

A schematic cross section of the HBT is shown in Figure 1.12. Key device features are: (1) Elevated extrinsic base regions self-aligned to the emitter window resulting in low extrinsic base resistance R_{Bx}. (2) Formation of the whole HBT structure in one active area without STI between emitter and collector contacts resulting in low collector resistance and small collector-substrate junction areas. (3) Device isolation without deep trenches resulting in reduced process complexity and improved heat dissipation.

Different stages of the HBT process are illustrated in Figure 1.13. The fabrication of the HBT module begins with the formation of the collector regions by high-dose ion implantation. The collector regions are laterally confined by shallow trench regions without introducing additional deep trenches [Hei02]. Active collector regions are defined by deposition and patterning

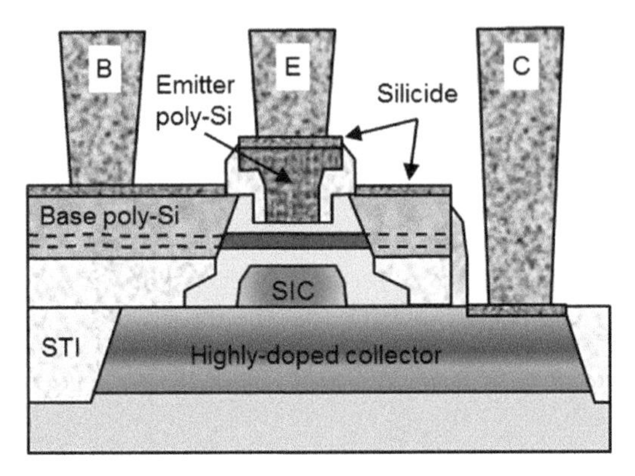

Figure 1.12 Schematic cross section of an NSEG HBT with elevated extrinsic base regions.

an oxide layer. The opened windows in this isolation oxide layer are filled by SEG of un-doped Si on the exposed collector areas. Next, selectively implanted collector (SIC) regions are formed via a patterned resist mask. A cross section of the HBT at this stage of the process is shown in Figure 1.13(a). Now, the non-selective growth of the base is performed. The layer stack consists of a Si buffer layer, the SiGe:C base layer, and a Si cap layer. It grows mono-crystalline in active collector regions and poly-crystalline on top of the isolation oxide as indicated in Figure 1.13(b). After epitaxy, an oxide/nitride/oxide layer stack is deposited and emitter windows

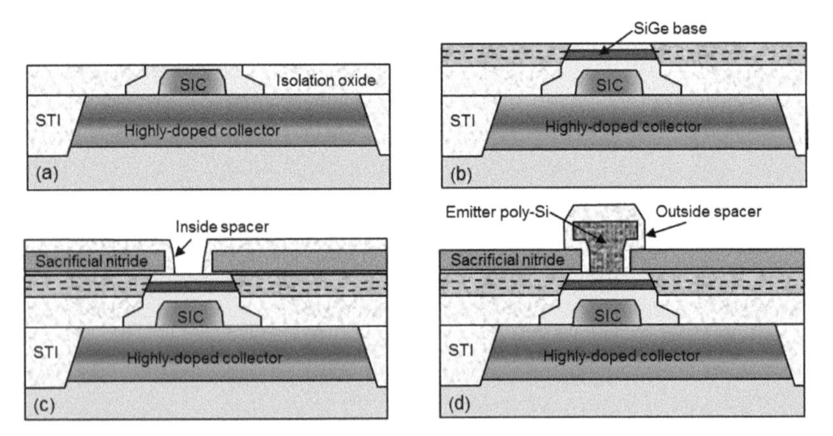

Figure 1.13 Process sequence for the NSEG HBT with elevated base regions: (a) after SIC formation, (b) after non-selective growth of the base, (c) before emitter deposition, (d) after emitter structuring.

are structured. Additional inside spacers are formed before depositing and structuring the As-doped emitter. Figure 1.13(c) shows the device cross section before emitter deposition. The emitter is capped with a dielectric layer and structured via a patterned resist mask. Outside spacers are formed on the emitter resulting in the device structure shown in Figure 1.13(d). Next, the sacrificial nitride layer is removed by wet etching followed by the selective growth of the B-doped elevated extrinsic base regions. The fabrication of the HBT module is continued with the patterning of the base poly-Si layer via a further resist mask. After the described process sequence for HBT structuring, the devices are exposed to a final rapid thermal processing (RTP) step which is used in the BiCMOS flow for the activation of source and drain regions. In the reference technology SG13G2 a spike anneal at 1,050°C is applied for this purpose. Finally, cobalt salicidation is performed on all contact areas and the aluminum metallization is processed. A schematic cross section of the final HBT structure is depicted in Figure 1.12.

The TEM cross section in Figure 1.14 shows an NSEG HBT with elevated extrinsic base regions from the SG13G2 BiCMOS process. The geometrical width of the emitter window of the final device is 120 nm. The elevated extrinsic base regions are separated from the emitter window by about 25 nm wide oxide spacers. This device construction facilitates the realization of very low extrinsic base resistances R_{Bx} due to the self-aligned positioning of the elevated extrinsic base to the emitter window and the high conductivity of the crystalline regions of the extrinsic base near the emitter. However, it has to be noted that the NSEG flow presented here exhibits a significantly lower degree of self-alignment than the DPSA-SEG and EBL process flows discussed in the section "Selective Epitaxial Growth of the Base."

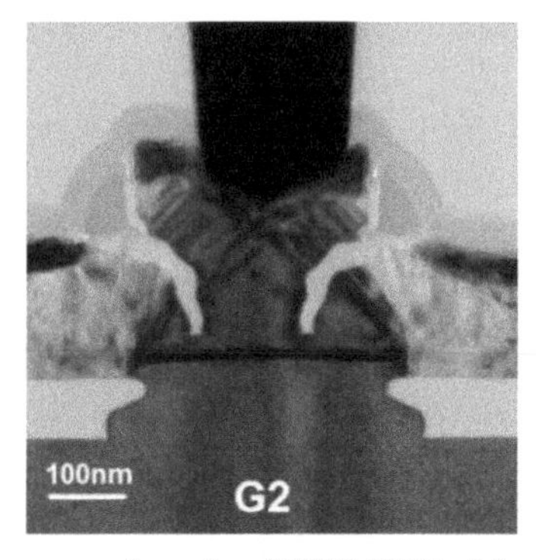

Figure 1.14 TEM cross section of an NSEG HBT of the technology SG13G2.

In particular, the collector window, the SIC implantation, the emitter window, and the emitter poly overhang are not self-aligned to each other. Their relative alignment is defined by the alignment accuracy of the respective lithographic mask steps. This sensitivity to the accuracy of the lithographic alignment can impose severe limitation for further scaling of lateral device dimensions.

Regardless of the above-mentioned limitations of the described NSEG HBT process with respect to self-alignment and scalability, it served as workhorse for optimizing the HBT performance by IHP within the projects DOTFIVE and DOTSEVEN. This arose from the greater experience and familiarity with the NSEG HBT compared to concepts with selective epitaxy discussed above. The fabrication of the NSEG HBT in the SG13G2 BiCMOS process resulted from intensive optimization of this transistor concept within the DOTFIVE project [Hei10]. This raises the question if and by what means further potential performance improvements could be achieved. According to device simulation, there are still respectable reserves for speed increase [Sch11a, Sch11b]. In particular, lateral scaling should help to increase f_{MAX} further. In addition, an appropriate vertical scaling of the doping and Ge profile is required for the desired objective of balanced high f_{T} and f_{MAX} values.

1.4 Optimization of the Vertical Doping Profile

Optimized device architectures as well as lateral and vertical scaling contributed to noticeable progress for f_{MAX} over the last years. In contrast, the potential for improving f_{T} seemed to be limited, in particular, if high f_{MAX} values have to be retained. All successful attempts to push the peak f_{T} of SiGe HBTs beyond 350 GHz delivered rather low f_{MAX} values. For example, the SiGe HBT which demonstrated the highest f_{T} until 2015 showed a peak f_{T} value of 410 GHz together with a peak f_{MAX} value of 190 GHz [Gey08].

Within the DOTSEVEN project, we considered two directions toward HBT performance optimization. First, we focused on aggressive scaling of vertical HBT doping and Ge profiles for increased f_{T}. Second, a device architecture and process flow with a balanced f_{T}-f_{MAX} design at highest performance level was aimed for. In the following, we describe the main results of the experiments for f_{T} optimization.

Various vertical doping and Ge profiles were investigated in a simplified HBT flow described in [Kor15]. In these experiments, a reduced thermal budget of the HBT process was utilized for limiting dopant diffusion. Lateral device dimensions were relaxed with respect to the reference process

SG13G2 in order to reduce process complexity. Measured f_T and f_{MAX} vs. collector current density are plotted in Figure 1.15 for the optimized vertical profile. The peak f_T values could be increased to 430 GHz together with peak f_{MAX} values of 315 GHz [Kor15]. These results were achieved for an HBT with a BEC layout configuration sketched in Figure 1.16(a). This layout corresponds to the standard HBT configuration in the SG13G2 reference process. It was optimized for the special collector construction without STI between collector contact and active emitter area. For better comparability with the results of 2D device simulations presented below, we have also investigated devices with CBEBC layout configuration sketched in Figure 1.16(b).

A further objective of these investigations in the DOTSEVEN project was the assessment of the accuracy of theoretical performance predictions from state-of-the-art device simulations based on the comparison of simulated and

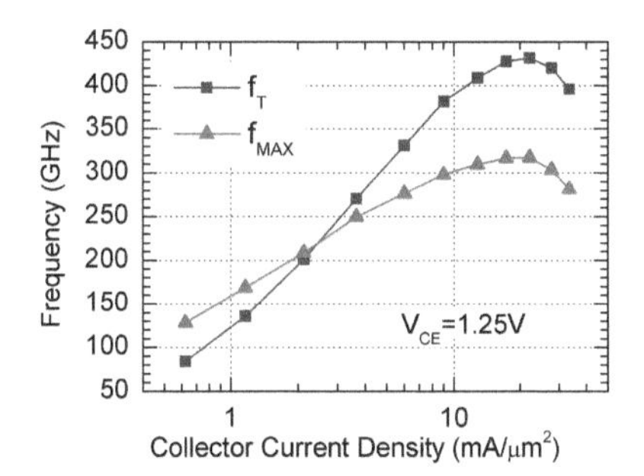

Figure 1.15 f_T and f_{MAX} vs. collector current density for an HBT with BEC layout configuration. Eight HBTs in parallel with individual emitter areas of $0.17 \times 1.01\ \mu m^2$ were measured [Kor15].

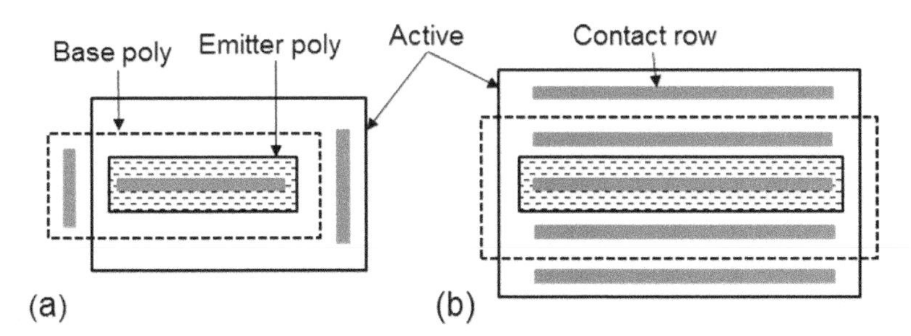

Figure 1.16 Layout configurations: (a) BEC configuration with base and collector contacts at the ends of the emitter line, (b) CBEBC configuration with base and collector contact rows parallel to the emitter line.

measured electrical data. For this purpose, it is essential to determine doping profiles, material compositions, and device geometries of the experimental reference structures as accurately as possible. Below, we summarize the results of the experimental profile characterization.

A combination of various analytic techniques was used to characterize the vertical profiles including secondary ion mass spectroscopy (SIMS), X-ray diffraction (XRD), and energy dispersive X-ray spectroscopy (EDX). The impact of the thermal budget of the fabrication process on the B and Ge profiles of the HBT is shown in Figure 1.17(a). The major part of the observed profile broadening occurred during the final spike rapid thermal processing (RTP) with 1,030°C peak temperature. The B profile experienced only a moderate broadening due to the suppression of B diffusion by Ge and C. However, significant diffusion is observed for the Ge profile itself resulting in a reduction of the peak Ge concentration from 32 at % in the as-grown layer to 28 at % in the final structure.

The accurate determination of doping and Ge profiles in actual HBT structures represents an additional challenge. Width and doping concentrations of epitaxial layers depend in general on the size of the exposed Si area. However, SIMS measurements require dimensions which are much larger than typical active HBT areas. We have performed EDX measurements of the Ge depth profiles in typical HBT structures and in large windows of 600 μm × 400 μm which were also used for SIMS measurements (Figure 1.17(b)). The width of the epitaxial SiGe layer was found to be 14%

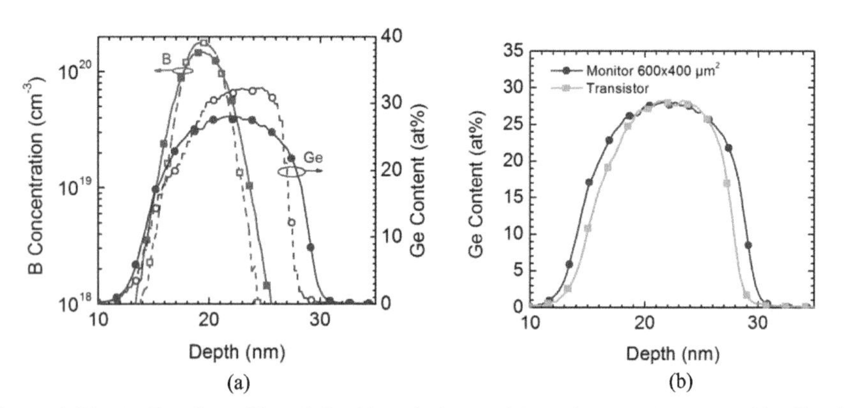

Figure 1.17 (a) Depth profiles of Ge (blue circles) and B (red squares) measured by SIMS. Open symbols are as-grown profiles. Filled symbols are profiles after the full fabrication process. (b) Ge depth profiles measured by EDX in a 600 μm × 400 μm window (blue) and in a typical HBT structure (green).

smaller in the small HBT window while about the same peak Ge concentrations were measured in both structures. The Ge profile of the small window can be obtained from the Ge profile in the SIMS monitor by shrinking the depth scale by 14%. We assume that the same shrink of the depth scale applies to the B profile resulting in a 14% thinner profile in the small HBT window.

The measured emitter, base, and collector profiles of the final HBT structure are plotted in Figure 1.18. The theoretically proposed doping profile corresponding to generation N3 of [Sch17] was included in Figure 1.18 for comparison. This profile N3 was proposed for an HBT generation with peak f_T values of about 500 GHz. The measured and the theoretical N3 profiles show similar widths of the base and of the EB and BC junctions. The steep increase of the theoretically proposed collector profile toward 10^{20} cm^{-3} was not realized in the experiment due to limitations in the formation of low-defective high-dose SIC profiles.

The extracted doping and Ge profiles were used as input for 1D device simulations with the Boltzmann transport equation (BTE) solver [Hon09] and for 2D simulations with the hydrodynamic (HD) transport model [Kor17]. For a quantitative comparison between simulation and experiment, all relevant features of the HBT must be represented adequately by the 2D structure. To accomplish this task, the widths of the EB and BC depletion regions were adjusted to meet the measured area-specific capacitances C_{BEj} and C_{BCj}. The extent of boron diffusion from the external base as well as the lateral extent of the SIC were tailored in such a way that the measured edge capacitance C_{BEe} and C_{BCe} are reproduced [Kor17]. The 2D geometry of the simulated device was adjusted to the TEM cross section of the measured device.

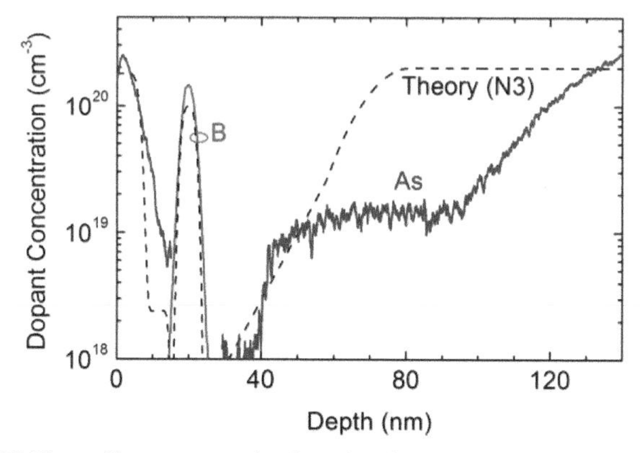

Figure 1.18 SIMS profiles measured after the final annealing step. The theoretically proposed doping profile N3 of [Sch17] is shown for comparison.

Measured and simulated transit frequencies f_T are plotted in Figure 1.19. At low and medium current densities, the simulated f_T values agree well with the measurement. However, at high I_C, the simulation markedly deviates from the measured curve. The degradation of f_T starts at lower I_C and is less abrupt in the simulation, leading to an about 11% smaller peak f_T. Further investigations are needed to clarify this deviation. Additionally, 1D HD and BTE simulations of the inner transistor were performed in [Kor17]. The simulated peak f_T of the 1D transistor is about 80% higher than the corresponding 2D value due to the absence of peripheral capacitances and resistances as well as self-heating. These simulation results indicate that a further enhancement of f_T can be expected for the given vertical profile when contributions of the device edges to the EB and BC capacitances and parasitic resistances are reduced.

1.5 Optimization towards 700 GHz f_{MAX}

In this section, we review attempts in the DOTSEVEN project to optimize the device structure and the fabrication process of the NSEG HBT for highest RF performance. The starting point for this optimization was the SG13G2 technology. The investigated process modifications addressed the reduction of device parasitics by reducing lateral device dimensions as well as by

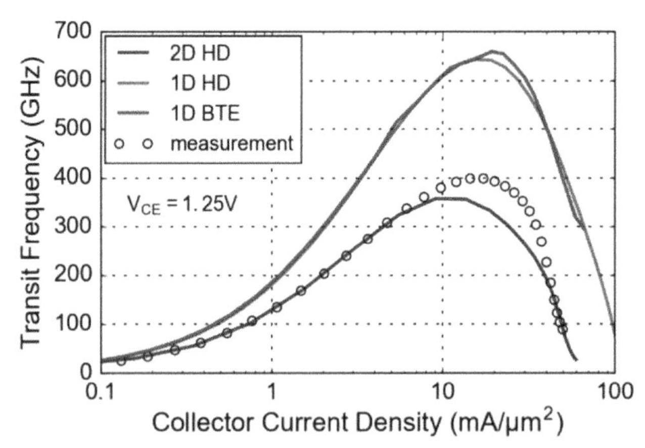

Figure 1.19 Measured and simulated f_T vs. collector current density [Kor17]. Simulations were performed with the hydrodynamic model in 2D and 1D. Results obtained from the 1D Boltzmann transport equation are shown for comparison. The measured device has an emitter area of 0.28 µm × 5.0 µm and the CBEBC layout corresponding to Figure 1.16(b).

improving the control of the doping profile and the conductivity of critical device regions.

The possibilities for lateral scaling of the emitter window by lithographic measures were already largely exhausted in the SG13G2 technology due to the resolution limits of the DUV tool at IHP. A further challenge for down-scaling of the emitter window width is the fabrication of conformal inside spacers in narrow emitter windows. This concerns in particular the deposition of homogenous dielectric layers with good step coverage and the reactive ion etching with minimum damage of the Si surface. Finally, we had to learn from a series of development loops that the room for a well-controlled down-scaling of the emitter width was very limited within the current process flow. Starting from a value of 120 nm in SG13G2, the emitter width was reduced to 100 nm in an intermediate process variant (split CR2). For the final device optimization, an emitter width of 105 nm was realized.

First, we discuss the process stage that was used for circuit fabrication in the DOTSEVEN project. After using the SG13G2 technology for a first circuit fabrication run, an HBT process with improved RF performance was targeted for a second circuit fabrication run (CR2). The following process changes were addressed in this split: smaller emitter–base spacers and smaller emitter window widths were formed by modifying the corresponding deposition and etching processes. An emitter deposition process with enhanced As concentration previously explored in [Hei10] was introduced. The doping concentration of the epitaxially elevated base link regions was enhanced and a new base profile was applied. In addition to these measures which are compatible with the SG13G2 BiCMOS process, we explored for further optimization of the HBT performance process changes which are in conflict with the reference BiCMOS flow. The thickness of the cobalt silicide was increased and the thickness of the silicide blocking spacer at the sidewall of the emitter poly-Si was reduced to minimize the external base resistance. The peak temperature of the final RTP step was reduced in order to minimize diffusion broadening of the base and consequently the base transit time. The HBT cross sections depicted in Figure 1.20 indicate the decreased emitter window width, the reduced widths of the base–emitter and silicide blocking spacers, and the enhanced $CoSi_2$ thickness of the split CR2 with respect to the reference process SG13G2 (G2).

Electrical device parameters for the investigated process variants are summarized in Table 1.3. The process developments introduced in the CR2 split resulted in significant improvements of f_T, f_{MAX}, and the CML ring oscillator gate delay compared to the reference process G2 (Table 1.3).

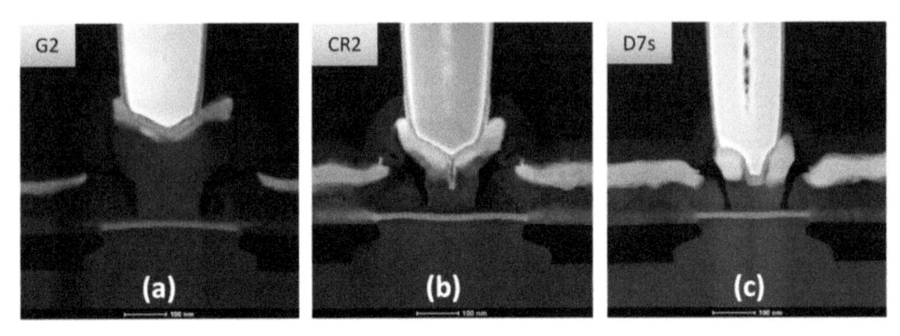

Figure 1.20 TEM cross sections of HBTs from the process splits G2 (a), CR2 (b), and D7s (c).

Table 1.3 HBT parameters of different process splits

	Unit	Measuring Condition	G2	CR2	G2N	G2NF	D7	D7s
No. emitters			8	8	8	8	8	8
w_E	μm		0.12	0.10	0.12	0.12	0.105	0.105
l_E	μm		1.02	1	1.02	1.02	1	1
f_T	GHz		314	351	325	331	498	505
f_{MAX}	GHz	$V_{CB} = 0.25$ V, $T = 300$ K	414	526	461	510	671	720
j_C @ peak f_T	mA/μm^2		20.3	26.3	19	20.6	31.9	34.4
C_{BC}	fF	*s*-parameter	14.7	15.0	14.8	14.5	15.7	15.1
C_{BE}	fF	*s*-parameter	20.6	19.2	21	20.8	20.3	20.4

Reduced emitter and base resistances of CR2 devices were facilitated by enhanced dopant concentration of the revised deposition processes for the emitter and the extrinsic base. For this analysis, we extracted the emitter resistance (R_E) from simple fly-back measurements [Get78]. The impact of the different process splits on base resistance (R_B) is assessed on the basis of $R_B + R_E$ values extracted from circle fits of s_{11}. The modified SiGe base epitaxy is an additional source for the improved RF parameters of the CR2 split. The increase of the Ge content enabled higher collector current densities leading to higher f_T values. A smaller base–emitter junction width and a reduced spike temperature of the final RTP created a more aggressive vertical profile compared to G2. Higher f_T values were obtained but at the cost of a decreased base–emitter breakdown voltage BV_{EB0}. The reduced base sheet resistance R_{SBi} of the split CR2 supported a further reduction of the base resistance and higher f_{MAX} values. Figure 1.21 shows a comparison of f_T and f_{MAX} as a function of collector current density for the splits G2 and CR2.

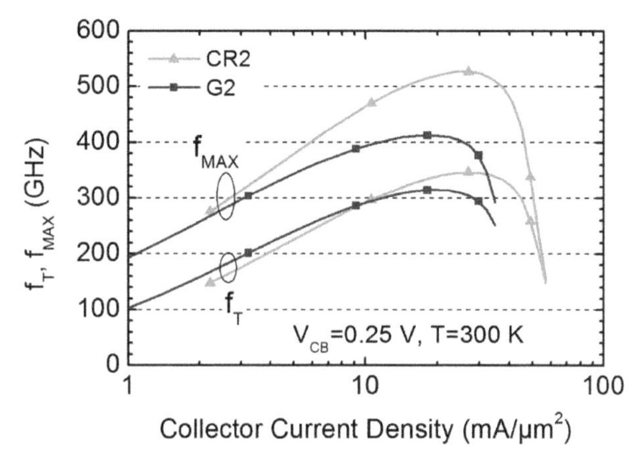

Figure 1.21 Transit frequency f_T and maximum oscillation frequency f_{MAX} vs. collector current density for devices of the split CR2 (second circuit fabrication run in DOTSEVEN) compared to the reference process G2. Device dimensions are given in Table 1.3.

Next, we investigated the potential of enhanced dopant activation by millisecond annealing and low-temperature BEOL processing for further performance improvement. Their benefit for SiGe HBTs has already been pointed out in [Liu14]. Millisecond annealing with laser or flash lamp techniques facilitates a very high level of dopant activation with almost no diffusion. In order to take full advantage of this high activation level, subsequent process steps have to be kept at sufficiently low temperatures to avoid dopant deactivation. Within the DOTSEVEN project, we exploited a non-commercial flash lamp annealing tool at the Helmholtz-Zentrum Dresden-Rossendorf, Germany. Peak temperatures beyond 1,200°C could be realized. A modified contact formation and nickel silicidation were applied to decrease the maximum temperature after flash annealing below 500°C. Cobalt silicide formation does not meet this requirement. In contrast, the nickel silicide fabrication widely used in advanced CMOS nodes does not need annealing above 500°C.

The impact of these processes on HBT performance was investigated in [Hei16] by two process splits of the G2 process. For the split G2N (Table 1.3), the original cobalt silicide process was replaced by a nickel silicide process and the temperature of the contact formation process was reduced. In addition, a flash annealing step was introduced before silicidation in the split G2NF. Figure 1.22 demonstrates the impact of these two process modifications on f_T and f_{MAX}. Markedly enhanced f_{MAX} values by 15%

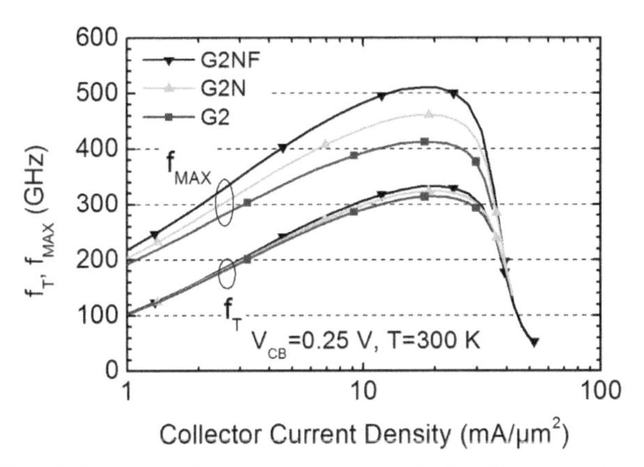

Figure 1.22 Transit frequency f_{T} and maximum oscillation frequency f_{MAX} vs. collector current density for devices of the split G2N (nickel silicide) and G2NF (nickel silicide and flash annealing) compared to the reference process G2 [Hei16]. Device dimensions are given in Table 1.3.

and 23% compared to G2 were obtained for the G2N and G2NF process splits, respectively. These improvements are mainly attributed to reduced base and emitter resistances due to the high level of dopant activation supported by reduced dopant deactivation during silicidation and BEOL processing and by enhanced activation due to flash annealing. Breakdown voltages and junction capacitances are hardly affected by these process modifications (see Table 1.3).

The final optimization stage performed in the DOTSEVEN project [Hein16] is represented by the split D7 in Table 1.3. The scaled device D7s corresponds to the same process flow. There, the width of the collector window is reduced by 17% and the width of the emitter poly-Si is reduced by 33% with respect to device D7 resulting in somewhat lower R_{B} and C_{BC}. A cross section of the final NSEG HBT is presented in Figure 1.20(c). Regarding the emitter-window and emitter–base spacer width, the situation is very similar to the interim case CR2. For this HBT, an emitter width of 105 nm was determined. The geometric dimensions of the revised versions also suggest that limits were reached for further decrease of the width of the EB spacers and of the emitter-poly overlap to the emitter window.

The D7 process split combines the above introduced process modifications of an improved EB spacer process, an extrinsic base with enhanced conductivity, nickel silicidation, and flash annealing with the following additional amendments. A further optimized SiGe base epitaxy is applied which closely resembles the B and Ge profiles of the f_T-optimized device described in the section "Optimization of the Vertical Doping Profile." The thicknesses of the lower doped parts of the emitter–base and base–collector junctions are reduced with respect to split CR2. The emitter utilizes the enhanced doping level but now with reduced layer thickness. Furthermore, the fabrication process for the SIC was revised. In the G2 HBT flow, the SIC was formed with the help of a patterned resist mask. In the D7 split, this process sequence was replaced by a hard mask with inside spacers. By this means, we gained additional flexibility in matching the lateral dimensions of the SIC and the emitter window. The implantation dose of the SIC was doubled in the split D7 with respect to the reference process G2.

Figure 1.23 illustrates the measurement procedure for the extraction of f_T and f_{MAX} for the device D7s. The small-signal current gain h_{21} and the unilateral gain U were derived from s-parameter measurements up to 50 GHz and plotted as a function of frequency. The f_T and f_{MAX} values are obtained from averaged values around an extrapolation frequency of 40 GHz assuming a gain decay of –20 dB per frequency decade. The quality of the measurement procedure was confirmed by independent measurements at IHP and at Infineon as described in [Hei16].

Figure 1.24 shows f_T and f_{MAX} values of the devices D7 and D7s as a function of collector current density. Data for the previous device generations G2 and CR2 are included for comparison. The optimized D7 process reveals

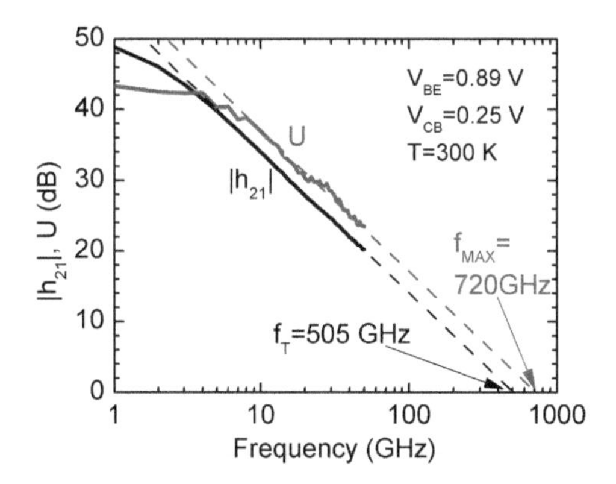

Figure 1.23 De-embedded small-signal current gain h_{21} and unilateral gain U vs. frequency of the device D7s used for extraction of transit frequency f_T and maximum oscillation frequency f_{MAX} with −20 dB decay per frequency decade [Hei16]. The emitter area is $8 \times (0.105 \times 1.0)\ \mu m^2$.

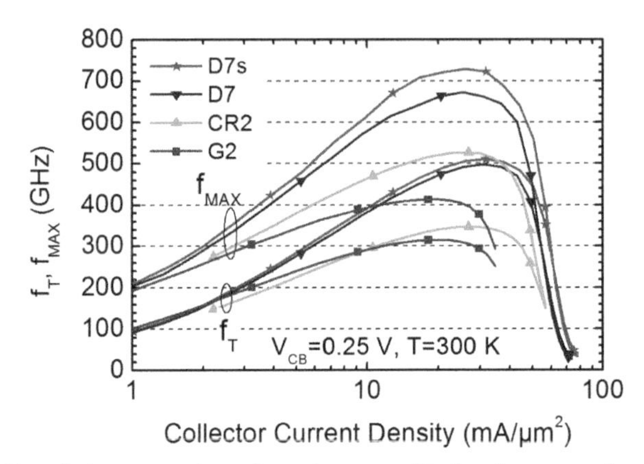

Figure 1.24 Transit frequency f_T and maximum oscillation frequency f_MAX vs. collector current density for two device geometries (D7 and D7s) of the latest process status of DOTSEVEN compared to the reference process G2 and the split CR2 [Hei16]. Device dimensions are given in Table 1.3.

a strong enhancement of both f_T and f_MAX. The obtained high f_T values are supported by reduced transit times for the aggressively scaled vertical doping profile and by strongly reduced R_E values which are accompanied by a strong enhancement of the transconductance g_m in the high current regime. Highest peak $f_\mathrm{T}/f_\mathrm{MAX}$ values of 505 GHz/720 GHz were measured for the scaled device D7s [Hei16]. Both of these values represent the state of the art for SiGe HBTs.

CML ring oscillator gate delays are plotted in Figure 1.25 as a function of current per gate. The oscillators consist of 31 stages and a 1:16 frequency divider. Currents per stage were adjusted to a single-ended voltage swing ΔV of 300 mV at a supply voltage V_EE of –2.5 V. The circuits use conventional resistive loads and do not apply special circuit techniques such as inductive peaking. The data plotted in Figure 1.25 for four of the device splits discussed above demonstrate that the improvement of f_T and f_MAX for the devices D7 and D7s is associated with significantly reduced gate delay times. The minimum gate delay of 1.34 ps for device D7s represents the shortest gate delay that has been reported so far for a SiGe HBT technology [Hei16]. Until now, shorter gate delays have not been reported for any other integrated circuit technology.

The aggressive lateral and vertical scaling for the D7 split resulted in reduced base–emitter and base–collector breakdown voltages as indicated in Table 1.3. In addition, Figure 1.26 illustrates a degradation of the base

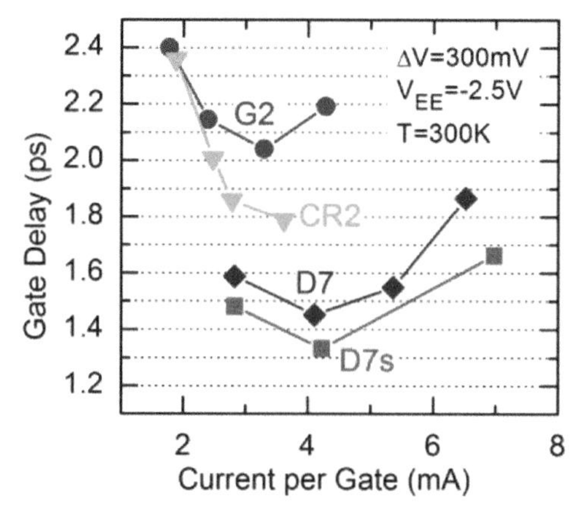

Figure 1.25 CML ring oscillator gate delays vs. current per gate for oscillators consisting of 31 stages with single-emitter HBTs for the splits G2 (A_E = 0.12 μm × 1.02 μm), CR2 (A_E = 0.1 μm × 1.0 μm), D7, and D7s (A_E = 0.105 μm × 1.02 μm) [Hei16].

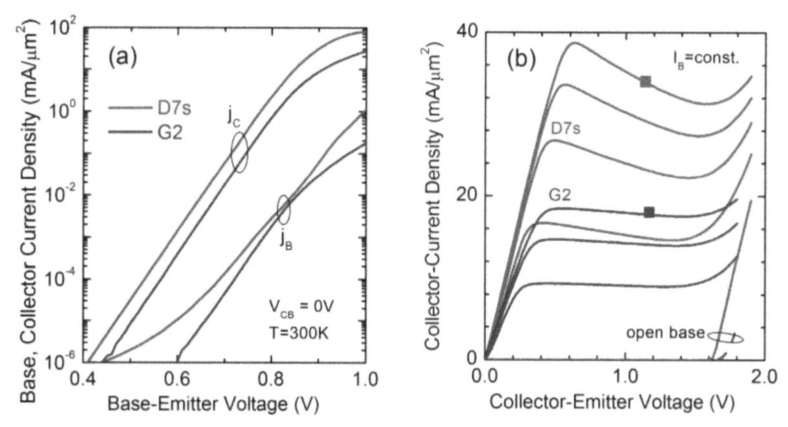

Figure 1.26 Gummel characteristics (a) and base-current forced output characteristics (b) for the splits D7s and G2. Symbols in (b) indicate the bias points for peak f_T. The emitter areas are 8 × (0.12 × 1.02) μm^2 for G2 and 8 × (0.105 × 1.0) μm^2 for D7s [Hei16].

current ideality factor n_{IB} and stronger self-heating for the device D7s. We suppose that a further optimization of the EB doping profile and process advancement for an improved alignment of the SIC to the collector window will weaken some of these drawbacks in future. In fact, excellent DC characteristics have already been demonstrated for devices with nearly 600 GHz f_{MAX} in [Hei16].

1.6 Summary

Due to the intensive work in the projects DOTFIVE and DOTSEVEN, a new high-speed performance level of SiGe HBTs has become a reality. Most valuable is the fact that not only f_{MAX} and the ring oscillator gate delay are improved significantly, but also a new f_{T} record is demonstrated for the same transistor. The balanced increase of f_{MAX} and f_{T} toward 700 GHz and 500 GHz, respectively, makes these devices attractive for an even wider range of RF, mm-wave and sub-mm-wave applications.

These improvements were achieved for the most part by optimization of the vertical profile, the SIC formation, and the reduction of the external resistances. A lower thermal budget in the BEOL processing or its combination with millisecond annealing can produce an extra performance increase due to an enhanced level of dopant activation. Apart from variations of the collector window or the emitter poly-Si overlap, the potential of a comprehensive lateral scaling was not exhausted. Besides adequate lithographic capabilities, this requires advanced conformal deposition and damage-less etching techniques as well as well-controlled epitaxial processes for the SiGe base in small windows.

The modifications applied here for pushing the HBT performance toward 700 GHz f_{MAX} did not consider the compatibility to any frozen CMOS or BiCMOS process. Additionally, we did not shrink back from additional processing steps as long as noticeable performance enhancement seemed possible and process safety was ensured. Consequently, it is concluded from the present results that the feasibility of the DOTSEVEN device targets is demonstrated but the challenging task to implement these capabilities in a next BiCMOS production technology has still to be done. It is an enormous challenge to integrate these HBTs in an advanced CMOS process with its tight constraints on thermal budget and device topology while reaching a similar HBT performance level and fulfilling simultaneously the industrial needs of simplicity, robustness, and high yield.

It was the intention of IHP and Infineon in the DOTSEVEN project to investigate whether the HBT module with EBL could be a promising candidate for this task. As a result, two major directions for next steps could be derived: (a) Transfer of performance enhancing modifications which were tested successfully in the NSEG concept to the SEG EBL flow and (b) revision of the flow to enable further down scaling of the emitter window and to improve process safety under a production-like BiCMOS environment.

The demonstration of a new SiGe HBT performance level should stimulate device engineers and technology developers to create further ideas for cost-effective process flows with best performance potential. In this context, technological challenges related to non-selective or selective base epitaxy and partial or full self-alignment of the HBT layers need to be reinvestigated under the process constraints of advanced CMOS technology nodes.

References

[Ave05] Avenier, G., Schwartzmann, T., Chevalier, P., Vandelle, B., Rubaldo, L., Dutartre, Boissonnet, L., et al. (2005). "A self-aligned vertical HBT for thin SOI SiGeC BiCMOS," in *Proceedings of the Bipolar/BiCMOS Circuits and Technology Meeting, 2005*, (Santa Barbara, CA: IEEE), 128–131.

[Ave09] Avenier, G., Diop, M., Chevalier, P., Troillard, G., Loubet, N., Bouvier, J., et al. (2009). 0.13 μm SiGe BiCMOS technology fully dedicated to mm-wave applications. *IEEE J. Solid State Circuits* 44, 2312–2321.

[Boe04] Böck, J., Schäfer, H., Aufinger, K., Stengl, R., Boguth, S., Schreiter, R., et al. (2004). "SiGe bipolar technology for automotive radar applications," in *BCTM Proceedings* (Montreal, QC: IEEE), 84–87.

[Boe15] Böck, J., Aufinger, K., Boguth, S., Dahl, C., Knapp, H., Liebl, W., et al. (2015). "SiGe HBT and BiCMOS process integration optimization within the DOTSEVEN project," in *BCTM Proceedings* (Boston, MA: IEEE), 121–124.

[Can12] Canderle, E., Chevalier, P., Montagné, A., Moynet, L., Avenier, G., Boulenc, P., et al. (2012). "Extrinsic base resistance optimization in DPSA-SEG SiGe:C HBTs," in *BCTM Proceeding* (Portland, OR: IEEE), 149–152.

[Che11] Chevalier, P., Meister, T. F., Heinemann, B., Van Huylenbroeck, S., Liebl, W., Fox, A., et al. (2011). "Towards THz SiGe HBTs", in *BCTM Proceedings* (Atlanta, GA: IEEE), 57–65.

[Che14] Chevalier, P., Avenier, G., Ribes, G., Montagné, A., Canderle, E., Céli, D., et al. (2014). "A 55 nm triple gate oxide 9 metal layers SiGe BiCMOS technology featuring 320 GHz f_T/370 GHz f_{MAX} HBT and High-Q Millimeter-Wave Passives", in *Proceedings of the IEDM Technical Digest*, (San Francisco, CA: IEEE), 77–79.

[Che15] Chevalier, P., Avenier, G., Canderle, E., Montagné, A., Ribes, G., and Vu, V. T. (2015). "Nanoscale SiGe BiCMOS technologies:

from 55 nm reality to 14 nm opportunities and challenges," in *BCTM Proceedings* (Boston, MA: IEEE), 80–87.

[Don07] Donkers, J. J. T. M., Kramer, M. C. J. C. M., Van Huylenbroeck, S., Cho, L. J., Meunier-Beillard, P., Sibaja-Hemnandez, A., et al. (2007). "A novel fully self-aligned SiGe:C HBT architecture featuring a single-step epitaxial collector-base process", in *Proceedings of the IEDM Technical Digest* (Washington, DC: IEEE), 655–658.

[Fox08] Fox, A., Heinemann, B., Barth, R., Bolze, D., Drews, J., Haak, U., et al. (2008). "SiGe HBT module with 2.5 ps gate delay," in *Proceedings of the IEDM Technical Digest* (San Francisco, CA: IEEE), 731–734.

[Fox11] Fox, A., Heinemann, B., Barth, R., Marschmeyer, S., Wipf, C., Yamamoto, Y. (2011). "SiGe:C HBT architecture with epitaxial external base," in *Proceedings of the BCTM*, (Atlanta, GA: IEEE), 70–73.

[Fox15] Fox, A., Heinemann, B., Rücker, H., Barth, R., Fischer, G. G., Wipf, C., et al. (2015). "Advanced Heterojunction Bipolar Transistor for Half-THz SiGe BiCMOS Technology," *IEEE Electron. Device Lett.* 36, 642–644.

[Gey08] Geynet, B., Chevalier, P., Vandelle, B., Brossard, F., Zerounian, N., Buczko, M., et al. (2008). "SiGe HBTs featuring f_T >400 GHz at room temperature", in *Proceedings of the BCTM*, (Monteray, CA: IEEE), 121–124.

[Get78] Getreu, I. E. (1978). *Modeling the Bipolar Transistor*. New York, NY: Elsevier.

[Has14] Hashimoto, T., Tokunaga, K., Fukumoto, K., Yoshida, Y., Satoh, H., Kubo, M., et al. (2014). SiGe HBT technology based on a 0.13-μm process featuring an f_{MAX} of 325 GHz. *IEEE J. Electron Device Soc.* 2, 50–58.

[Hei02] Heinemann, B., Rücker, H., Barth, R., Bauer, J., Bolze, D., Bugiel, E., et al. (2002). "Novel collector design for high-speed SiGe:C HBTs", in *Proceedings of the Technical Digest IEDM*, (San Francisco, CA: IEEE), 775–778.

[Hei07] Heinemann, B., Barth, R., Knoll, D., Rücker, H., Tillack, B., and Winkler, W. (2007). High-performance BiCMOS technologies without epitaxially-buried subcollectors and deep trenches. *Semicond. Sci. Technol.* 22, 153.

[Hei10] Heinemann, B., Barth, R., Bolze, D., Drews, J., Fischer, G. G., Fox, A., et al. (2010). "SiGe HBT technology with f_T/f_{max} of 300 GHz/500 GHz and 2.0 ps CML gate delay," in *Proceedings of the IEDM Technical Digest* (San Francisco, CA: IEEE), 30.5.1–30.5.4.

[Hei16] Heinemann, B., Rücker, H., Barth, R., Bärwolf, F., Drews, J., Fischer, G. G., et al. (2016). "SiGe HBT with f_T/f_{max} of 505 GHz/720 GHz," in *Proceedings of the IEDM Technical Digest* (San Francisco, CA: IEEE), 3.1.1–3.1.4.

[Hon09] Hong, S.-M., and Jungemann, C. (2009). A fully coupled scheme for a Boltzmann-Poisson equation solver based on spherical harmonics expansion. *J. Comput. Electron.* 8, 225–241.

[Hue96] Hueting, R. J. E., Slotboom, J. W., Pruijmboom, A., de Boer, W. B., Timmering, C. E., and Cowern, N. E. B. On the optimization of SiGe-base bipolar transistors. *IEEE Trans. Electron Device* 43, 1518–1524.

[Huy11] Van Huylenbroeck, S., Sibaja-Hernandez, A., Venegas, R., You1, S., Vleugels, F., Radisic, D., et al. (2011). "Pedestal collector optimization for high speed SiGe:C HBT," in *Proceedings of the BCTM* (Atlanta, GA: IEEE), 66–69.

[Jag02] Jagannathan, B., Khater, M., Pagette, F., Rieh, J.-S., Angell, D., Chen, H., et al. (2002). Self-aligned SiGe NPN transistor with 285 GHz fmax and 207 GHz f_T in a manufacturable technology. *IEEE Electron Device Lett.*, 23, 258–260.

[Joh07] John, J. P., Kirchgessner, J., Morgan, D., Hildreth, J., Dawdy, M., Reuter, R., et al., "Novel collector structure enabling low-cost millimeter-wave SiGe:C BiCMOS technology," in *Proceedings of the RFIC*, (Honolulu, HI: IEEE), 559–562.

[Joh14] John, J. P., Trivedi, V. P., Kirchgessner, J., Morgan, D., To, I., Welch, P. (2014). "An enhanced 180 nm millimeter-wave SiGe BiCMOS technology with f_T/f_{max} of 260/350 GHz for reduced power consumption automotive radar IC's," in *Proceedings of the BCTM*, (San Diego, CA: IEEE), 88–91.

[Kno04] Knoll, D., Heinemann, B., Barth, R., Blum, K., Borngraber, J., Drews, J., Ehwald, K.-E., et al. (2004). "A modular, low-cost SiGe:C BiCMOS process featuring high-f_T and high-BV_{CEO} transistors," in *Proceedings BCTM*, (Montreal, QC: IEEE), 241–244.

[Kor15] Korn, J., Rücker, H., Heinemann, B., Pawlak, A., Wedel, G., and, Schröter, M. (2015). "Experimental and Theoretical Study of f_T for SiGe HBTs with a Scaled Vertical Doping Profile," in *Proceedings BCTM*, (Boston, MA: IEEE), 117–120.

[Kor17] Korn, J., Rücker, H., and Heinemann, B. (2017). "Experimental verification of TCAD simulation for high-performance SiGe HBTs," in *Proceedings of the 16th Topical Meeting on Silicon Monolithic Integrated Circuits in RF Systems (SiRF)* (Phoenix, AZ: IEEE), 94–96.

[Lan96] Lanzerotti, L. D., Sturm, J. C., Stach, E., Hull, R., Buyuklimanli, T., and Magee, C. (1996). "Suppression of Boron Outdiffusion in SiGe HBTs by Carbon Incorporation," in *Proceedings of the IEDM Technical Digest*, (Piscataway, NJ: IEEE), 249–252.

[Lie16] Liebl, W., Boeck, J., Aufinger, K., Manger, D., Hartner, W., Heinemann, B., et al. (2016). SiGe applications in automotive radars. *ECS Trans.* 75, 91–102.

[Liu14] Liu, Q. Z., Adkisson, J., Jain, V., Camillo-Castillo, R., Khater, M., Gray, P., et al. (2014). SiGe HBTs in 90 nm BiCMOS technology demonstrating f_T/f_{MAX} 285 GHz/475 GHz through simultaneous reduction of base resistance and extrinsic collector capacitance. *ECS Trans.* 64, 285–294.

[Mei95] Meister, T. F., Schäfer, H., Franosch, M., Molzer, W., Aufinger, K., Scheler, U., et al. (1995) "SiGe base bipolar technology with 74 GHz f_{max} and 11 ps gate delay," in *Proceedings of the IEDM Technical Digest*, IEEE, 739–742.

[Orn03] Orner, B. A., Liu, Q. Z., Rainey, B., Stricker, A., Geiss, P., Gray, P., et al. (2003). "A 0.13 μm BiCMOS technology featuring a 200/280 GHz (f_T/f_{max}) SiGe HBT," in *Proceedings of the BCTM*, IEEE, 203–206.

[Ost97] Osten, H. J., Lippert, G., Knoll, D., Barth, R., Heinemann, B., Rucker, H., et al. (1997). "The effect of carbon incorporation on SiGe heterobipolar transistor performance and process margin," in *Proceedings of the IEDM Technical Digest*, IEEE, 803–806.

[Pek16] Pekarik, J. J., Adkisson, J., Gray, P., Liu, Q., Camillo-Castillo, R., Khater, M., et al. (2014). "A 90 nm SiGe BiCMOS technology for mm-wave and high-performance analog applications," in *Proceedings BCTM*, IEEE, 92–95.

[Pru95] Pruijmboom, A., Terpstra, D., Timmering, C. E., de Boer, W. B., Theunissen, M. J. J., Slotboom, J. W., et al. (1995). "Selective-epitaxial base technology with 14 ps ECL-gate delay, for low power wide-band communication systems," in *Proceedings of the IEDM Technical Digest,* IEEE, 747–750.

[Pre11] Preisler, E., et al. (2011). "A millimeter-wave capable SiGe BiCMOS process with 270 GHz f_{max} HBTs designed for high volume manufacturing," in *BCTM Proceedings*, 74–77.

[Rue99] Rücker, H., Heinemann, B., Bolze, D., Knoll, D., Krüger, D., Kurps, R., et al. (1999). "Dopant diffusion in C-doped Si and SiGe: physical model and experimental verification," in *Proceedings of the IEDM Technical Digest,* IEEE, 345–348.

[Rue03] Rücker, H., Heinemann, B., Barth, R., Bolze, D., Drews, J., Haak, U., et al. (2003). "SiGe:C BiCMOS Technology with 3.6 ps Gate Delay," in *Proceedings of the IEDM Technical Digest*, IEEE, 121–124.

[Rue04] Rücker, H., Heinemann, B., Barth, R., Bolze, D., Drews, J., Fursenko, O., et al. (2004). "Integration of High-Performance SiGe:C HBTs with Thin-Film SOI CMOS", in *Proceedings of the IEDM Technical*, IEEE, 239–242.

[Rue10] Rücker, H., Heinemann, B., Winkler, W., Barth, R., Borngräber, J., Drews, J., et al. (2010). "A 0.13μm SiGe BiCMOS Technology Featuring f_T/f_{max} of 240/330 GHz and Gate Delays below 3 ps," *IEEE J. Solid State Circuits* 45, 1678–1686.

[Rue12] Rücker, H., Heinemann, B., and Fox, A. (2012). "Half-Terahertz SiGe BiCMOS technology", in *Proceedings of the 11th Topical Meeting on Silicon Monolithic Integrated Circuits in RF Systems (SiRF)*, IEEE, 133–136.

[Rus09] Russo, S., d'Alessandro, V., La Spina, L., Rinaldi, N., and Nanver, L. K. (2009). "Evaluating the self-heating thermal resistance of bipolar transistors by DC measurements: a critical review and update," in *Proceedings BCTM*, IEEE, 95–98.

[Sat92] Sato, F., Hashimoto, T., Tatsumi, T., and Tashiro, T. (1992). "Sub-20 ps ECL circuits with high-performance super self-aligned selectively grown SiGe base (SSSB) bipolar transistors," in *Proceedings of the IEDM Technical Digest*, IEEE, 483–488.

[Sch11a] Schröter, M., Wedel, G., Heinemann, B., Jungemann, C., Krause, J., Chevalier, P., et al. (2011). Physical and electrical performance limits of high-speed SiGeC HBTs – part i: Vertical scaling. *IEEE Transactions on Electron Devices*, 58, 3687–3696.

[Sch11b] Schröter, M., Krause, J., Rinaldi, N., Wedel, G., Heinemann, B., Chevalier, P., et al. (2011). Physical and electrical performance limits of high-speed SiGeC HBTs – part ii: Lateral scaling. *IEEE Trans. Electron Devices* 58, 3697–3706.

[Sch17] Schröter, M., Rosenbaum, T., Chevalier, P., Heinemann, B., Voinigescu, S. P., Preisler, E., et al. (2017). SiGe HBT technology: Future trends and TCAD based roadmap. *Proc. IEEE* 105, 1068–1086.

[Tau98] Taur, Y., and Ning, T. H. (1998). *Fundamentals of Modern VLSI Devices*. Cambridge: Cambridge University Press.

[Thi13] Thibeault, T., Preisler, E., Zheng, J., Dong, L., Chaudhry, S., Jordan, S., et al. (2013). "A study of ultra-high performance SiGe HBT devices on SOI," in *Proceedings of the BCTM*, IEEE, 235–238.

[Tri16] Trivedi, V. P., John, J. P., Young, J., Dao, T., Morgan, D., To, I., et al. (2016). "A 90 nm BiCMOS Technology featuring 400 GHz f_{MAX} SiGe:C HBT," in *BCTM Proceedings*, IEEE, 60–63.

[Was00] Washio, K., Ohue, E., Shimamotot, H., Oda, K., and Hayami, R. (2000). "A 0.2-μm, 180 GHz-fT, 6.71ps-ECL SOI/HRS self-aligned SEG SiGe HBT/CMOS technology for microwave and high-speed digital applications," in *Proceedings of the IEDM Technical Digest*, IEEE, 741–744.

[Was03] Washio, K. (2003). SiGe HBT and BiCMOS technologies for optical transmission and wireless communication systems. *IEEE Trans. Electron Devices* 50, 656–668.

2

Device Simulation

M. Schröter[1,2], G. Wedel[1], N. Rinaldi[3] and C. Jungemann[4]

[1]Chair for Electron Devices and Integrated Circuits, Technische Universität Dresden, Germany
[2]Department of Electrical and Computer Engineering, University of California at San Diego, USA
[3]Department of Electrical Engineering and Information Technology, University Federico II, Italy
[4]RWTH Aachen University, Germany

2.1 Numerical Simulation

G. Wedel and M. Schröter

Numerical simulation (including TCAD) tools were heavily used during DOTSEVEN, aiding the device design and optimization during process development and the physical understanding of the HBT operation so as to support compact modeling. In addition, based on earlier work during DOTFIVE on exploring the physical limitations of SiGe HBTs [Sch11], various numerical simulation tools were used to create the first comprehensive and detailed roadmap in cooperation with the radio-frequency/analog mixed-signal committee of the International Technology Roadmap of Semiconductors (ITRS)[1] consortium in 2014. The flowchart displayed in Figure 2.1 for finding the vertical and lateral HBT structure of a major ITRS technology node is an example for the large variety of simulation tools that were

[1]In the course of the semiconductor industry consolidation only very few companies have remained that have the financial means for developing advanced CMOS technologies. Pursuing partially diverse manufacturing approaches for the associated MOSFET structures, the common basis for technology development, which initially led to the ITRS consortium, is gradually disappearing and rather results in a competition. Hence, the ITRS consortium has been abandoned by the CMOS industry in 2015. The performance tables of the ITRS have been taken over by the IRDS consortium.

developed and employed in DOTFIVE and in DOTSEVEN, since the same approach has been used to define the DOTSEVEN target HBT structure. As already mentioned in-house tools for carrier transport have been based on the DD, HD and BTE approach. In particular, a new deterministic BTE solver was developed and applied to SiGe HBTs, which is described in Section 2.2.2 In addition, simulation tools were developed for the analysis of thermal and parasitic effects in advanced HBTs.

A three-dimensional (3D) thermal solver was used for investigating the temperature distribution within the device structure and its impact on the HF characteristics. A more detailed analysis was performed and insights into the microscopic effects of self-heating were gained with a new Boltzmann solver for phonon transport which was then coupled to the already existing spherical harmonics expansion based BTE solver for charge carrier transport. This approach, which allows deeper insights into device reliability, is described in Section 2.3.

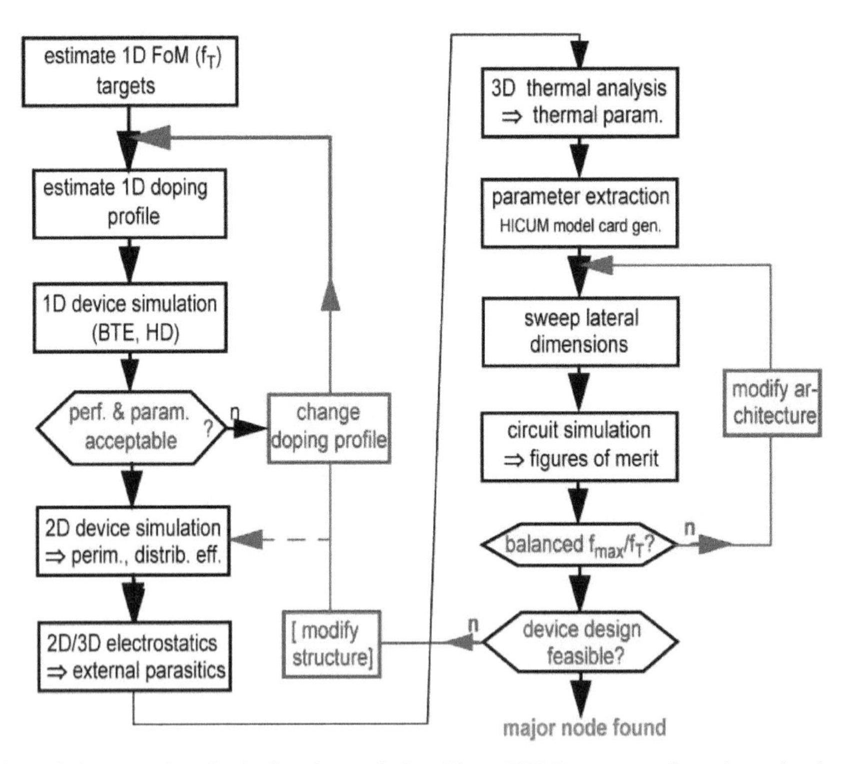

Figure 2.1 Flowchart for finding the vertical and lateral HBT structure of a major technology node [Sch17].

In order to obtain a realistic estimate for the HF device performance, a 3D Poisson solver has been used to calculate the various parasitic capacitances within an actual 3D transistor structure.

2.2 Device Simulation

2.2.1 TCAD Device Optimization

The device design of advanced SiGe HBTs demands an as accurate as possible prediction of the electrical behavior in order to shorten the time from process development of new devices to product and to identify promising device structures during the early stage of process development. Therefore, technology computer aided design (TCAD) software offers a relatively fast and cost effective approach, compared to fabricating and measuring test devices. Of course, the predictive capability of TCAD strongly depends on the accuracy of the employed physical models, such as those for carrier transport, and their parameters.

The most widely used description for carrier transport is the so called drift-diffusion (DD) model, which has been the workhorse in industry for over 40 years. The DD transport model is derived from the Boltzmann transport equation (BTE) by taking the first moments [Jun03][Sel84][Lun00] and consists of the

- Poisson equation,
- hole and electron continuity equation and
- carrier transport equations.

Its major advantage is the low computational cost in terms of both memory requirements and simulation time. However, these advantages are obtained at the cost of physical accuracy, especially for today's SiGe HBTs. For example, the DD transport model significantly underestimates the HF performance of advanced SiGe HBTs (i.e. too low transit frequencies f_T) and overestimates the impact of the Avalanche effect (e.g. too low BVCEO).

For improving the accuracy while maintaining reasonable simulation times, the hydrodynamic (HD) and the energy transport (ET) model were introduced. These models can also be derived from the BTE [Jun03][Lun00] and include a description of energy transport. The ET model neglects the momentum conservation and is thus a subset of HD model. The latter consists of the

- Poisson equation,
- hole and electron continuity equation,

- hole and electron transport equations,
- hole and electron energy balance equations and
- hole and electron energy flux equations.

Compared to DD, the HD transport model considers two additional equations for each carrier type. In addition and due to the two additional equations, new physical quantities, like the carrier energy relaxation time (for the energy balance equation) and the thermal conductivity involved in the energy flux description are introduced. Another point in terms of the HD transport model is the impact of the carrier temperature (solution variable of the energy balance equation as a representation of the kinetic carrier energy) on the carrier transport equation. For the transport equation, the impact of the carrier temperature, especially its gradient acting as additional driving force, needs to be rather heuristically adjusted than set by physics-based considerations in order to obtain reasonable results w.r.t. measurements or BTE simulation data [Wed10]. Once the above mentioned issues have been clarified, a reasonable agreement between HD and BTE can be obtained in terms of terminal characteristics, like transfer currents or transit frequencies. In addition, the HD simulation times are only slightly longer (by a factor of 1.5 up to 3.5) than those of DD simulation. However, BTE results or, if available, measurement data are needed for adjusting the HD transport model. Thus for the design of advanced devices and realistic estimations of their performance, a BTE solver is inevitable.

For solving the BTE, two kinds of methods exist: stochastic solvers based on the Monte-Carlo (MC) method [Jun03][Tom93][Jac83] and deterministic solvers [Hon11][Gal05][Wed16], which solve the BTE directly based on discretized equations just like in the DD and HD case. The MC method is more widely used, since it offers a relatively low implementation effort. In addition, the memory consumption is low compared to deterministic solvers. However, the major drawbacks of the MC method are long simulation times (since MC is inherently a transient method), noisy results (due to its stochastic nature) and a insufficient resolution of minority carrier densities, e.g. electrons in the base region of an HBT. These drawbacks and the advances in computer performance (faster CPUs and cheaper memory) gave rise to the development of deterministic solvers. These solvers enable the calculation of stationary solutions, which are smooth and noise free even for minority carriers. However, these advantages are obtained at the cost of a high memory consumption and especially a much more elaborate mathematical preprocessing and implementation effort. Despite the strong reduction in simulation time compared to the MC method, the computational effort still is significantly higher compared to DD and HD. Thus, the use of even deterministic BTE solvers is practically not feasible for direct device design optimization, but remains restricted to the 1D carrier transport and serves mostly as a reference solution for advanced vertical HBT structures.

2.2.2 Deterministic BTE Solvers

The BTE is a seven-dimensional integro-differential equation defined over the three real space dimensions (*x,y,z*), the three dimensions over the reciprocal space (k_x, k_y, k_z) and time *t*. In conservative form, the BTE for electrons reads [Jun03][Lun00][Honll][Gal05][Wedl6]

$$\frac{\partial f^v}{\partial t} + \nabla_{\vec{r}} \bullet (\vec{\upsilon}_g^v f^v) - \frac{q}{\hbar} \nabla_{k\vec{\upsilon}} \bullet (\vec{E}_{\text{eff}}^v f^v) = C, \tag{2.1}$$

with the particle distribution function (PDF) f^v as the solution variable of the BTE (with $0 < f^v \leq 1$), the carrier group velocity $\vec{\upsilon}_g^v$ and the collision term *C*. The effective field $\vec{E}_{\text{eff}}^v$ is defined as [Hon11][Wed16]

$$\vec{E}_{\text{eff}}^v = \nabla_{\vec{r}} \left(-\psi + \frac{E_{\text{C},0}^v}{q} + \frac{\varepsilon^v}{q} \right) \tag{2.2}$$

with the electrostatic potential ψ obtained by the Poisson equation, the material/composition dependent band edge $E_{\text{C},0}^v$ and the kinetic energy ε^v. The variable *v* denotes the observed valley within the first Brillouin-zone of the reciprocal space. For the sake of readability, the dependencies of the quantities involved have been omitted and are here shortly summarized:

- f^v is a function of $\vec{r}$, $\vec{k}^v$ and time *t*;
- $\vec{\upsilon}_g^v$ is a function of $\vec{r}$ and $\vec{k}^v$;
- ψ is a function of $\vec{r}$ and time *t*;
- $E_{\text{C},0}^v$ is a function of $\vec{r}$;
- ε^v is a function of $\vec{r}$ and $\vec{k}^v$.

The collision term *C* describes the carrier interaction (scattering) due to lattice vibrations (phonon scattering), impurities/alloys and other carriers.

For each considered scattering mechanism, a transition rate S^{X} (X is a placeholder for the considered scattering mechanism) is obtained by Fermi's Golden Rule and the collision term becomes (see e.g. [Jun03][Lun00])

$$C = \sum_{\text{X}} \sum_{\vec{k}^{v'}} S^{\text{X}}(\vec{k}^{v'} \to \vec{k}^v) f^{v'} (1 - f^v) - S^{\text{X}}(\vec{k}^v \to \vec{k}^{v'}) f^v (1 - f^{v'}), \tag{2.3}$$

where the summation over $\vec{k}^{v'}$ discards the spin degeneracy [Lun00]. The first term within the sums describes the in-scattering and the second one the out-scattering of particles, respectively. Equation (2.3) sums first over all possible $\vec{k}^{v'}$ (defined by the transition rate (S^{X})) from where in- or out-scattering might occur and sums these results up for each scattering mechanism. The formulation of the collision term (2.3) contains terms $(1-f^{v,v'})$, which measures the vacancy of the state $\vec{k}^{v/v'}$, respectively. Thus, in- or out-scattering might be blocked due to a fully occupied state, which is called Pauli-exclusion principle [Hon11]. For low doping concentrations, where the Fermi-level is sufficiently far away from the conduction band edge (non-degenerate semiconductor), the PDFs $f^{v,v'}$ are much smaller than one. In this case, (2.3) can be simplified to

$$C \cong \sum_{\mathrm{X}} \sum_{\vec{k}^{v'}} S^{\mathrm{X}}(\vec{k}^{v'} \to \vec{k}^{v}) f^{v'} - S^{\mathrm{X}}(\vec{k}^{v} \to \vec{k}^{v'}) f^{v}. \tag{2.4}$$

However, in both Equations (2.3) and (2.4) it is summed over discrete final states $\vec{k}^{v'}$. Under the assumption that adjacent states are close enough, these states are assumed to be continuous and the sum is converted into an integral by the relation [Gal05][Wed16]

$$\sum_{\vec{k}^{v'}} \cdots \to \frac{\Omega}{(2\pi)^3} \int \cdots d\vec{k}^{v'}, \tag{2.5}$$

with the crystal volume Ω. Thus, with (2.5) the collision terms become [Hon11][Gal05][Wed16]

$$C = \frac{\Omega}{(2\pi)^3} \sum_{\mathrm{X}} \int (S^{\mathrm{X}}(\vec{k}^{v'} \to \vec{k}^{v}) f^{v'} (1 - f^{v}) - S^{\mathrm{X}}(\vec{k}^{v} \to \vec{k}^{v'}) \\ f^{v}(1 - f^{v'})) d\vec{k}^{v'} \tag{2.6}$$

for the degenerate case and

$$C = \frac{\Omega}{(2\pi)^3} \sum_{\mathrm{X}} \int (S^{\mathrm{X}}(\vec{k}^{v'} \to \vec{k}^{v}) f^{v'} - S^{\mathrm{X}}(\vec{k}^{v} \to \vec{k}^{v'}) f^{v}) d\vec{k}^{v'} \tag{2.7}$$

for the non-degenerate case.

Thus, the considered system of equations is set up by the BTE (2.1) with the effective field (2.2) and the collision terms (2.6) or (2.7). For the numerical treatment, it is more advantageous to express the vectors $\vec{k}^{v}$ and $\vec{k}^{v'}$ in terms of the kinetic energy ε^{v} and $\varepsilon^{v'}$, measured from the minimum of the valley v/v', respectively. For carrier scattering, both momentum and energy conservation has to be fulfilled (see e.g. [Lun00]). However, due to the complex shape of the phonon energy as function of the scattering vector $\vec{\beta} = \vec{k}^{v'} - \vec{k}^{v}$ and the demand of $\varepsilon^{v'}(\vec{k}^{v'}) = \varepsilon^{v}(\vec{k}^{v}) \pm \hbar\omega(\vec{\beta}) + (E^{v}_{\mathrm{C},0} - E^{v'}_{\mathrm{C},0})$ (energy conservation), approximations are employed for the phonon energy as function of the scattering vector, which relax the momentum conservation. In practice, modeling the scattering is mainly focused on the energy conservation, since $\hbar\omega(\vec{\beta})$ is approximated to be either zero (elastic scattering) or constant (inelastic scattering). With these simplifications, it is equivalent to consider the kinetic energies. Thus for the BTE, a coordinate transformation has to be performed and the valley dispersion relation $\varepsilon^{v}(\vec{k}^{v})$ needs to be a analytic and invertible function. The most commonly used dispersion relation is the non-parabolic one [Lun00][Tom93][Hon11][Gal05][Wed16]

$$\varepsilon v(1 + \alpha\varepsilon^{v}) = \frac{\hbar^2}{2m_0 m^*} |\vec{k}^{v}|^2, \tag{2.8}$$

with the effective mass m^* and the non-parabolicity factor α. This dispersion relation models equi-energy surfaces in the reciprocal space as spheres (due to $|\vec{k}^{v}|^2$), where α describes the increase of energy for increasing $|\vec{k}^{v}|$. For $\alpha = 0$, a parabolic dispersion relation is obtained, where the kinetic energy increases quadratically with $|\vec{k}^{v}|$. With (2.8), the vector $\vec{k}^{v}$ can be expressed in spherical coordinates [Hon11][Gal05][Wed16]

$$\vec{k}^{v} = \frac{\sqrt{2m_0 m^*}}{\hbar} \sqrt{\varepsilon^{v}(1 + \alpha\varepsilon^{v})} \begin{bmatrix} \mu \\ \sqrt{1-\mu^2}\cos(\varphi) \\ \sqrt{1-\mu^2}\sin(\varphi) \end{bmatrix}, \tag{2.9}$$

with μ as the cosine of the polar angle and the azimuthal angle φ. In (2.9), the spherical coordinate system is rotated to measure the polar angle μ w.r.t. the k^{v}_{x}-axis instead of the k^{v}_{z}-axis. This rotation is advantageous for 1D simulations in x-direction, since it allows to omit φ due to the rotational symmetry of the PDF. Thus in this case, only two dimensions (ε^{v} and μ) have to be discretized instead of the full $[k_x k_y k_z]$ space [Honll][Wedl6].

Nevertheless, the equi-energy surfaces do usually not exhibit a spherical shape within the first Brillouin-zone of the reciprocal space. For example in Si, ellipsoids aligned to the axis $[k_x k_y k_z]$ are found for the so called X-valleys, which mainly contribute to the carrier transport. Thus, the dispersion relation (2.8) becomes in this case

$$\varepsilon^v(1+\alpha\varepsilon^v) = \frac{\hbar}{2m_0}\left(\frac{(k_x^v)^2}{m_x^v} + \frac{(k_y^v)^2}{m_y^v} + \frac{(k_z^v)^2}{m_z^v}\right), \tag{2.10}$$

with the anisotropic effective masses m_x^v, m_y^v and m_z^v. In order to account for the valley anisotropy in (2.9), the Herring-Vogt transformation [Her56] is employed, which basically scales the axis $[k_x^v\ k_y^v\ k_z^v]$ in such a way that the ellipsoids are mapped to spheres. After the mapping, the considered k-space $[\widetilde{k}_x^v\ \widetilde{k}_y^v\ \widetilde{k}_z^v]$ is given by

$$\vec{\widetilde{k}}^v = [T^{\mathrm{HV},v}] \bullet \vec{k}^v, \tag{2.11}$$

$$\vec{\widetilde{k}}^v = \frac{\sqrt{2m_0 m_v^*}}{\hbar}\sqrt{\varepsilon^v(1+\alpha\varepsilon^v)}\begin{bmatrix}\mu \\ \sqrt{1-\mu^2}\cos(\varphi) \\ \sqrt{1-\mu^2}\sin(\varphi)\end{bmatrix}, \tag{2.12}$$

with the Herring-Vogt transformation matrix

$$[T^{\mathrm{HV},v}] = \begin{bmatrix}\sqrt{\frac{m_v^*}{m_x^v}} & 0 & 0 \\ 0 & \sqrt{\frac{m_v^*}{m_y^v}} & 0 \\ 0 & 0 & \sqrt{\frac{m_v^*}{m_z^v}}\end{bmatrix}, \tag{2.13}$$

and the effective mass $m_v^* = \sqrt[3]{m_x^v m_y^v m_z^v}$. Like for the isotropic dispersion relation (2.8) and its k-vectors (2.9), also for the anisotropic description (2.10) and the k-vectors (2.11)–(2.12), the φ-dependence can be omitted for 1D simulations in x-direction, as long as the Herring-Vogt matrix does not contain off-diagonal elements. This assumption holds for the X-valleys in Si/SiGe, but not for the L-valleys in some III–V materials, like GaAs. If one considers the 1D transport in x-direction, the effective field $\vec{E}^v_{\mathrm{eff}}$ consists only of a non-zero x-component. Thus, the PDF is only altered by the effective field in k_x^v-direction. If the equi-energy surfaces are spheres or ellipsoids, aligned to the $[k_x k_y k_z]$ axis, their rotational symmetry allow to omit φ. Otherwise, a change of the PDF in k_x^v-direction forces changes in the $\widetilde{k}_x^v$, $\widetilde{k}_y^v$

and $\widetilde{k}_z^v$ directions and the PDF does not exhibit a rotational symmetry on the equi-energy surfaces anymore. Focusing on Si/SiGe and considering the x-direction only, the BTE (1) simplifies to [Wed16]

$$\frac{\partial f^v}{\partial t} + \frac{d}{dx}(T_{1,1}^{\mathrm{HV},v} \widetilde{v}_g^{v,x} f^v) - \frac{q}{\hbar} \nabla_{\vec{\tilde{k}^v}} \bullet (T^{\mathrm{HV},v} \bullet \vec{E}_{\mathrm{eff}}^v f^v) = C, \tag{2.14}$$

with the carrier group-velocity

$$\widetilde{v}_g^{v,x} = \frac{1}{\hbar} \frac{\partial}{\partial \widetilde{k}_x^v} \varepsilon^v \left(\vec{\tilde{k}^v} \right) \tag{2.15}$$

after the Herring-Vogt transformation and $T_{1,1}^{\mathrm{HV},v}$ being the first main-diagonal element of the matrix (2.13). Assuming a spatial independent transformation matrix $T^{\mathrm{HV},v}$, the BTE (2.14) transforms into the ε^v/μ-space to

$$\frac{\partial f^v}{\partial t} + \frac{d}{dx}(a_x f^v) - \frac{q}{\hbar \mathrm{dos}^v(\varepsilon^v)} \left\{ \frac{\partial}{\partial \varepsilon^v}(a_{\varepsilon^v} f^v) + \frac{\partial}{\partial \mu}(a_\mu f^v) \right\} = C \tag{2.16}$$

with the flux coefficients

$$a_x = \frac{1}{\hbar} \frac{\mu T_{1,1}^{\mathrm{HV},v}}{\frac{\mathrm{d}}{\mathrm{d}\varepsilon^{\mathrm{V}}} |\vec{\tilde{k}^v}|}, \tag{2.17}$$

$$a_{\varepsilon^v} = \frac{\mu \mathrm{dos}^v(\varepsilon^v)}{\frac{\mathrm{d}}{\mathrm{d}\varepsilon^{\mathrm{V}}} |\vec{\tilde{k}^v}|} T_{1,1}^{\mathrm{HV},v}, E_{\mathrm{eff}}^{v,x}, \tag{2.18}$$

$$a_{\mu^v} = \frac{(1-\mu^2)\mathrm{dos}^v(\varepsilon^v)}{|\vec{\tilde{k}^v}|} T_{1,1}^{\mathrm{HV},v}, E_{\mathrm{eff}}^{v,x}, \tag{2.19}$$

the abbreviation

$$\frac{\mathrm{d}}{\mathrm{d}\varepsilon^{\mathrm{V}}} |\tilde{k}^v| = \frac{\sqrt{2 m_0 m_v^*}}{\hbar} \frac{1}{2} \frac{1 + 2\alpha\varepsilon^v}{\sqrt{\varepsilon^v (1 + \alpha\varepsilon^v)}} \tag{2.20}$$

and the density of states

$$\mathrm{dos}^v(\varepsilon^v) = \frac{1}{2} \left(\frac{\sqrt{2 m_0 m_v^*}}{\hbar} \right)^3 \sqrt{\varepsilon^v (1 + \alpha\varepsilon^v)} (1 + 2\alpha\varepsilon^v), \tag{2.21}$$

which can be viewed as the transformed infinitesimal volume element

$$\vec{d\tilde{k}^v} = \mathrm{dos}^v(\varepsilon^v) d\varepsilon^v d\mu d\varphi. \tag{2.22}$$

At this point, the analytic pre-considerations/pre-processing, necessary for the discretization, is almost done. There are various approaches, such as the spherical harmonics expansion (SHE) [Hon11] or BIM-WENO approach [Gal05][Wed16], which rely on the same underlying equations but differ in the numerical representation for f^v (PDF). The choice of the ansatz for f^v strongly affects the further treatment of the collision term. However, regardless of the SHE or BIM-WENO ansatz, the transformed BTE (2.16) is multiplied by the density of states (2.21) and integrated over a discretized control volume in the reciprocal space. After the numerical representation of the derivatives and integrals (collision term) involved, a set of equations (for each discretization point) is obtained, which finally results in a sparse matrix to be solved. The system to be solved also contains the Poisson equation, needed for the electrostatic potential involved in the effective field $\vec{E}^v_{\text{eff}}$ (2.2). The main burden of solving the BTE deterministically arises from the number of needed discretization points, especially for ε^v. The step size for ε^v has to be fine enough to capture all energy exchanges by phonons, since otherwise these scattering processes get smeared out by others and results in less accurate results.

2.2.3 Drift-diffusion and Hydrodynamic Transport Models

As mentioned at the beginning of this chapter, both the DD and the HD transport model can be derived from the BTE by the method of moments with some simplifications. The Poisson and the continuity equations are employed in both DD and HD analysis. The Poisson equation reads

$$\nabla_{\vec{r}} \cdot (\varepsilon_0 \varepsilon_r \nabla_{\vec{r}}(\psi)) = -q(p - n + N_{\text{D}} - N_{\text{A}}), \tag{2.23}$$

with the relative material permittivity ε_r, the elementary charge q, and the donor and acceptor doping concentration N_{D} and N_{A}, respectively. The carrier continuity equation reads

$$\nabla_{\vec{r}} \cdot (\vec{J}_c) = -\text{sgn}(c) q \left(R + \frac{\partial c}{\partial t} \right) \tag{2.24}$$

with the carrier density c, the carrier current density $\vec{J}_c$ and the net recombination rate R.

The first difference between the DD and HD transport models is found for the carrier transport equation [Syn15]

$$\begin{aligned} \vec{J}_c = {} & qc\mu_c \vec{E}_{c,\text{eff}} - q\text{sgn}(c)\mu_c V_{\text{T},c} \nabla_{\vec{r}} c \\ & - \text{sgn}(c)\mu_c c k_{\text{B}} \left[f^{\text{td}} + \log\left(\frac{N_{\text{X}}}{n_{\text{ir}}}\right) \right] \nabla_{\vec{r}} T_c \end{aligned} \tag{2.25}$$

where the first two terms on the r.h.s. of (2.25) represent drift and diffusion transport, and the last term considers the gradient of the carrier temperature T_c acting as additional driving force for the carrier transport process. Here, μ_c is the carrier mobility, $V_{\mathrm{T,c}}$ is the thermal voltage for the carrier temperature T_c, N_x is the effective density of states of electrons and holes (depending on c), n_{ir} is the reference intrinsic carrier density, and f^{td} is a HD transport model parameter. Finally, $\vec{E}_{\mathrm{c,eff}}$ is the effective field given by

$$\vec{E}_{\mathrm{c,eff}} = -\nabla_{\vec{r}}(\Psi - \mathrm{sgn}(c)V_{\mathrm{B,c}}) \tag{2.26}$$

with the carrier band potential $V_{\mathrm{B,c}}$ accounting for the impact of high-doping and material composition effects on the respective band edge. Equations (2.24)–(2.26) contain the function sgn, which is depends the carrier type:

$$\mathrm{sgn}(c) = \left\{ \begin{array}{r} -1, \text{ for electrons } (\mathrm{c} = \mathrm{n}), \\ 1, \text{ for holes } (\mathrm{c} = \mathrm{p}). \end{array} \right\}. \tag{2.27}$$

The DD transport model consists of Equations (2.23)–(2.26), with a carrier temperature equal to the lattice temperature ($T_{\mathrm{c}} = T_{\mathrm{L}}$). Thus, for isothermal DD simulations, the last term on the r.h.s. in (2.25) vanishes. In the case of HD simulations, the carrier temperature T_{c} is obtained by the energy balance equation

$$\frac{\partial}{\partial t} c\omega_c = \vec{J}_c \cdot \vec{E}_{\mathrm{c,eff}} - (\nabla_{\vec{r}} \cdot \vec{S}_c) - R\omega_c + c\left.\frac{\partial \omega_c}{\partial t}\right|_{\mathrm{coll.}}, \tag{2.28}$$

with $\omega_c = \frac{3}{2}k_{\mathrm{B}}T_C$ and the energy flux

$$\vec{S}_c = -\left(\frac{5}{2} + f^{\mathrm{tc}}\right)\left(\frac{k_{\mathrm{B}}}{q}\right)^2 qc\mu_c\nabla_{\vec{r}}T_c + \mathrm{sgn}(c)\left(\frac{5}{2} + f^{\mathrm{ec}}\right)\frac{k_{\mathrm{B}}T_c}{q}\vec{J}_c. \tag{2.29}$$

Here, the first term on the r.h.s. Represents contains the thermal conductivity after Wiedemann-Franz and the gradient of the carrier temperature (energy transport due to spatially different carrier temperatures), where the second term models the energy transport due to the carrier current density. Within (2.25) and (2.29), three parameters f^{td}, f^{tc} and f^{ec} are available in the HD case for adjusting the impact of the respective contributions. These parameters are usually not fixed and vary across technologies and generations in the same material [Wed10]. The parameters are usually obtained by adjusting the HD terminal behavior (such as transfer and output characteristics and transit frequency) to those obtained by BTE simulations or, if available, to measured results. The simplicity of the DD and HD transport models demands an elaborate set of physical material models for the

- carrier mobility (low and high field case, DD, HD),
- recombination and generation (DD, HD),
- band potential (DD, HD), and
- energy relaxation time (HD only).

Usually, the material models are developed and their parameters are adjusted to BTE simulations of bulk material, where a homogenous and infinitely large semiconductor is assumed.

2.2.4 Simulation Examples

The intention of this section is to give an rough impression about the applicability of DD and HD transport models compared to the BTE for SiGe HBTs. As an example, a one-dimensional (1D) SiGe HBT structure with $f_T \approx 630$ GHz [Paw09] is considered. The doping and Germanium profile is shown in Figure 2.2.

2.2.4.1 DD simulation

The DD transfer characteristic and the transit frequency for the considered device are shown in Figure 2.3 and compared with the results obtained by the deterministic BTE solver based on the spherical harmonics expansion (SHE) of the electron distribution function [Hon11].

Compared to the BTE results, a reasonable agreement for the transfer characteristic between DD transport and BTE is obtained up to the onset of collector high current effects. However, DD severely underestimates the

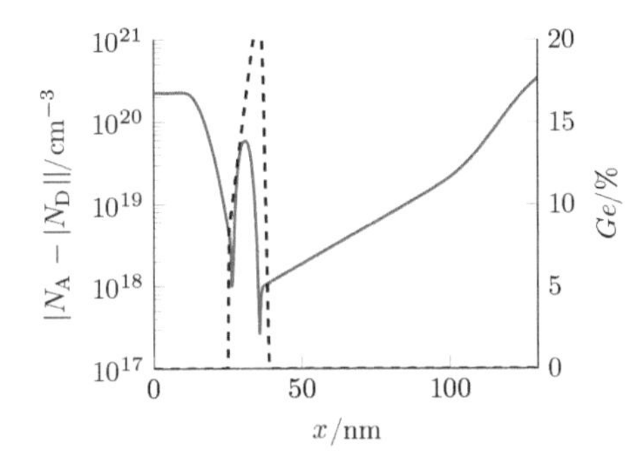

Figure 2.2 Net doping and Germanium profile of a SiGe HBT with f_T = 630 GHz.

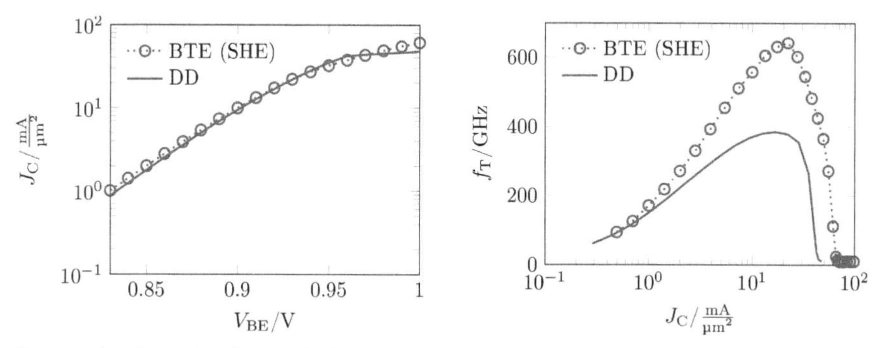

Figure 2.3 Transfer characteristic (left) and transit frequency (right) obtained from DD transport and BTE for the device of Figure 2.2. $V_{CE} = 1$ V.

transit frequency f_T. The main origin of the discrepancies is coming from the assumptions involved in the derivation of the DD transport model. For DD, it is assumed that the distribution function is in equilibrium with the lattice. Thus, any displacement of the distribution function (towards higher kinetic energies) is not directly taken into account, but indirectly by the electron velocity versus electric field model. Usually, a saturation drift velocity limits the carrier velocity although the velocity obtained by BTE might in some regions exceed that saturation limit (velocity overshoot) as shown in Figure 2.4(a) for the peak f_T range. Thus, DD predicts slower electrons and, for the same current, a higher electron density in the base-collector region (cf. Figure 2.4(b)). This results in a higher electron transit time and thus, compared to the BTE, a lower f_T. The constant DD electron density within

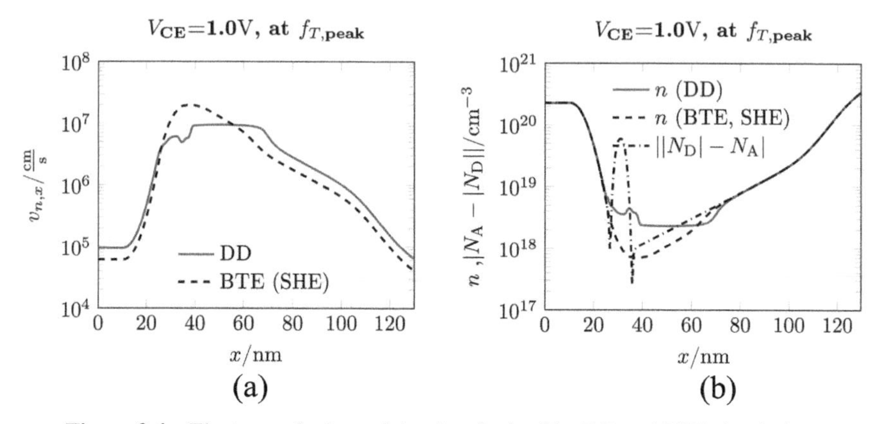

Figure 2.4 Electron velocity and density obtained by DD and BTE simulation.

the BC space charge region (SCR) is the result of the saturation drift velocity used in the DD simulation.

The discrepancies shown in Figures 2.3 and 2.4 are one example of the deficiency of DD transport models. Another one is the underestimation of the breakdown voltage.

2.2.4.2 HD simulation

Here, the most critical parameters are f^{td}, f^{tc} and f^{ec} introduced in Section 2.2.3. These parameters have a significant impact on the terminal characteristics, which can even show a non-physical behavior. Usually, a parameter constellations can be found that gives HD simulation results close to the ones obtained by the BTE. For the SiGe HBT in Figure 2.2, the impact of each of the above parameters is illustrated below. Figure 2.5 shows transfer, transit frequency and output characteristic for different values of f^{td}, while the remaining parameters are kept at zero.

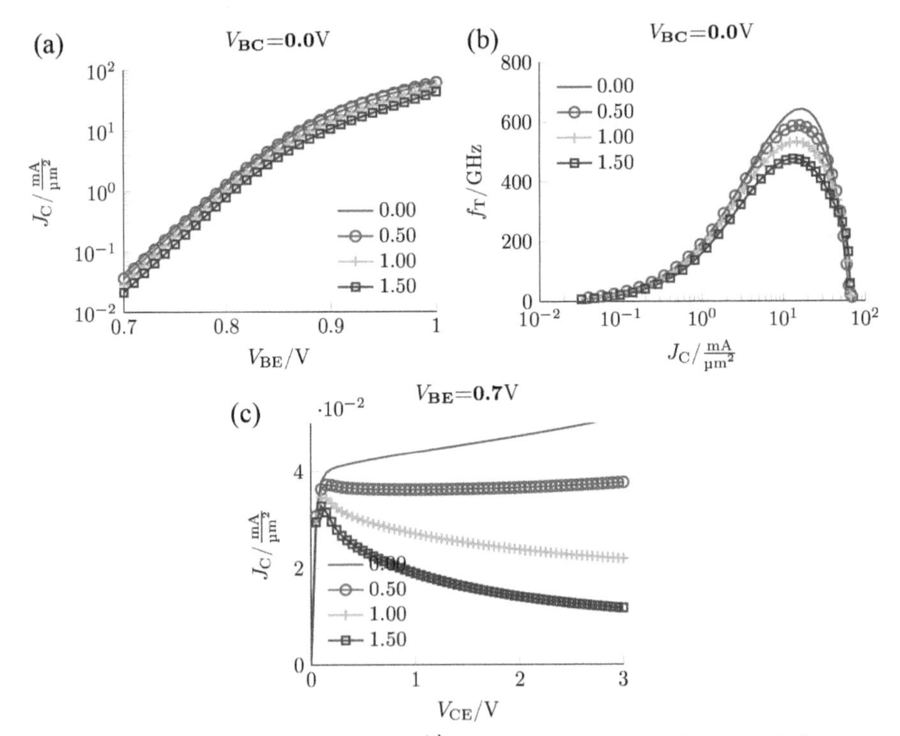

Figure 2.5 Illustration of the impact of f^{td} on the transfer characteristic, transit frequency and output characteristics for $f^{tc} = f^{ec} = 0$.

With increasing values of f^{td} both the collector current density and the transit frequency are decreasing, while the output conductance decreases and can even become negative and thus non-physical in this case (of constant V_{BE}). This is caused by a too strong gradient of the carrier temperature acting as driving force of the electron current. Thus, although f^{td} often requires larger values for adjusting $f_{\mathrm{T,peak}}$ it needs to be limited. Fortunately, with the parameter f^{tc} the impact of f^{td} on the output characteristics can be damped as shown in Figure 2.6. With f^{tc}, the thermal conductivity involved in the energy flux equation (2.29) is altered. The smaller f^{tc} the less energy is transported by the gradient of the carrier temperature. Thus, the carrier temperature profile becomes less smeared within the device, which in turn diminishes the impact of the carrier temperature gradient on the transport equation (2.25). With a increasing f^{tc}, the collector current densities are increased along with the output conductance.

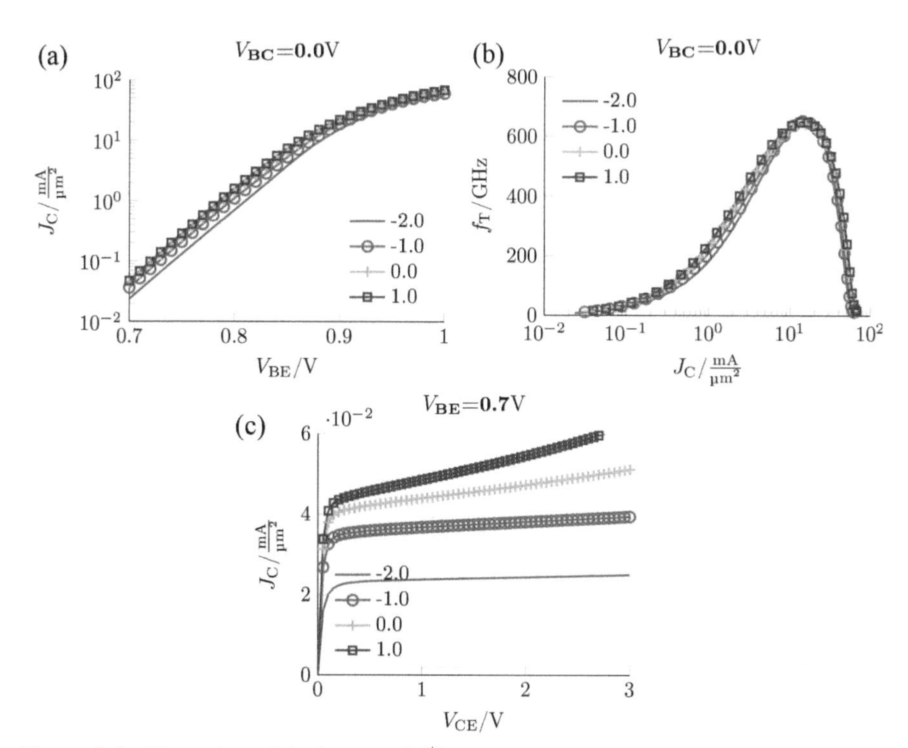

Figure 2.6 Illustration of the impact of f^{tc} on the transfer characteristic, transit frequency and output characteristics at $f^{\mathrm{td}} = f^{\mathrm{ec}} = 0$.

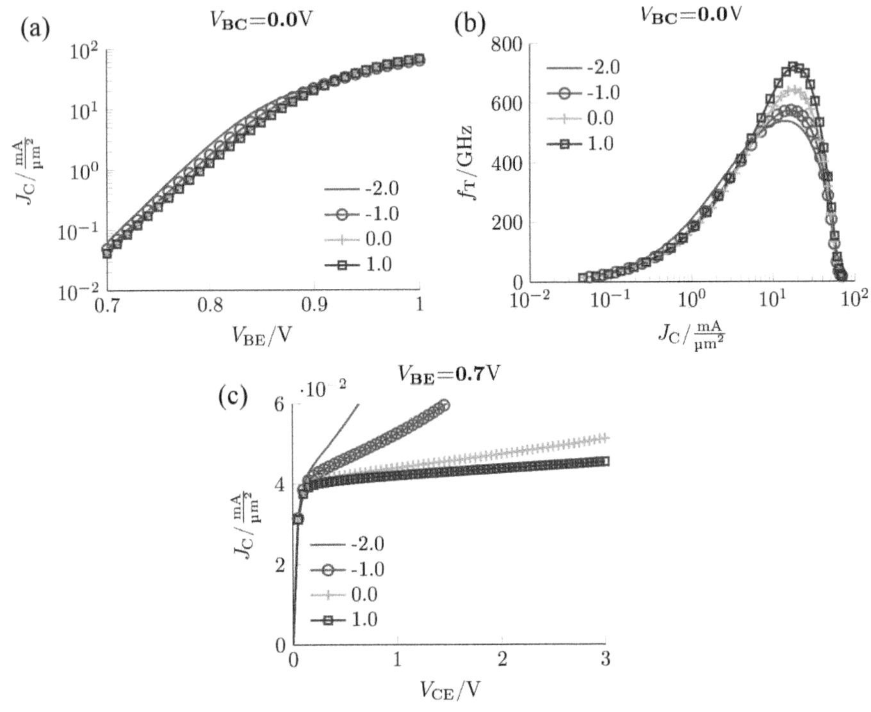

Figure 2.7 Exemplary illustration of the impact of f^{ec} (f^{td}= f^{tc} = 0) on the transfer characteristic, transit frequency and output characteristic.

The last HD transport model parameter f^{ec} scales the energy transport due to the current flow (see (2.29)). It has an opposite effect on the current density compared to f^{tc}, as shown in Figure 2.7. However, it allows to adjust the peak value of transit frequency ($f_{T,peak}$).

The main burden for meaningful HD simulations is the determination of a proper set of HD transport model parameters. The following adjustment strategy and range of values proved to be suitable for the simulation of advanced SiGe HBTs:

- f^{td} is used for adjusting $J_C(V_{BE})$ and $f_T(0.7 \leq f^{td} \leq 2)$;
- f^{tc} prevents a negative output conductance $(-2.25 \leq f^{td} \leq -1.75)$;
- f^{ec} is usually around zero $(-0.5 \leq f^{ec} \leq 0.5)$.

Figure 2.8 compares the HD results obtained by the default HD model parameter set taken from SDevice [Syn15] with those obtained after adjustment to BTE results. The DD results are also shown for comparison.

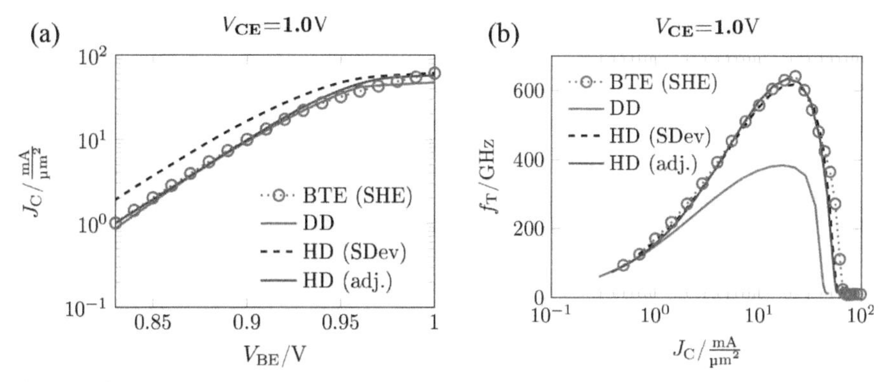

Figure 2.8 Transfer characteristic and transit frequency of the SiGe HBT in Figure 2.2 obtained from HD simulation with adjusted HD transport model parameters and SDevice defaults [Syn15], compared to BTE and DD simulation results.

Compared to DD, the HD transport model with the SDevice defaults gives already a good agreement for the transit frequency. However, the current densities are about twice as high as those obtained by the BTE and DD results. This discrepancy can be eliminated by adjusting the HD parameters to the BTE results. However, this agreement in the terminal characteristics is not reflected in the internal quantities (e.g. electron densities), which exhibit a different spatial dependence compared to the BTE results. In addition, there is no common set of HD parameters across technologies and generations in the same material. These parameters need to be readjusted for each major doping profile change in order to obtain reasonable results. Therefore, the computationally expensive BTE simulations are mandatory. In practice, it suffices though to simulate just the major technology nodes with the BTE and use those results to find the HD parameters for each node. With these parameters, HD simulations can then be applied for device optimization within a particular node [Wed10].

2.2.4.3 Effects beyond DD and HD transport

Due to the assumptions and simplifications involved in the derivation of the DD and HD transport models, some physical effects can not be captured by them compared to BTE solutions.

One assumption is the so-called single electron gas approximation [Blo70]. Here, the transport relevant electrons are assigned to one valley, which dominates the transport. In the case of silicon, the six valleys in Δ-direction (usually denoted by X-valleys) are combined to a single

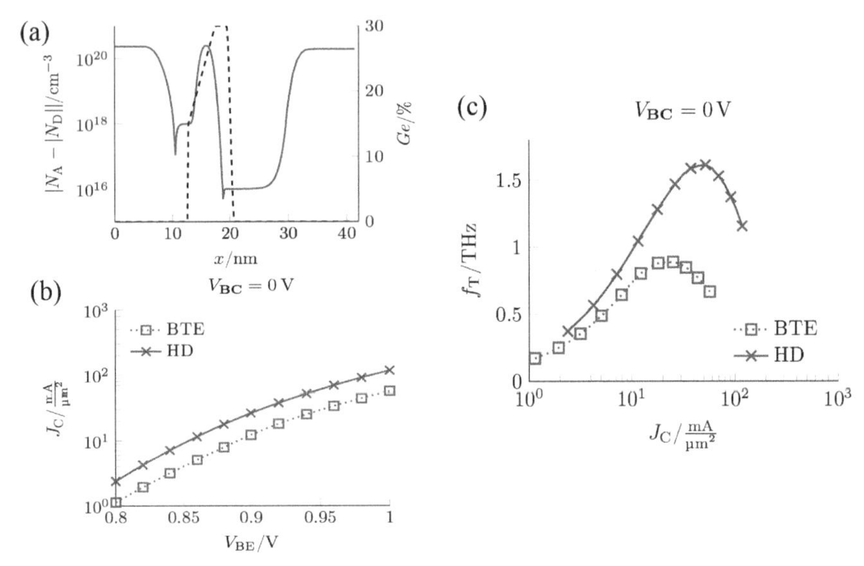

Figure 2.9 (a) Net doping and Ge profile of the SiGe HBT given in [Sch11] and the corresponding (b) transfer characteristic and (c) transit frequency obtained from BTE and HD simulation.

valley, which is energetically located at the conduction band edge. However, for SiGe and for material under biaxial-compressive strain, two of the six Δ-valleys are differently influenced with increasing Ge content [Hon11][Wed16]. In this case, two Δ-valleys are energetically shifted to higher potential energies, while the remaining four Δ-valleys undergo a downward shift in the potential energy. Hence, two conduction band edges are forming, where the lower conduction band edge is associated with the four and the higher with the remaining two valleys, respectively. In the case of DD or HD simulations, only the 4-fold lower conduction band edge is considered, neglecting the higher 2-fold conduction band edge. Figure 2.9 shows the impact of the neglected 2-fold conduction band for the SiGe HBT presented in [Sch11], which represents the presently known physical limit of SiGe HBT technology. According to Figure 2.9, the HD simulations overestimate both the collector current density and the peak transit frequency by about a factor of two. These discrepancies are caused by the abrupt rising edge of the Ge profile and the neglect of the second conduction band edge.

For clarification purposes, the band edges, valley occupancies and quasi-static electron densities at $f_{T,\,\text{peak}}$ as obtained by BTE simulations are shown in Figure 2.10. Due to the abrupt rising edge of the Germanium profile, the

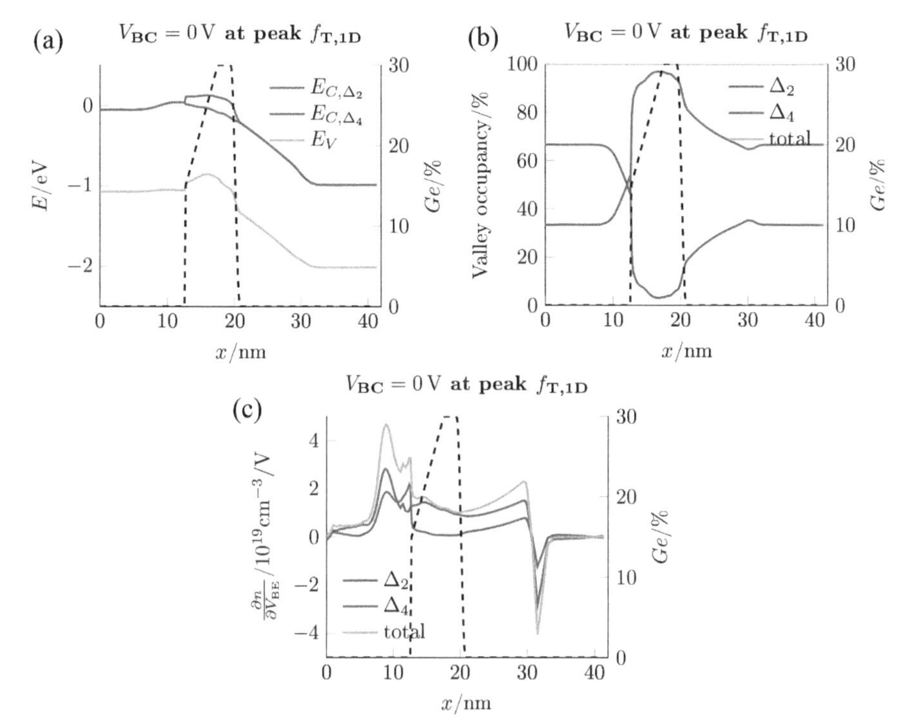

Figure 2.10 (a) Band edges and Ge profile, (b) valley occupancy, and (c) electron density of the SiGe HBT shown in Figure 2.9 obtained by BTE simulation at the operating point of peak transit frequency.

conduction band edge associated with the 2-fold Δ-valleys is energetically lifted up abruptly and thus forms an energy barrier (cf. Figure 2.10(a)). Only few high energetic electrons are able to overcome this barrier, while instead most of the electrons accumulate at the barrier (cf. Figure 2.10(b)). This leads to an additional charge, which reduces both the transit frequency and the transconductance [Hon11]. Obviously, the single electron gas approximation used in conventional DD and HD tools cannot capture this effect.

In fabrication, a step Ge profile is unrealistic and a graded profile rather occurs. The impact of a graded Ge profile is sketched in Figure 2.11. With the graded Ge profile (Figure 2.11(a)), the 2-fold Δ-valley is continuously shifted to higher energies so that the electrons are now able to gradually transfer to the 4-fold Δ-valley, preventing an abrupt charge accumulation (Figure 2.11(c)).

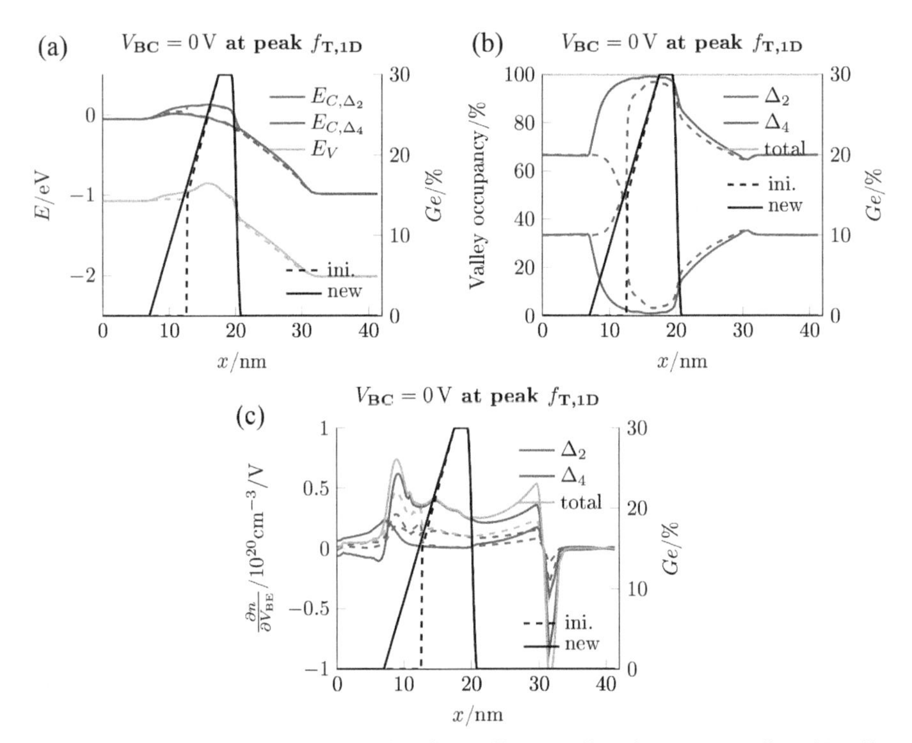

Figure 2.11 (a) Band edges and graded Ge profile as well as the corresponding (b) valley occupancy and (c) electron density obtained by BTE simulation at the operating point of peak transit frequency.

While the grading prevents a degradation of the transconductance, the higher quasi-static change of the electron density causes an increase of the total capacitance $\overline{C}_{\text{tot}}$ connected to the base terminal. According to Figure 2.12(a), $\overline{C}_{\text{tot}}$ increases by a factor of 1.76, whereas the transconductance $\overline{g}_{\text{m}}$ improves by a factor of 2.57 (Figure 2.12(b)). Overall this leads to an increase of f_{T}. According to

$$f_{\text{T}} = \frac{1}{2\pi}\frac{\overline{\overline{g}}_m}{\overline{C}_{\text{tot}}}, \tag{2.30}$$

the higher improvement of the transconductance is resulting in an increase of f_{T} by a factor of 1.46.

Figure 2.13 shows the improvements in the collector current (Figure 2.13(b)) and in f_{T} (Figure 2.13(c)) due to the graded Ge profile.

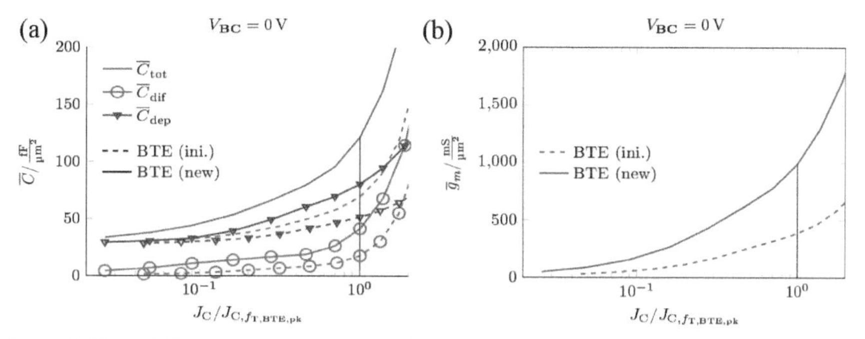

Figure 2.12 (a) Comparison of (a) the total 1D capacitance connected to the base node and its components and (b) the transconductance, obtained for the initial (abrupt) and the new (graded) SiGe HBT profile.

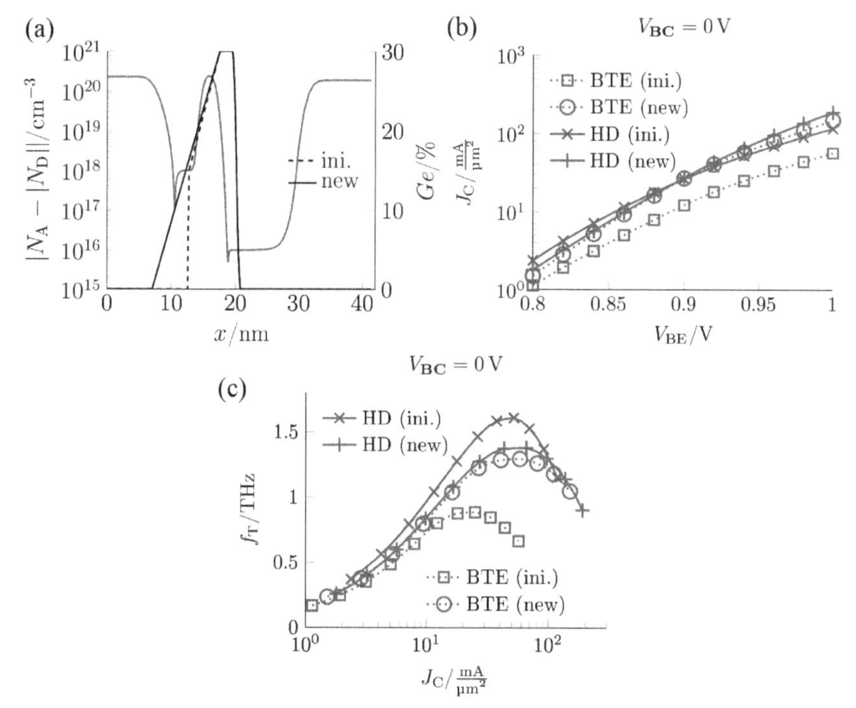

Figure 2.13 (a) Comparison of the initial and the new SiGe HBT profile with corresponding (b) transfer characteristic and (c) transit frequency, obtained by both BTE and HD simulation.

The good agreement of the HD results the with those of the BTE can only be obtained by readjusting the HD transport model parameters, since the Ge grading constitutes a major profile change.

Another limitation of the DD and HD transport model is the assumed shape of the distribution function for deriving them. Conventionally, a Maxwell-Boltzmann distribution is assumed over energy which, in the DD case, is assumed to be in equilibrium with the lattice or, in the HD case, has a modified decay over energy due the spatially dependent carrier temperature. However, in both cases a displacement of the distribution function off its equilibrium position is not considered. Contrary to DD and HD transport, the BTE is solved for the distribution function at each real-space point and over the reciprocal space and thus offers information about both the actual decay over energy and its displacement from the equilibrium position. Depending on the considered doping profile, different shapes of important characteristics can be obtained.

As an example, the SiGe HBT N3 in [Sch17] and shown in Figure 2.14(a) is considered. According to the transfer characteristics and transit frequency in Figure 2.14(b), (c), the deviations between HD and BTE occur despite

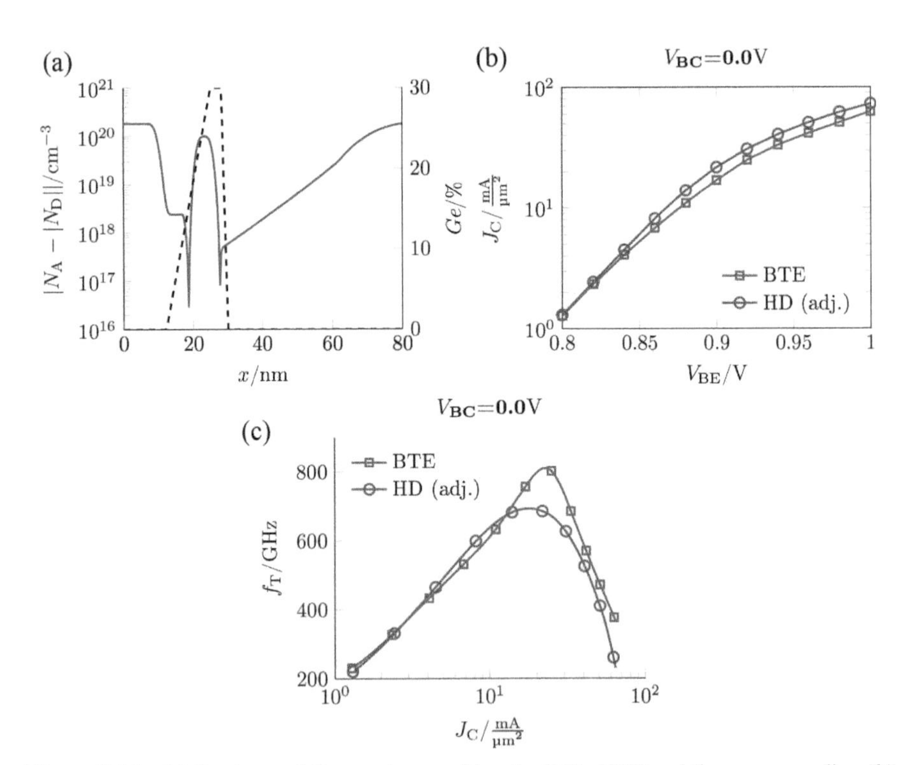

Figure 2.14 (a) Doping and Germanium profile of a SiGe HBT and the corresponding (b) transfer characteristic and (c) transit frequency obtained from BTE and HD simulation.

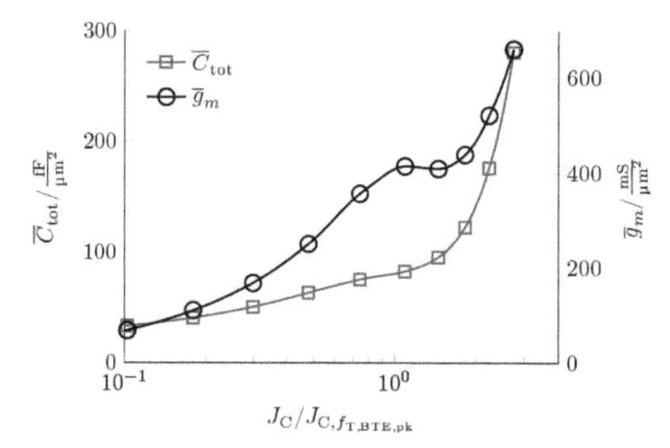

Figure 2.15 Transconductance and total capacitance of the N3 SiGe HBT.

adjusted HD transport model parameters. Especially for the transit frequency, the BTE predicts an spike-like increase to $f_{T,BTE,pk}$ compared to HD transport. As explained below, the deviations between HD and BTE originate from the doping profile in conjunction with the doping dependent description of the bandgap narrowing and the subsequent shift of the conduction band edge.

In Figure 2.15, the bias dependent total capacitance and the transconductance of the N3 HBT are shown. According to (2.30), the spike-like increase in f_T originates from the strong increase and dip of the transconductance.

Since the electron transport within the device is dominated by the peak of the conduction band edge. The conduction band edge of the 4-fold Δ-valley and the contours of the corresponding electron distribution function for three different operating points are shown in Figure 2.16. Since the Pauli exclusion principle is not considered, the distribution function exhibits values larger than one (i.e. > 0 in Figure 2.16) near the conduction band edge in the highly

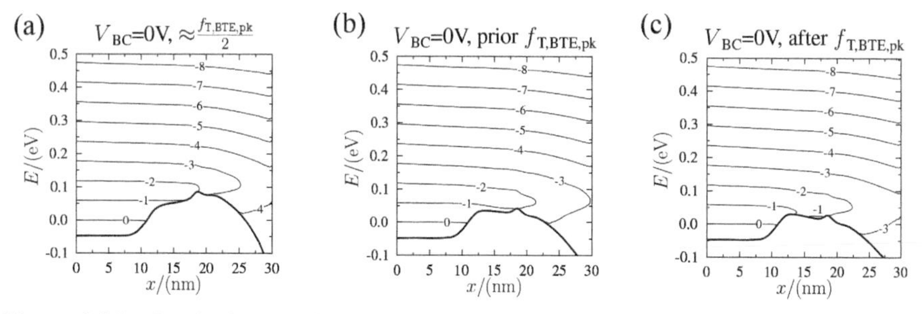

Figure 2.16 Conduction band edge versus location and superimposed contour lines of the (logarithm of the) electron distribution function of the 4-fold Δ-valley over energy within the emitter-base region for three operating points: (a) around $f_{T,BTE,pk}/2$, (b) just before $f_{T,BTE,pk}$, and (c) and just after $f_{T,BTE,pk}$.

doped regions. As discussed before, the 4-fold Δ-valley carries by far most of the electrons and thus it is sufficient to focus on this valley only.

According to Figure 2.16(a) and (b), the conduction band peak for the current range up to around $f_{\mathrm{T,BTE,pk}}$ is located at $x = 19$ nm, which coincides with the steep doping gradient at the BE junction (see Figure 2.14(a)). This decrease in conjunction with the commonly employed (doping dependent) bandgap narrowing model [Slo77] leads to a sudden increase of the bandgap and consequently to a bias independent conduction band barrier, which is for the considered doping profile approximately 2.5 nm wide with a barrier height of around 10 meV. Due to the bending of the conduction band at higher applied V_{BE}-voltages, a plateau like conduction band edge is seen prior to $f_{\mathrm{T,BTE,pk}}$ in Figure 2.16(b). At an operating point beyond $f_{\mathrm{T,BTE,pk}}$ (see Figure 2.16(c)), a potential well is forming between $x = 13$ nm, which corresponds to the transition from the low to the highly doped emitter region and the associated bandgap difference, and $x = 19$ nm.

At current densities below $f_{\mathrm{T,BTE,pk}}$, only high energetic electrons can overcome the conduction band peak at $x = 19$ nm, which corresponds to the classical function of the V_{BE} modulated conduction band barrier that blocks, in the absence of tunneling, low energetic electrons. With increasing V_{BE} this barrier decreases and is overcome by a larger fraction of electrons. The resulting higher average electron velocity increases the current and transconductance and thus f_{T}. Around $f_{\mathrm{T,BTE,pk}}$, the bias independent conduction band peak $x = 13$ nm starts to get exposed and leads to a wider barrier region with a potential well enclosed. Beyond $f_{\mathrm{T,BTE,pk}}$, this wider barrier is more efficient in blocking the low energetic carriers. In addition, carriers also accumulate in the potential well, thus causing a rapid decrease of f_{T}.

In the case of the HD transport model, the Boltzmann distribution function is altered in its spread by the carrier temperature. But the behavior of low and high energetic electrons can still not be separated and thus the blocking effect of low energetic electrons around $f_{\mathrm{T,BTE,pk}}$ can not be captured. Figure 2.17(b) compares the assumed HD distribution function with the one obtained by the BTE solver. Compared to the BTE result, HD overestimates the low energetic and underestimates the high energetic electron population. Therefore, the hill observed in $\overline{g}_{\mathrm{m}}$ in Figure 2.15 originating from the increased average electron velocity at the conduction band peak is not observed in the DD or HD results, which are compared in Figure 2.18(a). Thus, the spike-like increase in $f_{\mathrm{T,BTE}}$ (see Figure 2.14(c)) can neither be reproduced by DD nor HD.

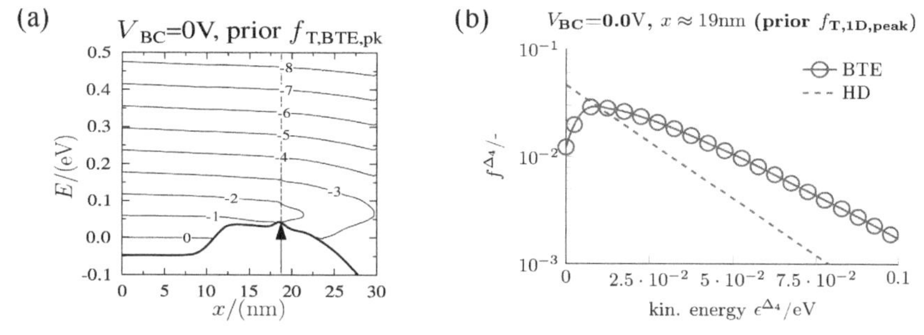

Figure 2.17 Electron distribution function within the 4-fold Δ-valley just below $f_{T,BTE,pk}$ within the emitter-base region. (a) Contours with the arrow marking the position of the doping induced conduction band barrier. (b) Comparison of HD and BTE distribution function at the barrier.

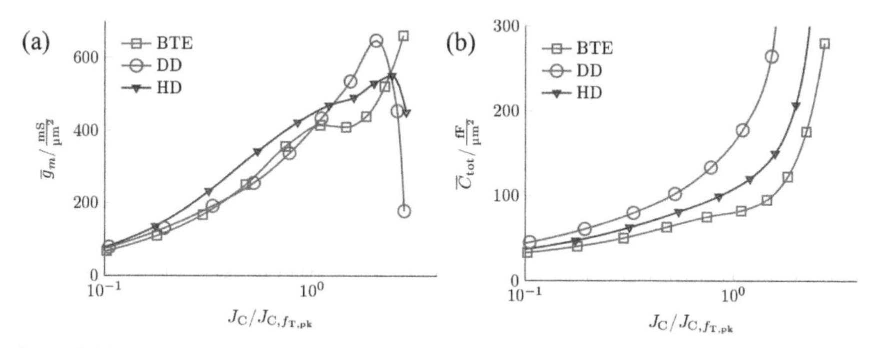

Figure 2.18 Comparison of (a) the transconductances and (b) the total capacitance obtained by DD, HD and BTE simulation results.

In addition and due to the missing energy separation, all DD or HD electrons participate to the charging of the capacitance, contrary to the BTE where the blocked low energetic carriers do not contribute. Thus, the DD and HD transport models are overestimating the total capacitance, as shown in Figure 2.18(b). A more detailed insight is given by Figure 2.19. Figure 2.19(a) compares the electron densities for the three considered operating points obtained by the BTE (lhs) and the HD (rhs) solver. Contrary to the HD results, only a slight variation of the electron densities within the lightly doped emitter is seen around $f_{T,BTE,pk}$ due to the blocking of low energetic carriers. Since these blocked electrons do not participate to the charging of the dynamic capacitances, the quasi-static electron densities (dn/dV_{BE}) in the lightly doped emitter region are lower compared to those obtained by HD (see Figure 2.19(b)).

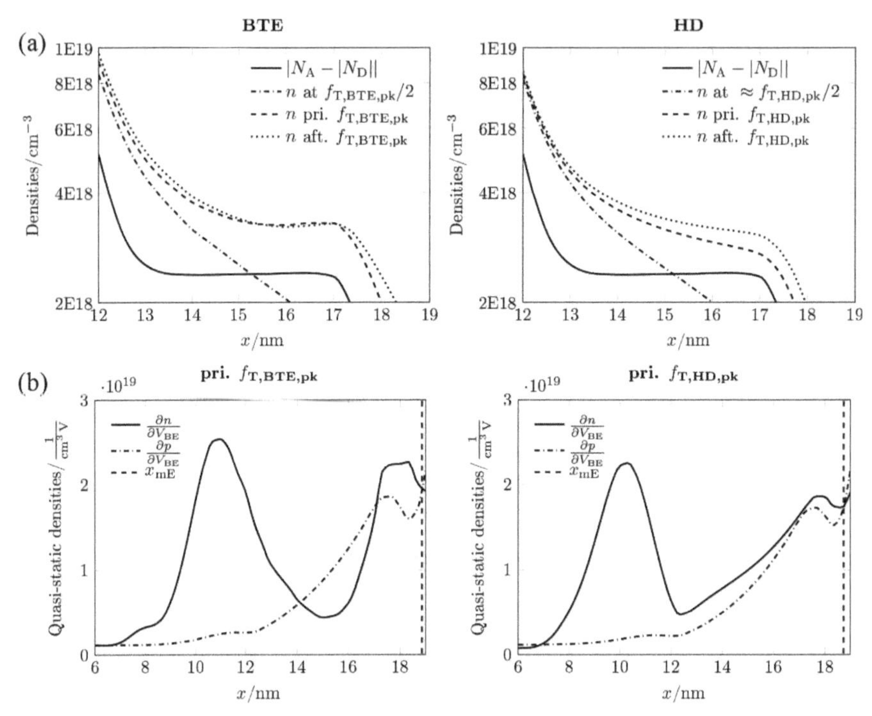

Figure 2.19 Comparison of the BTE and HD electron densities obtained by (a) an DC and (b) an quasi-static analysis. In (b), also the quasi-static hole densities are shown. For (b), a different axis intercept is used compared to (a) in order to visualize the contributions to the emitter junction capacitance $\overline{C}_{jEi}$.

In terms of the emitter depletion capacitance [Sch06]

$$\overline{C}_{jEi} = \int_0^{x_{mE}} \frac{\partial}{\partial V_{BE}}(n-p)\bigg|_{v_{BC}} dx, \tag{2.31}$$

where $x_{mE} = x(dn/dV_{BE}{=}dp/dV_{BE})$, the region of blocked electrons reduces the contributions to $\overline{C}_{jEi}$ from the regions before, due to the higher quasistatic hole density (see Figure 2.19(b), lhs). Therefore, around $f_{T,pk}$ the value of $\overline{C}_{jEi}$ is overestimated in the HD results compared to the BTE results, as shown in Figure 2.20.

The spike-like increase discussed above has not been measured on fabricated devices, possibly because this effect might either be weakened due to tunneling or be masked by the impact of peripheral and external elements. In addition, the fabricated doping concentrations so far just do not have such a

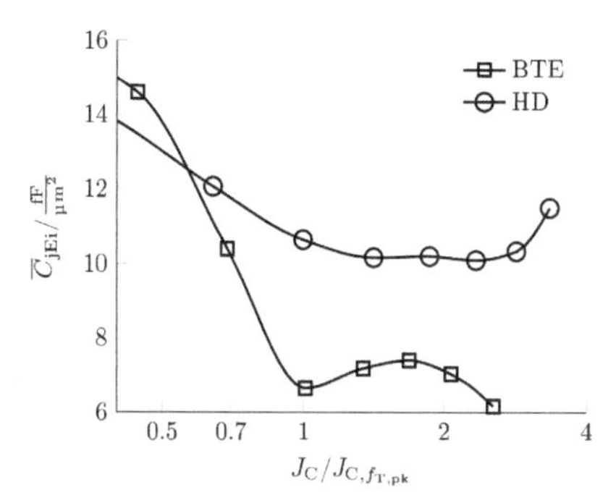

Figure 2.20 Comparison of $\overline{C}_{\mathrm{jEi}}$ obtained by BTE and HD simulations.

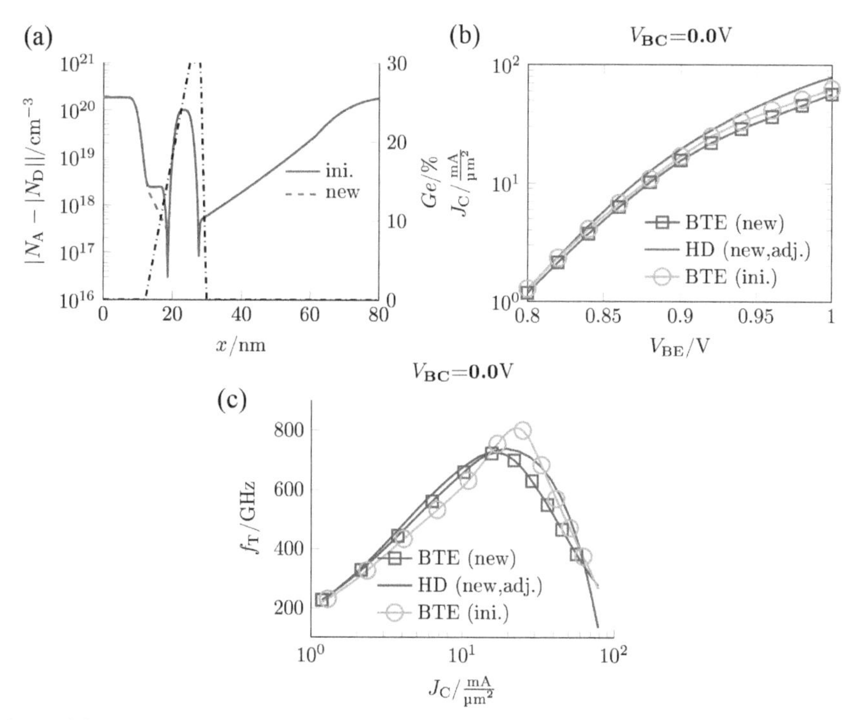

Figure 2.21 (a) Doping profile with smoothed high to low transition in the emitter (new) and previous step-like profile (ini.). Corresponding terminal characteristics: (a) transfer current and (b) transit frequency.

steep gradient. The latter hypothesis has been tested in Figure 2.21 showing a smoothed transition in the emitter doping (Figure 2.21(a)). This results in the disappearance of the previously observed peak f_T overshoot for the BTE

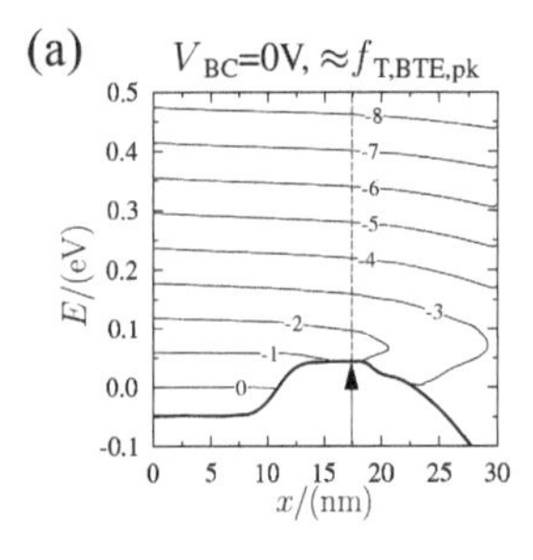

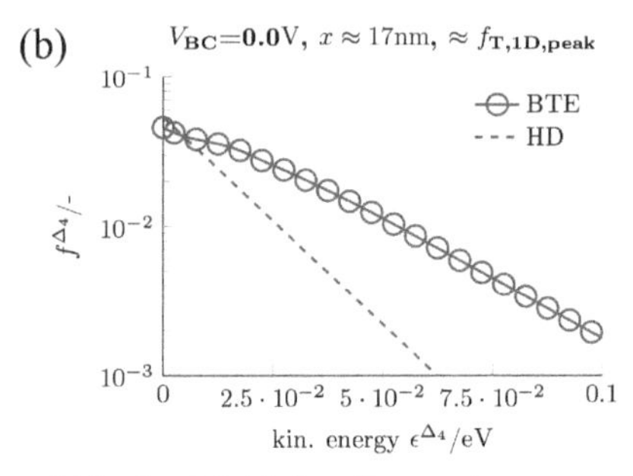

Figure 2.22 Electron distribution function within the 4-fold Δ-valley at $f_{T,BTE,pk}$ within the emitter-base region. (a) Contours with the arrow marking the position of the doping induced conduction band barrier. (b) Comparison of HD and BTE distribution function at the barrier peak.

solver and terminal characteristics that are fairly close to the HD results (after adjusting though the HD transport model parameters (f^{td}, f^{tc} and f^{ec}).

In Figure 2.22(a), the conduction band edge and the contour lines of the electron distribution function for the smoothed doping profile are shown within the BE region. The corresponding distribution function in Figure 2.22(b) at the conduction band maximum at $x = 17$ nm shows no blocking of low energetic electrons and thus the HD and BTE results are approaching each other.

2.2.4.4 Comparison with experimental data

In the course of the DOTSEVEN project, a variety of experimental data of fabricated SiGe HBTs were evaluated. In order to evaluate the predictive capability of the TCAD infrastructure employed within the project, measured terminal characteristics, such as transfer current and f_T characteristics, were compared with simulation results. Here, a SiGe HBT fabricated by IHP and shown in Figure 2.23(a) is considered. For the comparison with the 1D simulation results, the measured data were deembedded by the external elements of the actual 3D transistor structure. This was accomplished by using the physics-based and geometry scalable properties of the HICUM/L2 compact model and its parameters, which were extracted from measured data of transistors and special test structures [Paw17][Kor15]. A comparison between the respective 1D measurements with the HD and BTE simulation results is given in Figure 2.23(b), (c) for the transfer current and transit frequency in the absence of self-heating, the impact of which has already been accounted for during the parameter extraction.

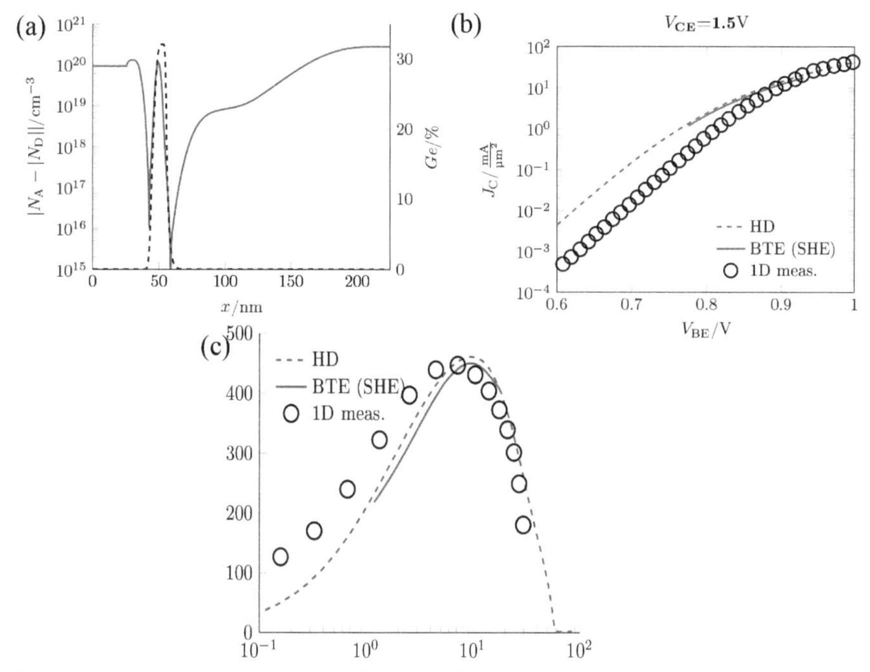

Figure 2.23 (a) Doping and Ge profile of a SiGe HBT (fabricated by IHP) with corresponding (b) transfer current, and (c) transit frequency. Comparison HD and BTE simulation results with 1D measurement data.

According to Figure 2.23(c), both HD and BTE simulation results agree well with the measured data around peak f_T and beyond. Below peak f_T, discrepancies exist which may be attributed to (i) too strong deembedding of parasitic or external junction capacitances in the measured data, (ii) an incorrect doping and Ge profile resulting in lower junction capacitances or doping and Ge dependent bandgap for the device simulation, (iii) the neglect of Carbon in the base region, or (iv) a significant difference in the doping, Ge and C dependent bandgap modeling in the simulation. The observed discrepancy in the transfer current (cf Figure 2.23(b)) indicates the latter as a major cause for the differences at low current densities.

According to Figure 2.24, the experimentally determined bandgap narrowing is indicating the presence of metastable strain, as reported in [Bea92]. For a first estimation of the impact of different bandgap narrowing values as function of Germanium on the terminal characteristics, a simple linear model ("lin." in Figure 2.24) is employed for DD and HD simulations. The corresponding terminal characteristics are shown in Figure 2.25.

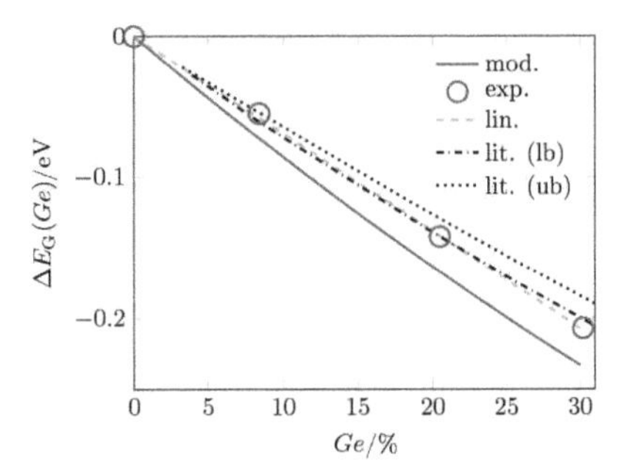

Figure 2.24 Comparison of the Ge concentration induced bandgap narrowing from experimental data (exp.) and device simulation model (mod.). In addition, the lower and upper boundary (lb and ub) for bandgap narrowing as function of Ge the presence of metastable strain [Bea92] is shown.

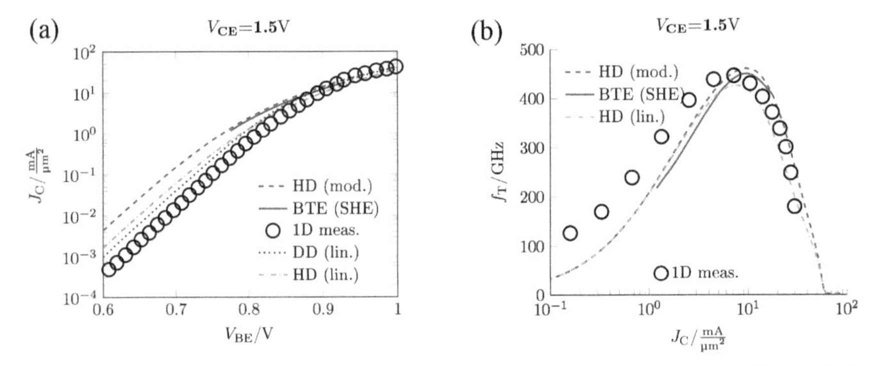

Figure 2.25 Comparison of the 1D measurement data with DD, HD and BTE simulation results. For the DD and HD simulation, the linear model indicated in Figure 2.24 is used. The transit frequency obtained by DD is not shown, since its parameters have not been adjusted.

According to Figure 2.25(a), the linear bandgap narrowing model improves the agreement between the simulated and experimental transfer characteristics at low current densities, but there is little improvement for f_T (cf. Figure 2.25(b)). Since advanced SiGe HBTs exhibit a significant carbon content, which is has not been considered in the simulations, further investigations based on experimental data (variation of the Germanium and carbon contents) are needed to clarify the origin of the discrepancies.

Nevertheless and focusing on trends only, the TCAD infrastructure employed in the DOTSEVEN project is capable of predicting performance trends correctly. Figure 2.26 displays the peak values of the measured 1D transit frequency for four different SiGe HBTs obtained from a process split. Here, both the quantitative and qualitative trend of the measurements is well

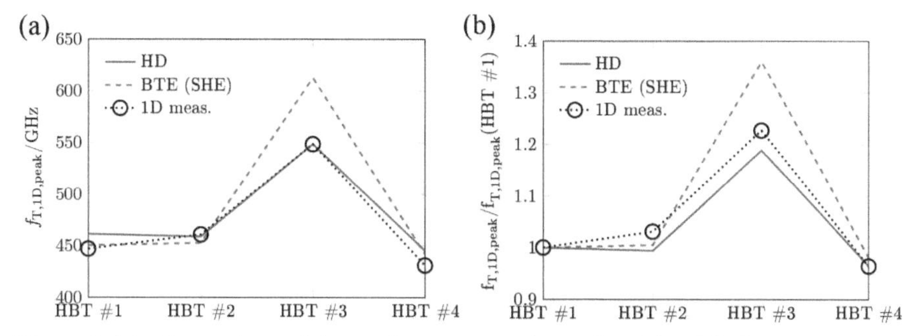

Figure 2.26 Comparison of the performance trends predicted by TCAD (HD and BTE simulation) with 1D measurement results (three samples) for a process split with four different SiGe HBTs (fabricated by HP).

captured by HD and BTE simulation. For the SiGe HBT labeled by HBT #3, the BTE results are overestimating the peak transit frequency. However, as explained before, this overestimation is based on an improper doping profile description at the metallurgical emitter-base junction and thus due to an underestimation of low energetic electrons.

2.3 Advanced Electro-thermal Simulation

C. Jungemann and N. Rinaldi

2.3.1 Carrier–Phonon System in SiGe HBTs

Charge carriers are accelerated under high electric fields in semiconductor devices and gain high kinetic energies. Carriers with energies higher than 60 meV scatter mainly with optical phonons. The optical phonons, which have a negligible group velocity, cannot participate in heat transport. Instead, they decay into long-wavelength acoustic phonons, which determine heat conduction in semiconductors. Since this decay is relatively slow, compared to the carrier–phonon interactions, a bottleneck for energy dissipation can occur, which results in a large number of hot optical phonons in high-field domains. The carrier–phonon and phonon–phonon interaction processes with their corresponding scattering time constants are illustrated in Figure 2.27 [Pop06].

In order to investigate self-heating in ultra-scaled bipolar transistors precisely, we have to consider a coupled system of transport equations for electrons, holes, and phonons. For this goal, the coupling terms, which describe carrier–phonon interactions, must be modeled accurately. In fact, carriers gain energy from the electric field and diffuse several mean free paths

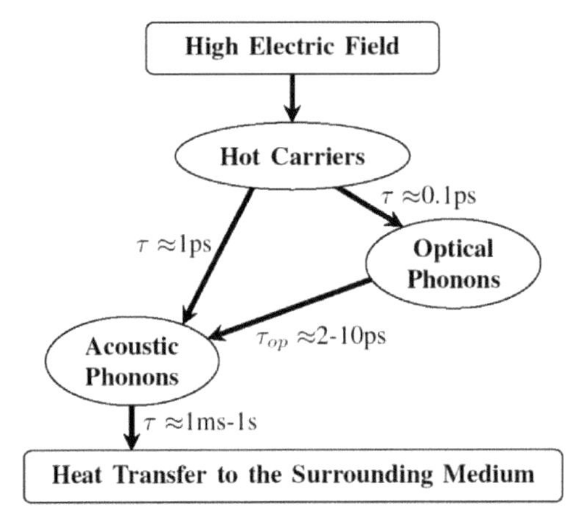

Figure 2.27 Thermal energy transport diagram in semiconductor devices.

(tens of nanometers) before they lose their energy to the lattice. Therefore, the so-called Joule-heating term, which represents the energy that carriers receive from the electric field, is not appropriate to capture the spatial distribution of heat generation in submicron devices. In order to tackle this problem, an advanced hydrodynamic model has been proposed [Mus08] to describe heat generation and transport in sub-micron silicon devices. However, this approach is still based on a single averaged carrier temperature within the relaxation time approximation and does not account for the spectral information regarding the emitted phonons. Since inelastic carrier–phonon scattering, which is described in detail by the scattering integral of the Boltzmann transport equation (BTE), causes heat generation, solving the BTE is the most accurate approach to study carrier–phonon interactions [Pop10].

The Monte Carlo (MC) method has been widely used to solve the BTE for electrons coupled with heat transport equations. The Fourier heat equation as the most elementary approach to consider heat conduction was used in [Zeb06, Sad10], which is not valid for sub-micron devices. In a CPU-efficient approach, a system of energy balance equations for both optical and acoustic phonons was extracted from the phonon BTEs. This system of equations coupled with an electron MC simulator, which was used to study self-heating in silicon-on-insulator devices, can describe the phonon bottleneck in thermal energy transport by distinguishing between optical and acoustic phonon temperatures [Ral08, Vas09]; however, it cannot capture all microscopic effects of non-equilibrium phonon transport due to the averaged phonon temperatures. Nghiem et al. [Ngh14] introduced recently an electrothermal simulator, which solves the BTE for both electrons and phonons self-consistently. However, in their system of equations, the feedback to the electron system is the effective temperature extracted from the phonon distribution function.

Despite great advances in understanding the physics of phonon transport using the MC method, the stochastic basis of this method hinders calculation of parameters with very small or slow variations. Alternatively, a deterministic approach based on spherical harmonics expansion (SHE) can be used to solve the BTE [Gnu93]. In this regard, Ramonas et al. [Ram15] presented a deterministic solution of a non-equilibrium bulk electron–phonon system for noise calculations.

Most recently, Kamrani et al. [Kam17a] presented a SHE method for the coupled BTEs of electrons, holes, and phonons under stationary conditions in a SiGe HBT. Since it has been shown that carriers in SiGe lose their energy mainly by scattering with longitudinal optical (LO) phonons [Pop05, Ni12], they solved the phonon BTE only for the LO phonon mode, and used energy balance equations for the other optical and acoustic phonon modes. In addition, the reduction of the thermal conductivity in a SiGe HBT was accounted for by analytical models for the lattice thermal conductivity in a way that is consistent with empirical data.

2.3.2 Deterministic and Self-consistent Electrothermal Simulation Approach

In the framework of semi-classical transport theory, the kinetics of a non-equilibrium carrier–phonon system under stationary conditions is described by a coupled set of BTEs for the distribution functions of carriers (electrons/holes) $f^{e/h}(\vec{r}, \vec{k})$ and phonons $n(\vec{r}, \vec{q})$, defined on the position vector $\vec{r}$, carrier and phonon wave vectors $\vec{k}$ and $\vec{q}$, respectively. To investigate the impact of hot LO phonons on carrier transport, the non-equilibrium distribution function of LO phonons must be obtained, while for the other phonon modes, equilibrium distribution functions, which are evaluated at averaged phonon temperatures of the optical T_{op} and acoustic T_{ac} phonon branches, can be assumed. In this case, the BTE for the charge carriers is written as:

$$L\{f\} = Q^{LO}\{f, n\} + S\{f, T_{op}, T_{ac}\} \tag{2.32}$$

where $L\{f\}$ is the free-streaming operator, $Q^{LO}\{f, n\}$ is the scattering operator for inelastic interactions of carriers with non-equilibrium LO phonons, and $S\{f, T_{op}, T_{ac}\}$ denotes the scattering operator of all other scattering mechanisms. The scattering term for non-equilibrium LO phonons is expressed as:

$$Q^{LO}\{f,n\} = \frac{V_0}{(2\pi)^3}\sum_{v'}\int\left[W^{v,v'}\left(\vec{r},\vec{k},\vec{k}',n\right)f^{v'}\left(\vec{r},\vec{k}'\right)\right.$$
$$\left.-W^{v',v}\left(\vec{r},\vec{k}',\vec{k},n\right)f^{v}\left(\vec{r},\vec{k}\right)\right]d^3k' \tag{2.33}$$

where V_0 is the system volume, and $W^{v,v'}\left(\vec{r},\vec{k},\vec{k}',n\right)$ is the transition rate from the initial state $\left(v',\vec{k}'\right)$ into the final state $\left(v,\vec{k}\right)$ given by:

$$W^{v,v'}\left(\vec{r},\vec{k},\vec{k}',n\right) = C_0\left(\vec{r}\right)\left[n\left(\vec{r},\vec{q}\right)+\tfrac{1}{2}\mp\tfrac{1}{2}\right]$$
$$\delta\left(\varepsilon\left(\vec{k}\right)-\varepsilon\left(\vec{k}'\right)\mp\hbar\omega_{op}\right) \tag{2.34}$$

where $C_0\left(\vec{r}\right)$ is the interaction constant, $\hbar\omega_{\mathrm{op}}$ is the constant energy for the dispersionless LO phonons, and the upper sign refers to phonon absorption and the lower to phonon emission.

The BTE for non-equilibrium LO phonons is written as:

$$\frac{n\left(\vec{r},\vec{q}\right)-n_{eq}\left(T_{\mathrm{L}}\right)}{\tau_{op}} + G^c\{n,f\} = 0 \tag{2.35}$$

where the first term represents interactions between optical and acoustic phonons within the relaxation time (τ_{op}) approximation, $n_{eq}\left(T_{\mathrm{L}}\right)$ is the equilibrium phonon distribution function, which follows the Bose–Einstein statistics, T_{L} is the lattice temperature which is equivalent to the averaged acoustic phonon temperature, and $G^c\{n,\ f\}$ is the phonon generation term given by:

$$G^c\{n,f\} = \frac{2V_0}{(2\pi)^3}\sum_{v}\int\left[W_{ab}^{v,v}\left(\vec{r},\vec{k},\vec{q},n\right)\right.$$
$$\left.-W_{em}^{v,v}\left(\vec{r},\vec{k},\vec{q},n\right)\right]f^{v}\left(\vec{r},\vec{k}\right)d^3k \tag{2.36}$$

where $W_{ab}^{v,v}$ and $W_{em}^{v,v}$ refer to the transition rate for phonon absorption and emission, respectively.

To solve this coupled system of BTEs, all the terms have to be expanded into spherical harmonics, and the spherical coordinates of the q-space

(q, θ_q, φ_q) can be expressed based on the modulus of the wave vector and the angles of the initial and final carrier states by using the momentum conservation rule, $\overrightarrow{k'} = \overrightarrow{k} \pm \overrightarrow{q}$.

In this simulation approach, non-equilibrium effects for the other phonon modes are accounted for by a coupled set of energy balance equations for the optical and acoustic phonon branches, where the energy loss rate due to inelastic carrier–phonon scattering is used as the heat generation term [Kam15]. Furthermore, the effects of Ge content [Pal04], doping profile [Lee12], and boundary scattering [Vas10] in the reduction of the lattice thermal conductivity, as the main parameter that models heat conduction by acoustic phonons, were considered via analytical models.

2.3.3 Hot Phonon Effects in a Calibrated System

To investigate non-equilibrium effects for the carrier–phonon system in bipolar transistors, a state-of-the-art toward-THz SiGe HBT fabricated by Infineon Technologies AG within the framework of the European DOTFIVE project with an emitter width of W_{E} = 0.13 μm and a length of L_{E} = 2.73 μm, and belonging to a technology development stage referred to as set #3 in [d'Al14], was used to extract the thermal resistance based on simple DC measurements. The extracted thermal resistance from measurements R_{TH} = 6,800 K/W [d'Al16] results in a junction temperature increase of ΔT_{j} = 38.5 K at V_{BE} = 0.9 V and V_{CE} = 1 V with I_{C} = 5.66 mA.

To study self-heating in this device, 2-D electrothermal simulations were performed for the structure, which is partly shown in Figure 2.2, the doping profiles of which were extracted by secondary ion mass spectrometry (SIMS). For these simulations, a self-consistent steady-state solution of the BTEs for electrons, holes, and LO phonons coupled with the energy balance equations was obtained. Due to minor uncertainty in the extracted Ge profile, a calibration of the Ge content by a few percent is used to reproduce the measured I_{C} at T_{B} = 300 K by simulation. Figure 2.28 (top) depicts the thermal conductivity over the 2-D SiGe HBT by considering the effect of Ge content, doping concentration, and boundary scattering at 300 K.

Since the SHE solution of the BTEs for a 3-D real space is too CPU intensive, only a 2-D real space was used and some important parts of the structure for thermal conduction, such as metal layers, were neglected. To mimic the 3-D nature of the heat propagation in 2-D simulation, the thermal boundary conditions are adjusted to match the simulated average lattice temperature increase over the space-charge region of the base–emitter junction

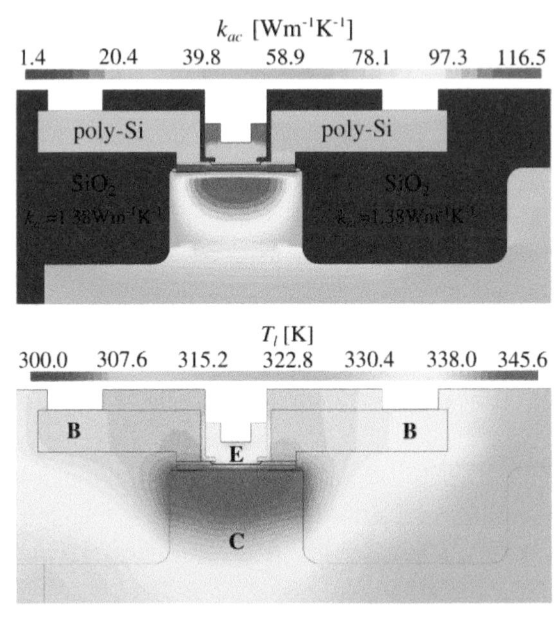

Figure 2.28 (Top) Thermal conductivity in the 2-D SiGe HBT structure by taking into account the effect of Ge content, doping concentration, and boundary scattering at 300 K. (Bottom) Self-consistent lattice temperature at $V_{\mathrm{BE}} = 0.9$ V and $V_{\mathrm{CE}} = 1$ V.

equal to the extracted junction temperature from measurements [d'Al02]. Hence, a convective boundary condition at the emitter contact is considered, and the heat transfer coefficient is calibrated to obtain an average junction temperature increase of 38.5 K from the temperature distribution of the simulation at the corresponding bias conditions. In this simulation, Neumann (adiabatic) boundary conditions are considered for artificial boundaries, the base and collector contacts, whereas the bulk contact at the bottom of the substrate is set to a constant temperature of 300 K. Figure 2.8 (bottom) shows the self-consistent lattice temperature obtained from the energy balance equations.

The spatial distribution of the Joule-heating term and the energy loss rate due to in-elastic carrier–phonon scattering are shown along the symmetry axis of the device in Figure 2.29. This figure depicts the distance that carriers have to travel before releasing their energy to the lattice; therefore, carriers receive energy from the high electric field at the collector–base junction, while they lose their energy via net phonon generation deep in the collector region. Moreover, Figure 2.29 shows that carriers are mainly scattered by

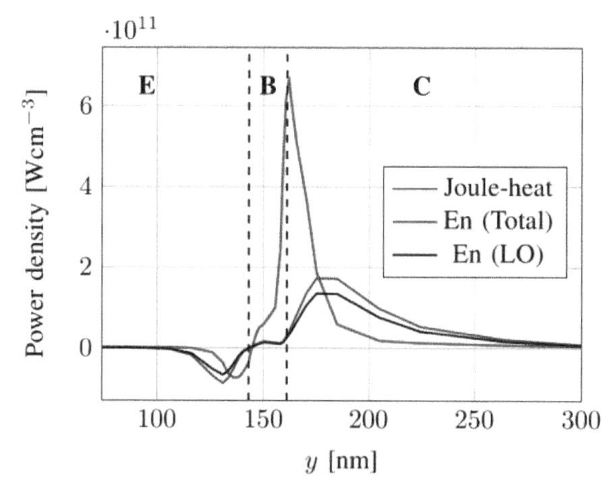

Figure 2.29 Profiles of the power densities received and dissipated by carriers which are calculated from Joule-heating and energy loss rate due to inelastic phonon scattering, respectively, along the symmetry axis of the HBT at V_{BE} = 0.9 V and V_{CE} = 1 V.

LO phonons which can lead to a strong deviation in the LO phonon distribution function with respect to the equilibrium value evaluated at the lattice temperature in the collector region which is shown in Figure 2.30 (top).

In order to investigate the effect of hot LO phonons and to make a comparison with the lattice temperature, an effective temperature is extracted from the non-equilibrium LO phonon distribution function which is shown in Figure 2.30 (bottom). The higher value of the effective temperature for LO phonons with respect to the lattice temperature, in the collector region, refers to the so-called phonon energy bottleneck in thermal conduction obtained for τ_{op} = 2 ps [Pop10]. The equality of T_{L} and T_{eff} around the base–emitter junction reveals the negligible effect of hot LO phonons on the collector current, because the temperature at this junction dominantly determines the impact of self-heating on the collector current increase. However, the large difference between these two temperatures in the collector region might influence some electrical phenomena, such as impact ionization (II) due to hot electrons, which occurs mainly deep in the collector region. To examine this possibility, the injected current due to electron II (I_{II}) was calculated in a simulation with high collector–base voltage V_{CB} = 2 V. A stronger phonon scattering due to hot LO phonons obtained from the full electrothermal simulation leads to a lower number of hot electrons. However, the reduction of the I_{II} at the same collector current due to temperature increase is just a few percent. As a result, the impact of hot LO phonons on electron II is not very strong in this case.

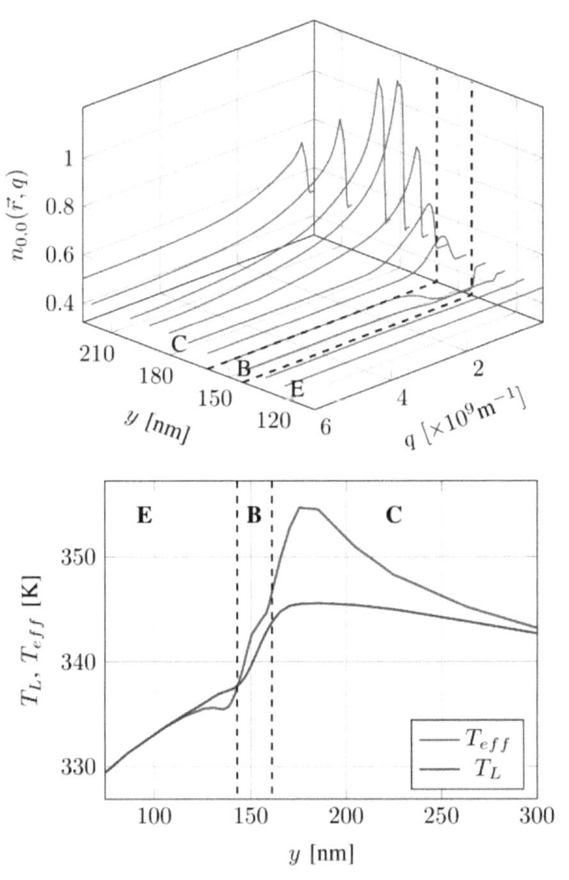

Figure 2.30 (Top) The LO phonon distribution function (zeroth-order harmonic), and (bottom) lattice temperature and effective temperature for LO phonons, along the symmetry axis of the investigated HBT at V_{BE} = 0.9 V and V_{CE} = 1 V.

2.3.4 Thermal Resistance Extraction from the Simulated DC Characteristics

To extract the thermal resistance by an approach similar to the experimental extraction method [d'Al14], the required DC characteristics were calculated and compared with measurement data (Figure 2.31). Figure 2.31 (top) displays the $I_{\mathrm{C}} - V_{\mathrm{BE}}$ characteristics of the HBT at different homogeneous temperatures, compared to experimental data measured under DC conditions at various thermo-chuck temperatures. In these voltage/current ranges, self-heating can be safely disregarded and the results are used for calculating the

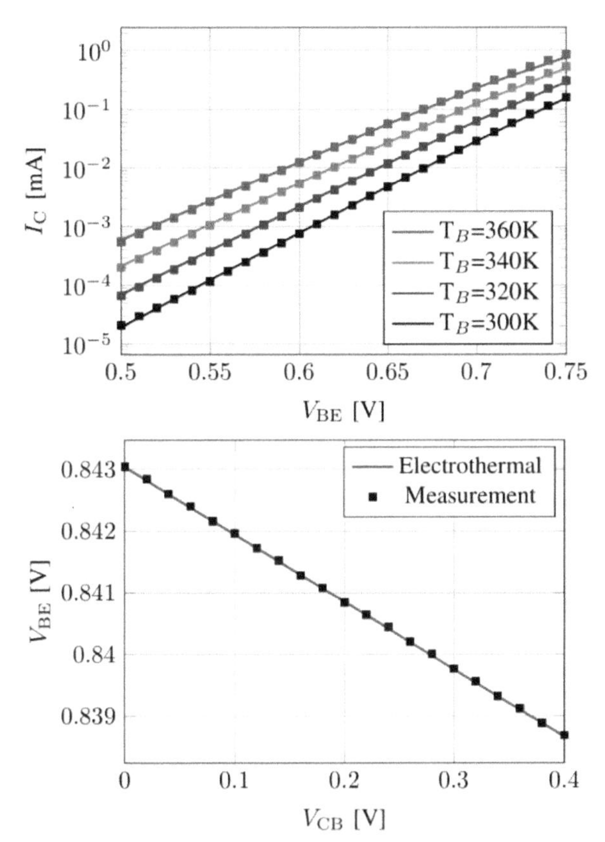

Figure 2.31 (Top) $I_C - V_{BE}$ characteristics for different homogeneous temperatures at $V_{CE} = 0.6$ V. Solid lines show the isothermal simulation results at $T_B = 300, 320, 340, 360$ K and symbols show the corresponding measurement data. (Bottom) $V_{BE} - V_{CB}$ characteristics from electrothermal simulation and measurement at $I_E = 2$ mA.

temperature coefficient $\varphi = -(\partial V_{BE}/\partial T_B)\,|_{I_C}$. The extraction of the other parameter $\gamma = (\partial V_{BE}/\partial V_{CB})\,|_{I_E}$ can be troublesome with the MC method, because the variation of V_{BE} with respect to V_{CB} for a constant I_E is very small. Moreover, thermal parameters of the device determine the slope of the $V_{BE} - V_{CB}$ curve [Kam15]; consequently, a self-consistent electrothermal simulation is needed to evaluate the slope of this curve consistent with the lattice temperature distribution of the device. Figure 2.32 shows the $V_{BE} - V_{CB}$ characteristics obtained from the deterministic and self-consistent electrothermal simulator in comparison with the measurement data.

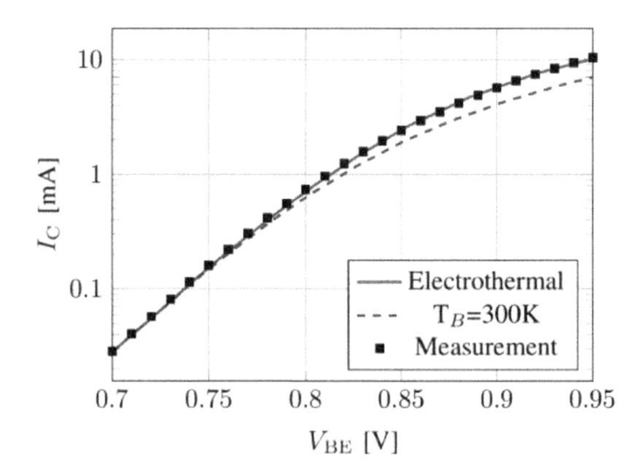

Figure 2.32 $I_C - V_{BE}$ characteristics with and without including self-heating compared to measurement data at V_{CE} = 1 V.

Since the simulated DC characteristics are in very good agreement with measurement results, the extracted thermal resistance from electrothermal simulations $R_{TH} = -\gamma/(I_E\varphi)$ matches the value obtained from measurements. This confirms the consistency of the extracted junction temperature from the simulated DC characteristics with the average lattice temperature over the base–emitter junction observed in the temperature profile shown in Figure 2.30 (bottom). Therefore, this result attests the accuracy of the analytical model on which the experimental procedure is based.

Figure 2.32 shows the effect of self-heating on the collector current increase observed in $I_C - V_{BE}$ characteristics of the SiGe HBT by electrothermal SHE simulations, which matches measurement results very well.

2.4 Microscopic Simulation of Hot-carrier Degradation

2.4.1 Physics of Hot-carrier Degradation

Due to inevitable trade-offs in the performance optimization of SiGe HBTs, these devices are operated closer and even beyond the classical safe-operating area (SOA) borders. Hot-carrier degradation (HCD) is the main reliability concern in bipolar transistors that strongly limits the lifetime of a device operated close to the SOA limit [Fis15]. This degradation happens due to trap states generated by hot-carriers along the oxide interfaces over time. In general, imperfections at the Si/SiO_2 interface lead to silicon dangling bonds, which can capture electrons or holes. Hence, these dangling bonds

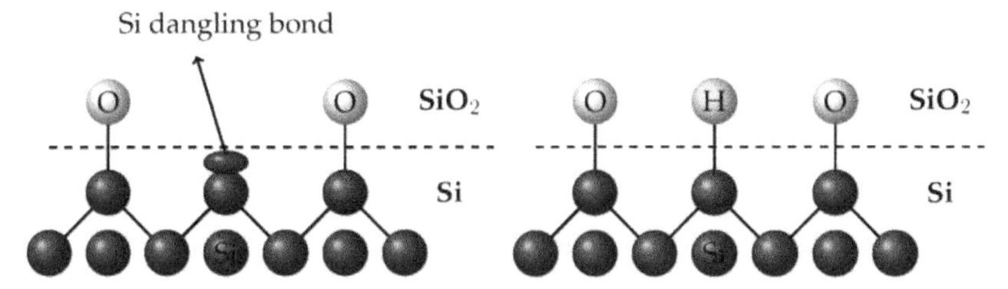

Figure 2.33 (Left) Creation of Si dangling bonds at the Si/SiO$_2$ interface. (Right) Passivation of the dangling bonds by incorporating hydrogen atoms.

are intentionally passivated by incorporating hydrogen atoms (Figure 2.33). However, hot carriers can supply enough energy to break the passivated Si–H bonds. A hot-carrier is a charge carrier which is accelerated under a high electric field inside the device and attains significant kinetic energy (higher than 1.5 eV) to break the bonds directly. Therefore, devices operating under bias conditions, which produce large electric fields, are susceptible to the HCD phenomenon.

In high-performance HBTs, shallow trench isolation (STI) and deep trench isolation (DTI) schemes together with the emitter–base (EB) spacer oxide are used to reduce parasitic capacitances and leakage currents [Mar09]. However, trap states resulting from the Si–H bond dissociation at the EB spacer and STI oxide interfaces produce excess non-ideal base current via Shockley–Read–Hall (SRH) recombination in the forward mode and reverse mode, respectively (Figure 2.34). As a result, traps generated due to hot carriers along the EB spacer oxide interface degrade the main parameters of

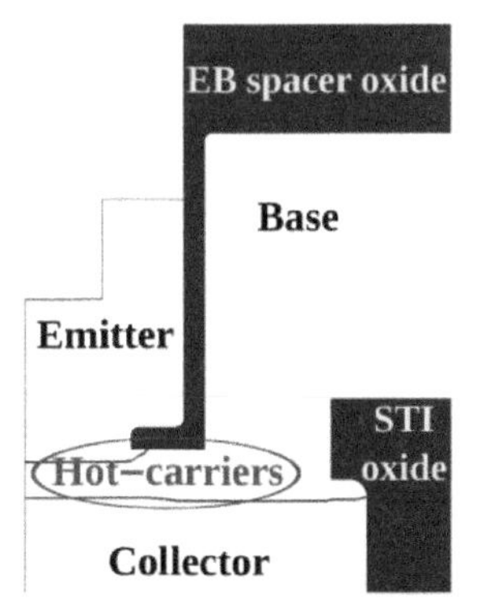

Figure 2.34 Schematic of a state-of-the-art SiGe HBT with the corresponding EB spacer and STI oxides.

the device such as current gain and noise figure [Cre04]. Hence, a profound knowledge of the microscopic mechanisms of the interface trap generation and annihilation as well as their impact on the electrical characteristics is essential.

Although conventional methods for a physics-based investigation of HCD in bipolar transistors, which are based on the lucky electron model, are electric field driven [Che09, Moe12, Wie16], it has been shown that the trap generation rate at the oxide interface is determined by the energy of the interacting charge carriers [DiM89, DiM01]. Hence, another quantity called the acceleration integral (AI), which is calculated from the carrier energy distribution function (EDF), has been introduced to describe accurately the spatial distribution of the interface traps obtained from charge pumping measurement data [Sta11, Sta12]. In consequence, an energy-driven paradigm based on the AI has been developed to model both single- and multiple-carrier processes of the bond dissociation in the degradation analysis of the n-channel MOSFETs [Bin14, Sha15, Tya16].

This model has been recently extended to include the effects of both hot electrons and hot holes for describing the underlying mechanisms of HCD in bipolar transistors [Kam16, Kam17b]. For this purpose, a coupled system of BTEs for electrons and holes, which accounts for II and SRH, has to be solved. Since stochastic algorithms such as the MC method impose an enormous computational burden to resolve the high-energy tail of the EDF, a deterministic approach based on a SHE was used to solve the BTEs including full band structure effects [Hon11].

The reaction-diffusion model has been widely used to represent the complex dynamics of the trap generation and subsequent annihilation in HCD of bipolar transistors and negative-bias temperature-instability degradation of MOSFETs [Moe12, Rag15, Kuf07]. Despite a very good matching for a wide range of experimental observations, it has been shown that the reaction-diffusion model is inconsistent with the measurement data at the microscopic level [17]. Therefore, in [Kam17b] a degradation model based on the AIs was used to calculate the dispersive bond-breakage rates associated with a reaction-limited model to describe HCD effects in a SiGe HBT.

2.4.2 Modeling of Hot-carrier Effects

In an energy-driven framework, the bond-breakage rate is modeled by considering the interaction of the incident charge carriers with the passivated Si–H bond. A Si–H bond is represented as a truncated harmonic oscillator

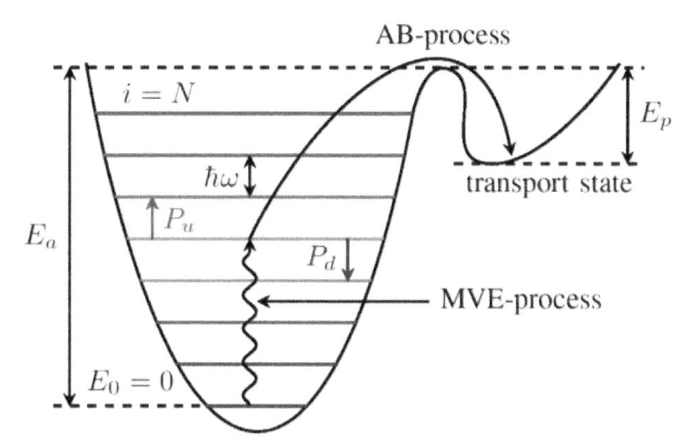

Figure 2.35 The energy configuration of the Si–H bond modeled as a truncated harmonic oscillator.

characterized by a system of eigenstate energies [Sto98], which is depicted in Figure 2.35.

Bond dissociation occurs via excitation of one of the bonding electrons to the transport state, which is known as an antibonding (AB) process. As a result, a repulsive force is induced, which detaches the hydrogen atom. The dissociation rate from the ith state of the Si–H bond with the energy E_i, triggered either by a hot electron or by a hot hole, is given by:

$$R_{AB,i} = I^e_{AB,i} + I^h_{AB,i} + v_r \exp\left(-(E_a - E_{ox}d - E_i)/k_B T_0\right) \quad (2.37)$$

where $I^e_{AB,i}$ and $I^h_{AB,i}$ are the AB acceleration integrals of electrons and holes, respectively, v_r is an attempt frequency, and E_a is the bond-breakage activation energy, which is reduced due to the interaction of the bond dipole moment d with the oxide electric field E_{ox}. Furthermore, to account for the fluctuations of the activation energy, a Gaussian distribution is considered with a mean value and standard deviation of $\langle E_a \rangle$ and σ_E, respectively. The AI for the AB process is given by [Bin14]

$$I^{e/h}_{AB,i} = \sigma_0^{AB} \int_{E_{th,i}}^{\infty} \left[f^{e/h}(E)\, z^{e/h}(E)\, v_g^{e/h}(E) \left[(E - E_{th,i})/E_{ref}\right]^p \right] dE \quad (2.38)$$

where $E_{th,i} = E_a - E_{ox}d - E_i$ is a threshold energy for the ith level, σ_0^{AB} is the AB reaction cross section, $f^{e/h}(E)$ is the carrier distribution function, $z^{e/h}(E)$ is the carrier density of states, $v_g^{e/h}(E)$ is the carrier group velocity, $p = 11$ is an empirical parameter, and $E_{ref} = 1\text{eV}$.

If a charge carrier does not provide enough energy to trigger the AB mechanism, it can still contribute to the bond-breakage procedure via multiple vibrational excitation (MVE) of the bond. In an accumulative consideration of the MVE and AB mechanisms, the bonding electron can be firstly excited by several colder particles to an intermediate energy level, and then dissociated by a carrier with a relatively high energy. The bond excitation and deexcitation rates triggered by either a cold electron or a cold hole, are given, respectively, by

$$P_u = I^e_{MVE} + I^h_{MVE} + \omega_e \exp\left(-\hbar\omega/k_B T_0\right) \tag{2.39}$$

$$P_d = I^e_{MVE} + I^h_{MVE} + \omega_e \tag{2.40}$$

where ω_e is the reciprocal phonon life-time and $\hbar\omega$ is the energy distance between the Si–H energy levels. The AI for the MVE process is defined as

$$I^{e/h}_{MVE} = \sigma_0^{MVE} \int_{\hbar\omega}^{\infty} \left[f^{e/h}(E)\, z^{e/h}(E)\, v_g^{e/h}(E) \left[(E - \hbar\omega)/E_{ref}\right] \right] dE \tag{2.41}$$

The cumulative bond-breakage rate, which accounts for all possible superpositions of the AB and MVE mechanisms, is calculated as

$$R_a = \frac{1}{k} \sum_i R_{AB,i} \left(\frac{P_u}{P_d}\right)^i \tag{2.42}$$

where k is a normalization prefactor defined as $k = \sum_i (P_u/P_d)^i$.

In the reaction-limited approach, the rate equation for the generation and recombination of the interface trap states is written as [Jep77]

$$\partial N_{it}/\partial t = (N_0 - N_{it})\, R_a - {N_{it}}^2 R_p \tag{2.43}$$

where N_{it} is the density of the generated interface traps, N_0 is the density of the primary passivated Si–H bonds, and R_p is the recovery rate.

In this approach, the dispersion of the bond-breakage energy determines the power-law time dependence of the HCD results. The dispersive effect of E_a is incorporated by discretizing an energy grid in the range of $[\langle E_a \rangle - 3\sigma_E, \langle E_a \rangle + 3\sigma_E]$ and evaluating Nit for each discretization point. The interface trap density profile, which is the combination of every single defect, is obtained by calculating the average of Nit at each energy point weighted by the Gaussian distribution [Bin14].

To obtain the required distribution functions of the carriers, a coupled system of the BTEs for electrons and holes has to be solved. The BTE for electrons in the stationary case is written as:

$$L\left\{f^e\right\} = S\left\{f^e\right\} + Q\left\{f^e, f^h\right\} - \Gamma^e\left\{f^e, f^h\right\} \tag{2.44}$$

where $S\{f^e\}$ is the scattering operator, which accounts for carrier–phonon scattering, impurity scattering, alloy scattering, and II scattering of primary particles, $Q\{f^e, f^h\}$ is the generation operator of secondary particles due to II [Jab14], and $\Gamma^e\{f^e, f^h\}$ is the SRH recombination operator [Jun07, Rup16] defined on the boundary of the oxide interface.

In this simulation approach, a full-band SHE simulator is used to obtain the carrier EDFs for a SiGe HBT under stress conditions. Then, the $N_{it}(\vec{r}, t)$ profile calculated at each stress time step is fed into the SHE solver to calculate the characteristics of the degraded device which change due to SRH recombination.

2.4.3 Simulation of SiGe HBTs under Stress Conditions Close to the SOA Limit

The conventional mixed-mode (MM) stress conditions, which are the concurrent applications of a high collector-base voltage and a high collector current density to accelerate the degradation procedure, set an upper limit for HCD of the SiGe HBTs during RF operation [Fis15]. However, as the main drawback they are far from typical operating conditions. Hence, to study the physics behind the long-term base current degradation under more practical operating conditions, three stress bias conditions P1, P2, and P3, along the border of the SOA, were selected to degrade the device up to 1,000 h at 300 K, and Gummel plots (V_{CB} = 0 V) were measured at certain stress time intervals [Jac15]. The corresponding bias voltage, current, and junction-to-ambient temperature increase obtained from the extracted thermal resistance R_{TH} = 2,850 K/W [d'Al14] are summarized in Table 2.1. Measurements showed that the examined npn SiGe HBT is negligibly affected by stress at P1, and P3 exhibits a higher base current degradation over time in comparison to P2.

For numerical simulations, a 2-D structure, the doping profiles of which were extracted from SIMS, was used. As a first step of this analysis, the simulator has to be calibrated to reproduce the measured Gummel plot and $I_{\mathrm{B}} - V_{\mathrm{CE}}$ characteristics of the fresh device. The base current reversal at

Table 2.1 Definition of the stress bias conditions P1, P2, and P3 and their corresponding junction temperatures

	P1	P2	P3
V_{CE} [V]	1 ($<BV_{\mathrm{CEO}}$)	2 ($>BV_{\mathrm{CEO}}$)	3 ($>BV_{\mathrm{CEO}}$)
J_{C} [mA/μm^2]	10	5	1
ΔT_{j} [K]	37	37	11

$V_{CE} > BV_{CEO}$ due to avalanche multiplication of carriers is used as a basis to determine the II rates initiated by primary electrons and holes. Figure 2.36 shows the II generation rates at P3 due to electrons and holes separately.

Under this stress condition, the high electric fields within the collector–base junction accelerate electrons to reach enough energy required for initiating avalanche generation of electron–hole pairs via II. Figure 2.36 shows that hot electrons responsible for II are deep in the collector region while hot holes are mainly found in the base region. To obtain a better understanding of the energy of the carriers which participate in the degradation process, Figure 2.11 depicts a cut of the EDFs for electrons and holes along the symmetry axis of the investigated SiGe HBT with respect to kinetic energies.

Electrons move toward the collector region and gain sufficiently high energies to initiate II, whereas the holes generated by II in the collector due to hot electrons [Figure 2.36 (top)] are accelerated toward the base and gain a lot of energy. Due to this high energy, some of the holes can shoot through

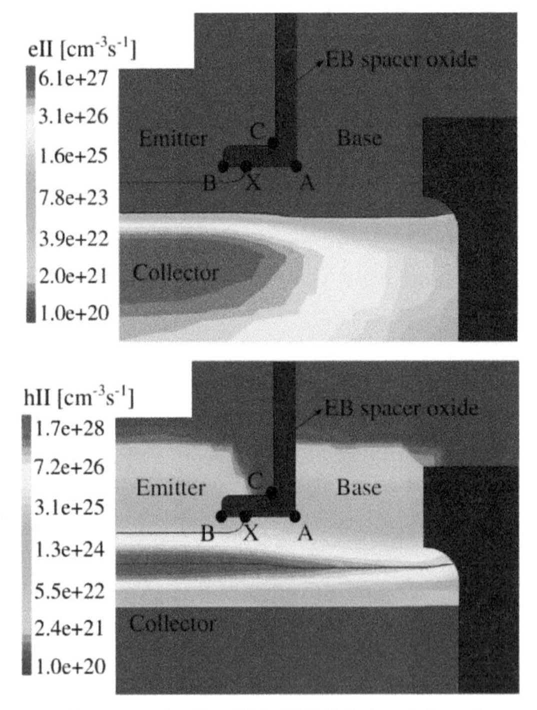

Figure 2.36 II generation rates in the SiGe HBT induced by electrons (top) and holes (bottom) at P3.

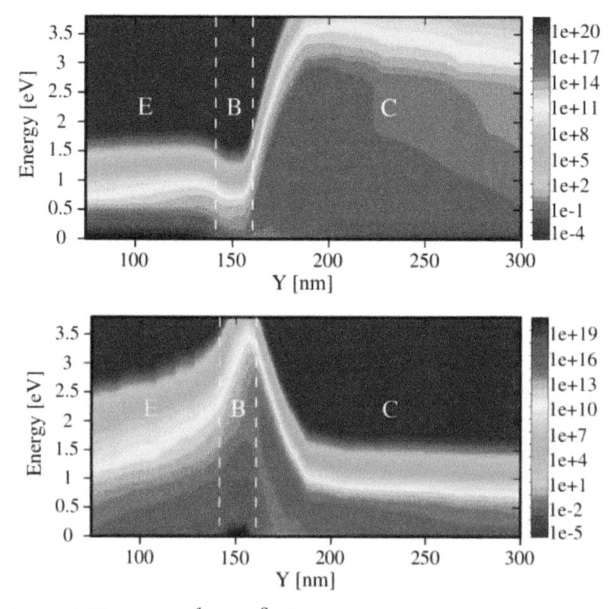

Figure 2.37 Cut of EDFs [eV^{-1} cm^{-3}] for electrons (top) and holes (bottom) along the symmetry axis of the HBT at P3.

the base into the emitter, where they still have a relatively large energy [Figure 2.37 (bottom)]. A certain fraction of these hot holes hit the EB spacer oxide interface, where they might break Si–H bonds.

This effect can only be captured by a model which resolves the dependence of the carriers on energy. Thus, it is not possible to directly describe the behavior of the hot holes with a drift-diffusion or a hydrodynamic model, in which the hole gas in the base is assumed to be close to equilibrium. Unavoidably, to perform physics-based degradation analysis relying on the classical models, the probability of hot carrier creation has to be calculated via the lucky electron model. This calculation based on the effective electric field shows inaccurately that hot holes are found at the collector–base junction. Subsequently, the possibility that a hot carrier reaches the oxide interface without any deflection has to be separately estimated [Moe12].

The interface traps located within the EB space-charge-region have the highest impact on the forward mode base current degradation via SRH recombination. Therefore, the EDFs at the intersection of the EB junction and the oxide interface are compared for different stress conditions in Figure 2.38 (top). The negligible role of electrons in the degradation process

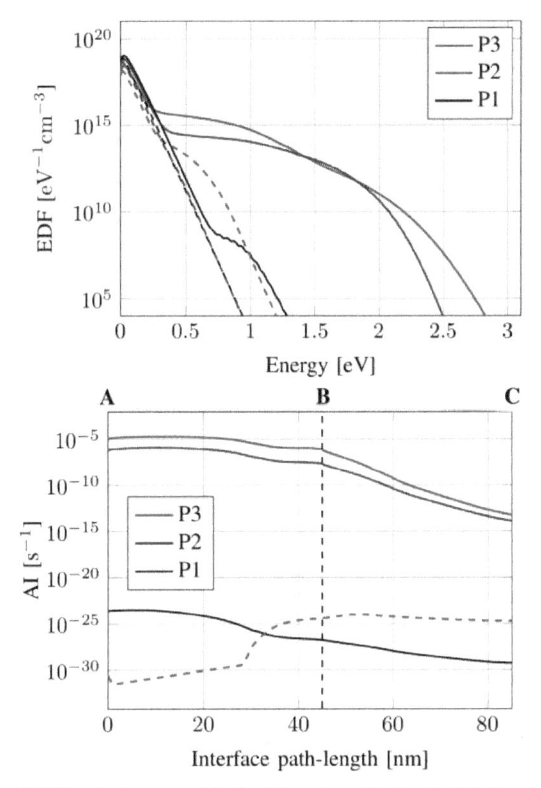

Figure 2.38 (Top) EDFs of electrons (dashed lines) and holes (solid lines) at the intersection of the EB spacer oxide interface and the EB junction [denoted by node X in Figure 2.36 (Bottom)]. Profiles of the AB AIs for electrons (dashed lines) and holes (solid lines) along the EB spacer oxide interface from node A to C denoted in Figure 2.36.

is concluded from the EDFs for electrons at P1 and P2, which exactly follow the equilibrium EDF, and at P3, with a low-energy hump. Because of small collector–emitter voltage in P1, even holes do not gain high energies, which explains the negligible degradation rate at P1 observed in measurements. Furthermore, the high-energy tails of the hole EDFs at P2 and P3 reveal the dominant role of hot holes in the degradation process under the stress conditions close to the SOA limit, which was also reported for conventional MM stress conditions in [Van06]. However, the deterministic solver provides the possibility to comprehensively describe the microscopic effects of hot carriers and accurately obtain the EDFs up to high energies in a realistic 2-D device structure and develop a practical physics-based degradation model to evaluate the resulting excess base current.

Although cold carriers can also participate in the MVE of the Si–H bonds, hot carriers with energies greater than 1.5 eV have a much higher chance to break the Si–H bonds directly [Tya16]. Moreover, the activation energy parameters determine the power-law time dependence and the dependence of the excess base current on the stress conditions. Hence, $\langle E_a \rangle$ = 1.6 eV and σ_E = 0.2 eV, which are in good agreement with those experimentally obtained [Ste96, Pob13], were considered for the activation energies of the bond-breakage to achieve good agreement with measurement data.

The AB AIs along the EB spacer oxide interface obtained from this calibration are depicted in Figure 2.38 (bottom). As an expected result, the equilibrium electron EDFs at P1 and P2 result in zero AB AIs along the oxide interface. Furthermore, due to the low-energy humps in the EDFs of electrons at P3 and holes at P1, their corresponding AIs are very small. In consequence, the measured base current degradations under P2 and P3 have to be ascribed to hot holes with relatively high AB rates, in which the bigger AI at P3 compared to P2 explicitly explains the bigger degradation current under this stress condition.

Figure 2.39 represents the trap densities along the EB spacer oxide interface for several stress time steps, which are significantly generated by the AB process due to hot holes. These interface trap densities, calculated for $N_0 = 10^{12}$ cm^{-2}, reveal that the large variation of the AB AIs along the oxide interface results in a strong non-uniformity in the spatial distribution of the N_{it} profile.

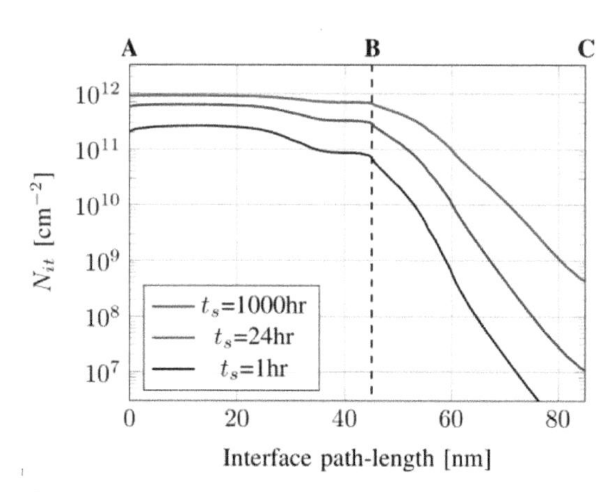

Figure 2.39 Interface trap densities generated at different stress time steps from node A to C denoted in Figure 2.36 at P3.

The generated traps at the EB spacer oxide interface cause a non-ideal increase in the forward mode base current via field-enhanced SRH recombination [Hur92]. Figure 2.14 (top) shows the Gummel characteristics of the fresh device and the degraded device after 1,000 h stress application at P3. The leakage currents observed for the fresh device due to packaging [Jac15] have no impact in the degradation analysis and are not taken into account. Since the recombination process has no considerable impact on the collector current, the resulting increase in the base current degrades the current gain of the transistor.

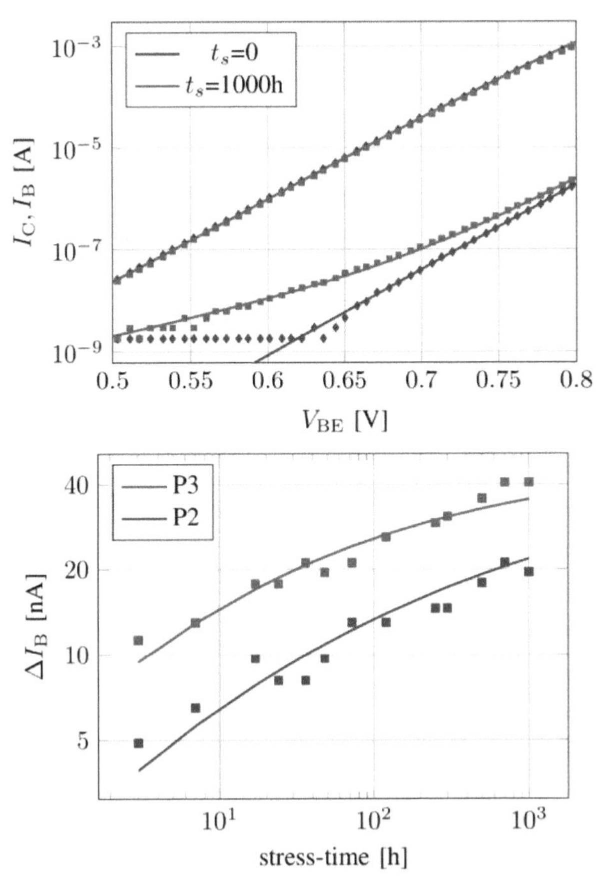

Figure 2.40 (Top) Gummel characteristics (V_{CB} = 0 V) of the fresh and degraded SiGe HBT after 1,000 h at P3 obtained from simulation (lines) and measurement (symbols). (Bottom) Excess base currents over the stress time obtained from simulation (lines) and measurement (symbols) at V_{BE} = 0.67 V and V_{CB} = 0 V.

In order to assess the time dependence of the HCD effects, the excess base current $\Delta I_{\mathrm{B}} = I_{\mathrm{B}}(t) - I_{\mathrm{B}}(0)$ is extracted over stress time [Figure 2.14 (bottom)]. Very good agreement between the simulation results and measurement data proves that the EDF-based degradation model can directly explain the time dynamics of the HCD results together with their dependence on the stress bias conditions.

References

[Bea92] J. C. Bean. Silicon-based semiconductor heterostructures: column IV bandgap engineering. Proceedings of IEEE, pp. 571–589, April 1992.

[Bin14] Bina, M., et al. (2014). Predictive hot-carrier modeling of n-channel MOSFETs. *IEEE Trans. Electron Devices* 61, 3103–3110.

[Blo70] K. Blotekjaer, Transport equations for electrons in two-valley semiconductors. *IEEE Transactions on Electron Devices*, vol. ED-17, no. 1, pp. 38–47, 1970.

[Che07] Cheng, P., et al. (2007). Understanding radiation-and hot carrier-induced damage processes in SiGe HBTs using mixed-mode electrical stress. *IEEE Trans. Nucleic Sci.* 54, 1938–1945.

[Che09] Cheng, P., Grens, C. M., and Cressler, J. D. (2009). Reliability of SiGe HBTs for power amplifiers-Part II: Underlying physics and damage modeling. *IEEE Trans. Electron Devices* 9, 440–448.

[Cre04] Cressler, J. D. (2004). Emerging SiGe HBT reliability issues for mixed-signal circuit applications. *IEEE Trans. Device Mater. Reliab.* 4, 222–236.

[d'Al02] d'Alessandro, V., and Rinaldi, N. (2002). A critical review of thermal models for electro-thermal simulation. *Solid State Electron.* 46, 487–496.

[d'Al14] d'Alessandro, V., Sasso, G., Rinaldi, N., and Aufinger, K. (2014). Influence of scaling and emitter layout on the thermal behavior of toward-THz SiGe: C HBTs. *IEEE Trans. Electron Devices* 61, 3386–3394.

[d'Al16] d'Alessandro, V., Magnani, A., Codecasa, L., Rinaldi, N., and Aufinger, K. (2016). Advanced thermal simulation of SiGe:C HBTs including back-end-of-line. *Microelectron. Reliab.* 67, 38–45.

[DiM01] D. J. DiMaria and J. H. Stathis, Anode hole injection, defect generation, and breakdown in ultrathin silicon dioxide films. *J. Appl. Phys.*, vol. 89, no. 9, pp. 5015–5024, May 2001.

[DiM89] DiMaria, D. J., and Stasiak, J. W. (1989). Trap creation in silicon dioxide produced by hot electrons. *J. Appl. Phys.*, vol. 65, no. 6, pp. 2342–2356, Mar. 1989.

[Fis15] G. G. Fischer and G. Sasso, Ageing and thermal recovery of advanced SiGe heterojunction bipolar transistors under long-term mixed-mode and reverse stress conditions. *Microelectron. Reliab.*, vol. 55, no. 3, pp. 498–507, Mar. 2015.

[Gal05] M. Galler. Mulitgroup Equations for the Description of the Particle Transport in Semiconductors, Vol. 70, World Scientific, 2005.

[Gnu93] A. Gnudi, D. Ventura, G. Baccarani, and F. Odeh, Two-dimensional MOSFET simulation by means of a multidimensional spherical harmonics expansion of the Boltzmann transport equation. *Solid State Electron.*, vol. 36, no. 4, pp. 575 – 581, Apr. 1993.

[Gra14] T. Grasser, K. Rott, H. Reisinger, M. Waltl, F. Schanovsky, and B. Kaczer, NBTI in nanoscale MOSFETs-the ultimate modeling benchmark. *IEEE Trans. Electron Devices*, vol. 61, no. 11, pp. 3586–3593, Nov. 2014.

[Her56] C. Herring and E. Vogt. Transport and deformation – potential theory for many – valley semiconductors with anisotropic scattering. *Physical Review*, Vol. 101, pp. 944–961, Feb. 1956.

[Hon11] S.-M. Hong, A.-T. Pham, and C. Jungemann. *Deterministic Solvers for the Boltzmann Transport Equation*, Springer, 2011.

[Hur92] G. Hurkx, D. Klaassen, and M. Knuvers, A new recombination model for device simulation including tunneling. *IEEE Trans. Electron Devices*, vol. 39, no. 2, pp. 331–338, Feb. 1992.

[Jab14] D. Jabs and C. Jungemann, "Avalanche breakdown of pn-junctions-Simulation by spherical harmonics expansion of the Boltzmann transport equation," in *Proc. Int. Conf. Simulation Semiconductor Process. Devices (SISPAD)*, Sep. 2014, pp. 173–176.

[Jac15] T. Jacquet *et al.*, Reliability of high-speed SiGe: C HBT under electrical stress close to the SOA limit. *Microelectron. Reliab.*, vol. 55, no. 9, pp. 1433–1437, Sep. 2015.

[Jac83] C. Jacoboni and L. Reggiani. The Monte Carlo method for the solution of charge transport in semiconductors with applications to covalent materials. *Reviews of Modern Physics*, vol. 55, pp. 645–705, July 1983.

[Jep77] K. O. Jeppson and C. M. Svensson, Negative bias stress of MOS devices at high electric fields and degradation of MNOS devices. *J. Appl. Phys.*, vol. 48, no. 5, pp. 2004–2014, May 1977.

[Jun03]C. Jungemann and B. Meinerzhagen. Hierarchical Device Simulation – The Monte-Carlo Perspective. Springer, 2003.

[Jun07] C. Jungemann, "A deterministic solver for the Langevin Boltzmann equation including the Pauli principle," *Proc. SPIE, Noise and Fluctuations in Circuits, Devices, and Materials*, vol. 6600, no. 1, p. 660007, May 2007.

[Kam15] H. Kamrani, T. Kochubey, D. Jabs, and C. Jungemann, "Electrothermal simulation of SiGe HBTs and investigation of experimental extraction methods for junction temperature," in *Proc. IEEE Int. Conf. Simulation Semiconductor Process. Devices (SISPAD)*, Sep. 2015, pp. 108–111.

[Kam16] H. Kamrani, D. Jabs, V. d'Alessandro, N. Rinaldi, and C. Jungemann, "Physics-based hot-carrier degradation model for SiGe HBTs," in *Proc. Int. Conf. Simulation Semiconductor Process. Devices (SISPAD)*, 2016, pp. 341–344.

[Kam17a] H. Kamrani, D. Jabs, V. d'Alessandro, N. Rinaldi, K. Aufinger, and C. Jungemann, "A deterministic and self-consistent solver for the coupled carrier-phonon system in SiGe HBTs," accepted in *IEEE Trans. Electron Devices*.

[Kam17b] H. Kamrani, *et al.*, "Microscopic hot-carrier degradation modeling of SiGe HBTs under stress conditions close to the SOA limit," submitted in *IEEE Trans. Electron Devices*.

[Kor15] J. Korn, H. Ruecker, B. Heinemann, A. Pawlak, G. Wedel, and M. Schroter. Experimental and Theoretical Study of fT for SiGe HBTs with a Scaled Vertical Doping Profile, Proc. IEEE BCTM, Boston, pp. 117–120, 2015.

[Kuf07] H. Kufluoglu and M. A. Alam, "A generalized reaction-diffusion model with explicit H-H2 dynamics for negative-bias temperature-instability (NBTI) degradation," *IEEE Trans. Electron Devices*, vol. 54, no. 5, pp. 1101–1107, 2007.

[Lee12] Y. Lee and G. S. Hwang, "Mechanism of thermal conductivity suppression in doped silicon studied with nonequilibrium molecular dynamics," *Phys. Rev. B*, vol. 86, no. 7, pp. 075 202-1–075 202-6, Aug. 2012.

[Lun00] M. Lundstrom. *Fundamentals of Carrier Transport*. Cambridge University Press, 2000.

[Mar09] I. Marano, V. d'Alessandro, and N. Rinaldi, "Analytical modeling and numerical simulations of the thermal behavior of trench-isolated bipolar transistors," *Solid State Electron.*, vol. 53, no. 3, pp. 297–307, Mar. 2009.

[Moe12] K. A. Moen, P. S. Chakraborty, U. S. Raghunathan, J. D. Cressler, and H. Yasuda, "Predictive physics-based TCAD modeling of the mixed-mode degradation mechanism in SiGe HBTs," *IEEE Trans. Electron Devices*, vol. 59, no. 11, pp. 2895–2901, Nov. 2012.

[Mus08] O . Muscato, V. Di Stefano, and C. Milazzo, "An improved hydrodynamic model describing heat generation and transport in submicron silicon devices," *J. Comput. Electron.*, vol. 7, no. 3, pp. 142–145, Sep. 2008.

[Ngh14] T. Thu Trang Nghiem, J. Saint-Martin, and P. Dollfus, "New insights into self-heating in double-gate transistors by solving boltzmann transport equations," *J. Appl. Phys.*, vol. 116, no. 7, pp. 074 514-1–074 514-9, Aug. 2014.

[Ni12] C. Ni, Z. Aksamija, J. Y. Murthy, and U. Ravaioli, "Coupled electrothermal simulation of MOSFETs," *J. Comput. Electron.*, vol. 11, no. 1, pp. 93–105, Mar. 2012.

[Pal04] V. Palankovski and R. Quay, *Analysis and simulation of heterostructure devices*. Vienna, Austria: Springer-Verlag, 2004.

[Paw09] A. Pawlak, M. Schröter, J. Krause, G. Wedel, and C. Jungemann. On the Feasibility of 500 GHz Silicon-Germanium HBTs, SISPAD 2009, San Diego, CA, 2009, pp. 1–4.

[Paw17] A. Pawlak, B. Heinemann, and M. Schröter, "Physics-based modeling of SiGe HBTs with f_T of 450 GHz with HICUM Level 2", IEEE BCTM, Miami , FL, p. 4, 2017.

[Pob13] G. Pobegen, S. Tyaginov, M. Nelhiebel, and T. Grasser, "Observation of normally distributed energies for interface trap recovery after hot-carrier degradation," *IEEE Electron Device Lett.*, vol. 34, no. 8, pp. 939–941, Aug. 2013.

[Pop05] E. Pop, R. W. Dutton, and K. E. Goodson, "Monte Carlo simulation of Joule heating in bulk and strained silicon," *Appl. Phys. Lett.*, vol. 86, no. 8, pp. 082 101-1–082 101-3, Feb. 2005.

[Pop06] E. Pop, S. Sinha, and K. E. Goodson, "Heat generation and transport in nanometer-scale transistors," *Proc. IEEE*, vol. 94, no. 8, pp. 1587–1601, Aug. 2006.

[Pop10] E. Pop, "Energy dissipation and transport in nanoscale devices," *Nano Research*, vol. 3, no. 3, pp. 147–169, Mar. 2010.

[Rag15] U. S. Raghunathan *et al.*, "Bias-and temperature-dependent accumulated stress modeling of mixed-mode damage in SiGe HBTs," *IEEE Trans. Electron Devices*, vol. 62, no. 7, pp. 2084–2091, Jul. 2015.

[Ral08] K. Raleva, D. Vasileska, S. M. Goodnick, and M. Nedjalkov, "Modeling thermal effects in nanodevices," *IEEE Trans. Electron Devices*, vol. 55, no. 6, pp. 1306–1316, Jun. 2008.

[Ram15] M. Ramonas and C. Jungemann, "A deterministic approach to noise in a non-equilibrium electron-phonon system based on the Boltzmann equation," *J. Comput. Electron.*, vol. 14, no. 1, pp. 43–50, Mar. 2015.

[Rup16] K. Rupp, C. Jungemann, S. M. Hong, M. Bina, T. Grasser, and A. Jungel, "A review of recent advances in the spherical harmonics expansion method for semiconductor device simulation," *J. Comput. electron.*, pp. 1–20, May 2016.

[Sad10] T. Sadi and R. W. Kelsall, "Monte Carlo study of the electrothermal phenomenon in silicon-on-insulator and silicon-germanium-on-insulator metal-oxide field-effect transistors," *J. Appl. Phys.*, vol. 107, no. 6, pp. 064 506-1–064 506-9, Mar. 2010.

[Sch06] M. Schröter, H. Tran. Charge-storage related parameter calculations for Si and SiGe bipolar transistors from device simulation, Proc. WCM, International NanoTech Meeting, Boston , MA, pp. 735–740, 2006.

[Sch11] M. Schröter, G.Wedel, B. Heinemann, C. Jungemnn, J. Krause, P. Chevalier, and A. Chantre. Physical and Electrical Performance Limits of High-Speed SiGeC HBTs-Part I: Vertical Scaling, IEEE Transactions on Electron Devices, Vol. 58, No. 11, pp. 3687–3696, Nov. 2011.

[Sch17] M. Schröter, T. Rosenbaum, P. Chevalier, B. Heinemann, S. Voinigescu, E. Preisler, J.Boeck, and A. Mukherjee. SiGe HBT technology: Future trends and TCAD based roadmap, Proc. of the IEEE, Vol. 105, No. 6, pp. 1068–1086, 2017.

[Sel84] S. Selberherr. Analysis and Simulation of Semiconductor Devices. Springer, 1984.

[Sha15] P. Sharma *et al.*, "Modeling of hot-carrier degradation in nLDMOS devices: different approaches to the solution of the Boltzmann transport equation," *IEEE Trans. Electron Devices*, vol. 62, no. 6, pp. 1811–1818, Jun. 2015.

[Slo77] J. W. Slotboom and H. C. de Graaff. Bandgap narrowing in silicon bipolar transistors. *IEEE Trans. Electron Devices*, vol. 24, no. 8, pp. 1123–1125, Aug. 1977.

[Sta11] I. Starkov *et al.*, "Hot-carrier degradation caused interface state profile- simulation versus experiment," *J. Vac. Sci. Technol. B*, vol. 29, no. 1, pp. 01AB091–01AB098, Jan. 2011.

[Sta12] I. Starkov, H. Enichlmair, S. Tyaginov, and T. Grasser, "Charge-pumping extraction techniques for hot-carrier induced interface and oxide trap spatial distributions in MOSFETs," in *Physical and Failure Analysis of Integrated Circuits (IPFA)*, July 2012, pp. 1–6.

[Ste96] A. Stesmans, "Revision of H2 passivation of pb interface defects in standard (111) Si/SiO2," *Appl. Phys. Lett.*, vol. 68, no. 19, pp. 2723–2725, Feb. 1996.

[Sto98] K. Stokbro *et al.*, "STM-induced hydrogen desorption via a hole resonance," *Phys. Rev. Lett.*, vol. 80, no. 12, pp. 2618–2621, Mar. 1998.

[Str62] R. Stratton. Diffusion of hot and cold electrons in semiconductor barriers. Physical Review, vol. 126, no. 6, pp. 2002–2014, 1962.

[Syn15] Synopsys, Inc., 690 E. Middlefield Road Mountain View, CA 94041. Sentaurus Device User Guide, version k-2015.06 edition, June 2015.

[Tom93] K. Tomizawa. Numerical Simulation of Submicron Semiconductor Devices. Artech House Boston, London, 1993.

[Tya16] S. Tyaginov, M. Jech, J. Franco, P. Sharma, B. Kaczer, and T. Grasser, "Understanding and modeling the temperature behavior of hot-carrier degradation in SiON n-MOSFETs," *IEEE Electron Device Lett.*, vol. 37, no. 1, pp. 84–87, Jan. 2016.

[Van] T. Vanhoucke *et al.*, "Physical description of the mixed-mode degradation mechanism for high performance bipolar transistors," in *Proc. IEEE Bipolar/BiCMOS Circuits Technol. Meeting*, 2006, pp. 25–28.

[Vas09] D. Vasileska, K. Raleva, and S. M. Goodnick, "Self-heating effects in nanoscale FD SOI devices: the role of the substrate, boundary conditions at various interfaces, and the dielectric material type for the box," *IEEE Trans. Electron Devices*, vol. 56, no. 12, pp. 3064–3071, Dec. 2009.

[Vas10] D. Vasileska, K. Raleva, and S. M. Goodnick, "Electrothermal studies of FD SOI devices that utilize a new theoretical model for the temperature and thickness dependence of the thermal conductivity," *IEEE Trans. Electron Devices*, vol. 57, no. 3, pp. 726–728, Mar. 2010.

[Wed10] G. Wedel and M. Schröter. Hydrodynamic simulations for advanced SiGe HBTs. IEEE Bipolar/BiCMOS Circuits and Technology Meeting (BCTM) 2010, pp. 237–244, Oct 2010.

[Wed16] G. Wedel. A Box-Integration/WENO solver for the Boltzmann Transport Equation and its Application to High-Speed Heterojunction Bipolar Transistors. PhD thesis, Technische Universitt Dresden, 2016.

[Wie16] B. R. Wier, K. Green, J. Kim, D. T. Zweidinger, and J. D. Cressler, "A physics-based circuit aging model for mixed-mode degradation in SiGe HBTs," *IEEE Trans. Electron Devices*, vol. 63, no. 8, pp. 2987–2993, Aug. 2016.

[Zeb06] M. Zebarjadi, A. Shakouri, and K. Esfarjani, "Thermoelectric transport perpendicular to thin-film heterostructures calculated using the Monte Carlo technique," *Phys. Rev. B*, vol. 74, no. 19, pp. 195 331-1–195 331-6, Nov. 2006.

3

SiGe HBT Compact Modeling

A. Pawlak[1], M. Schröter[1,2] and B. Ardouin[3]

[1]Chair for Electron Devices and Integrated Circuits, Technische Universität Dresden, Germany
[2]Department of Electrical and Computer Engineering, University of California at San Diego, USA
[3]XMOD Technologies, France

Abstract

Fabrication and circuit design are linked by compact device modeling; i.e., the electrical characteristics of the devices fabricated on a wafer are represented by sufficiently simple but preferably still physics-based models that are suitable for circuit simulation and optimization. The importance of modeling has been growing rapidly due to strongly increased device complexity, manufacturing cost, and fabrication time. There is an increased demand from industry for first-pass success of high-frequency (HF) analog circuits in order to stay competitive. For SiGeC HBT technologies, ranging from production to the most advanced process, this has been successfully addressed by the standard compact bipolar transistor model HICUM/L2 [Schr10].

For practical applications, a compact model (CM) itself is not sufficient though. Its model parameters need to be determined from measurements of terminal (current, voltage) characteristics, preferably making use of clever test structures and mathematical manipulations (so-called parameter extraction methods) in order to be able to separate the various, often superimposed, physical effects and their related parameters. Consistent physics-based parameter extraction methods that provide for a given process accurate geometry-scalable and statistical device models not only represent a key enabler to first-pass design success but also yield important information for process development. One objective of DOTSEVEN was the development of improved or even new parameter extraction methods and to provide a unified set of test structures.

3.1 Introduction

The predicted THz performance of SiGeC HBTs along with their integration with digital CMOS indicates a bright future of the corresponding BiCMOS technologies for serving HF applications [Sch11a, Sch11b, Sch16a]. These predictions and the rising demand for mm- and sub-mm-wave[1] applications have motivated the development of improved SiGeC process technologies, leading to the most recent results [Hei16] of $(f_{\mathrm{T}}, f_{\mathrm{max}}) = (505, 720)$ GHz fabricated with a 130 nm lithography within the DOTSEVEN project.

The large variety of HF applications requires a versatile and accurate representation of such process technologies within the design kits in order to enable circuit optimization and exploiting the performance limits of the technology. Thus, an important focus of DOTSEVEN was the development of suitable simulation and modeling tools as well as the verification of the new models [Sch16c]. The corresponding effort on compact modeling of high-speed SiGeC HBTs and the associated experimental results are presented in this chapter.

Compact models – which are also sometimes called *electrical models* or *SPICE models* – were introduced in the 1970s as a constitutive and inseparable part of simulators for electronic circuits. Based on a set of parameterized analytical equations, CMs are meant to provide a suitable representation of the electrical characteristics of electronic devices under given bias, frequency, and temperature conditions. The large variety of applications in the Si industry has led to a strong preference for physics-based CMs, in which (i) the formulations for current and charge are derived as simplified solutions of fundamental equations for carrier transport and electrostatics, (ii) most of the parameters retain a physical meaning, and (iii) the equivalent circuit corresponds to the physical structure of the device. Such physics-based models enable not only device sizing-based circuit optimization but also efficient modeling of statistical process variations and process debugging.

The determination of the model parameters from device measurements, typically called *parameter extraction*, includes the specification of measurement conditions and the mathematical procedure for data manipulation for obtaining the desired parameter values. This task is sometimes, especially in

[1] In this chapter, the designation HF will be used for all these frequency ranges.

the III–V community, called model extraction. This incorrectly implies the "construction" of the CM, which is actually not included. Similarly, often just the term *model* is used referring to both the CM (formulation) and its associated parameter set. Since for the vast majority of application cases the CM is given, the usually tedious and lengthy task of its development should be distinguished from the task of extracting its parameters.

While CMs are developed independently of circuit simulators, the availability of high-level description languages (such as Verilog-A, cf. e.g., [Muk16]) and associated model compilers has significantly sped up not only the model implementation but also its release for all commercial simulators in the form of a single (reference) model code.

The various tasks mentioned above eventually lead to a working CM. When evaluating its accuracy, different aspects come into play, which are sometimes a source of confusion. The necessary conditions for obtaining an accurate CM are listed below:

1. The intrinsic ability of a CM to describe a device, i.e., the versatility and accuracy of its constitutive equations and its equivalent circuit.
2. The model coding by its developers (e.g., in Verilog-A), e.g., correctly modeling capacitances by their respective charges, following from the integration of the capacitance between equilibrium and the controlling voltage(s).
3. A correct model implementation by the EDA vendors in their circuit simulator. Even when using appropriate tools, like model compilers, this step induces many non-trivial optimizations and customizations, which can be a source of error.
4. The accuracy of the model parameters, i.e., how accurate the individual electrical characteristics of the model agree with those of the measured device to be represented.

Only meeting all of the above criteria with sufficient quality will provide the desired CM accuracy for a given technology or process. No matter how accurate and versatile the CM itself may be, poorly determined parameters will ruin the effort spent not only on its development but also for process development and circuit design.

Compact modeling of high-speed SiGeC HBTs within DOTSEVEN addressed two main aspects. First, libraries with geometry scalable model parameters for the HBTs fabricated in the project were provided, enabling the design of mm-wave circuit building blocks and demonstrators. Second, CMs in combination with TCAD were employed for analyzing the process in terms of causes of performance reduction by separating 3D parasitic effects from 1D transport effects in the intrinsic device structure [Kor15]. The compact HBT modeling in the project was based on HICUM Level 2 (L2), which has been a CMC supported standard since 2004. Compared to other existing CMs, HICUM/L2 has been continuously developed for HF/high-speed HBT technologies and applications and offers various HF-specific features as well as high accuracy up to higher frequencies and over a wider bias range. This chapter highlights recent model extensions that are relevant for DOTSEVEN-like technologies. Furthermore, important steps during parameter extraction are presented with an emphasis on the model extensions and physics-based geometry scaling. In addition, most recent evaluations of methods for determining the series resistances are briefly discussed.

Section 3.7 gives an overview about publications showing comparisons of DC, AC and non-linear large-signal characteristics with experimental data.

3.2 Overview of HICUM Level 2

A detailed derivation of the formulations of HICUM/L2 can be found in [Sch10] and is beyond the scope of this book. Below, just a brief overview of the model components is given to provide a reference for further discussions.

The equivalent circuit of HICUM/L2 is displayed in Figure 3.1. The intrinsic (1D) transistor behavior is described by the controlled current source for the transfer current I_{T}, that is calculated based on the generalized integral charge-control relation (GICCR, cf. Section 3.3), the dynamic currents resulting from the time-dependent depletion charges (Q_{jEi}, Q_{jCi}) and diffusion charges (Q_{dEi}, Q_{dCi}), and diodes for the currents injected into the emitter (I_{jBEi}) and collector (I_{jBCi}). Collector impact ionization is represented by the current I_{avl}. As proved in [Sch16b], the complicated behavior of the (ohmic) intrinsic collector (epi) region can be accurately described within the GICCR framework and does not require a separate element with a typically complicated description as it is the case in some other CMs.

Laterally distributed effects in the internal base region are modeled by the bias-dependent internal base resistance R_{Bi}, which takes conductivity modulation and emitter current crowding into account. Dynamic emitter current crowding during small-signal operation is modeled by the parallel capacitance C_{RBi}.

The injection across the emitter perimeter junction is represented by Q_{jEp} and I_{jBEp}. The current source $I_{\mathrm{BET(i,p)}}$ models band-to-band (BTB) tunneling and can be connected either to the internal (B$'$) or to the perimeter (B*) base node, depending on the transistor architecture.

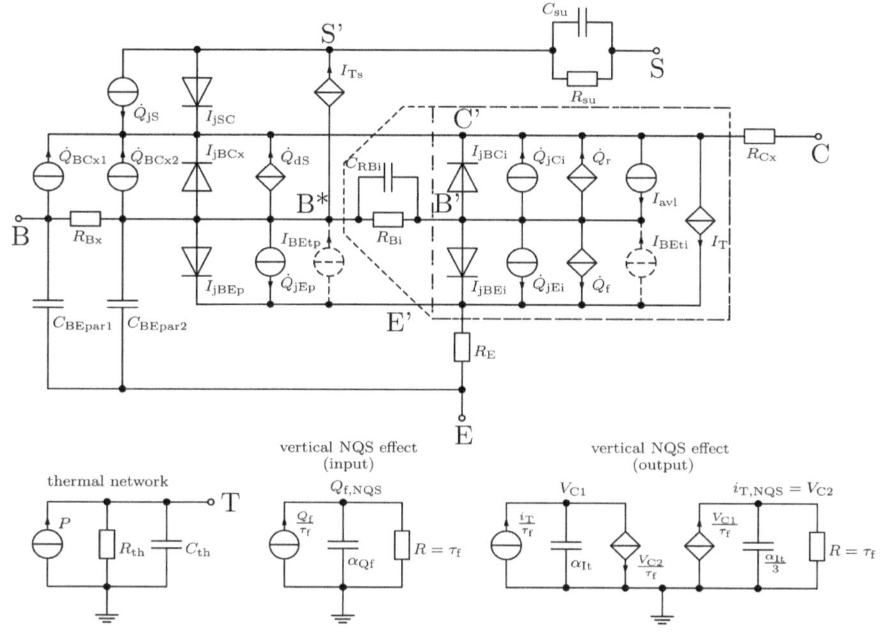

Figure 3.1 Equivalent circuit of HICUM/L2 including the adjunct networks for modeling electro-thermal effects and NQS effects. Not shown are the networks for modeling correlated noise [Her12, Sak15, Sch10]. The dash-dotted line defines the intrinsic (1D) transistor representation and the dashed line defines the internal transistor.

The external BC region is modeled by the junction current I_{jBCx} and the dynamic currents through the time-dependent charges $Q_{\text{jCx}'}$ and $Q_{\text{jCx}''}$. The latter include the depletion charge in the external base as well as the oxide capacitance of the shallow trench and contact region related parasitic capacitance between the base and the collector. The charges are split across the external base resistance to account for distributed lateral effects at high frequencies. The emitter contact and poly-silicon resistance are represented by R_{E}, while the external collector resistance R_{Cx} includes the collector contact, sinker, and buried layer contributions. The capacitances $C_{\text{BE,par1}}$ and $C_{\text{BE,par2}}$ include the BE spacer and parasitic poly-silicon contact region related capacitances between the base and the emitter.

The substrate transistor is modeled with simple expressions for the transfer current source I_{Ts} and the respective back-injection current I_{jSC}. The SC depletion charge is modeled by Q_{jS}, and a simple bias-independent storage time is used for describing the diffusion charge Q_{dS}.

Substrate coupling effects are described by a simple first-order frequency dependence with R_{Su} representing the finite resistance of the path between sub-collector and substrate contact and C_{Su} caused by the permittivity of the bulk substrate. An improvement of this simple equivalent circuit is presented in Section 3.5. Note that, based on the final circuit layout, any elaborate equivalent circuit can be connected to the substrate node.

Electro-thermal effects are taken into account by a simple first-order low-pass network consisting of the thermal resistance R_{th} and thermal capacitance C_{th}. The externally available temperature node allows the connection of both higher order and thermal coupling networks if required by the application [Leh14].

The extraordinary performance of SiGeC HBTs in DOTSEVEN has been achieved with changes in the transistor architecture. Changes in lateral directions result in a modification of both geometry scaling laws and the equivalent circuit. The most relevant aspects here will be discussed in Sections 3.5 and 3.6.3. Structural changes in the vertical direction impact the intrinsic transistor behavior. They have been accounted for in the formulations for the transfer current and stored charge, which will be discussed in the next two chapters. In versions of HICUM/L2 previous to 2.3, those effects were not considered explicitly, which led to increased effort for parameter extraction using conventional methods, potentially yielding non-physical model parameters.

3.3 Modeling of the Quasi-Static Transfer Current

3.3.1 Basics of the GICCR

As shown in [Sch10, Sch16b], the GICCR can be derived as closed-form solution from integrating the 3D drift-diffusion transport equation. This section summarizes the relevant features of the GICCR by focusing on just the vertical (1D) npn transistor structure. In this case, the 1D GICCR master equation reads

$$I_{\mathrm{T}} = \frac{(qA_{\mathrm{E}})^2 V_{\mathrm{T}} \overline{\mu_{\mathrm{nB}} n_{\mathrm{iB}^2}}}{Q_{\mathrm{ph}}} \left[\exp\left(\frac{V_{\mathrm{B'E'}}}{V_{\mathrm{T}}}\right) - \exp\left(\frac{V_{\mathrm{B'C'}}}{V_{\mathrm{T}}}\right) \right] = I_{\mathrm{Tf}} - I_{\mathrm{Tr}}, \tag{3.1}$$

with the elementary charge q, the emitter area A_{E}, the thermal voltage V_{T}, the electron mobility μ_{n}, the intrinsic carrier density n_{i}, and the voltages between the terminals of the 1D transistor $V_{\mathrm{B'E'}}$ and $V_{\mathrm{B'C'}}$. The denominator results

$$Q_{\mathrm{ph}} = A_{\mathrm{E}} q \int_{x'_{\mathrm{E}}}^{x_{\mathrm{C}'}} h(x) p(x) dx \tag{3.2}$$

from integrating the transport equation from the mono- to poly-silicon emitter interface, which defines the 1D emitter contact $x_{\mathrm{E}'}$, to the peak of the buried layer, which defines the 1D collector contact $x_{\mathrm{C}'} \cdot Q_{\mathrm{ph}}$ is a weighted hole charge, with p being the hole carrier density and the weight function $h(x)$

$$h(x) = h_{\mathrm{g}}(x) h_{\mathrm{J}}(x) h_{\mathrm{v}}(x). \tag{3.3}$$

Its first component,

$$h_{\mathrm{g}}(x) = \frac{\overline{\mu_{\mathrm{nB}} n_{\mathrm{iB}^2}}}{\mu_{\mathrm{n}}(x) n_{\mathrm{i}}^2(x)}, \tag{3.4}$$

accounts for all effects related to the field-dependent electron mobility and the spatially varying bandgap. The normalization factor $\overline{\mu_{\mathrm{nB}} n^2{}_{\mathrm{iB}}}$ is taken as average value over the neutral base. The second component,

$$h_{\mathrm{J}}(x) = -\frac{A_{\mathrm{E}} J_{\mathrm{n}}(x)}{I_{\mathrm{T}}}, \tag{3.5}$$

is, in the 1D case, related to volume recombination, while in the 3D case it also accounts for the lateral spreading of electron current density, typically in the collector. The third term

$$h_{\mathrm{v}}(x) = \exp\left(\frac{V_{\mathrm{B'E'}} - \varphi_{\mathrm{p}}(x)}{V_{\mathrm{T}}}\right), \tag{3.6}$$

takes into account the spatial variation of the hole Fermi-potential φ_{p} w.r.t. the chosen controlling (node) voltage. The deviation is expected to be negligible across the vertical neutral base region, but can become significant in the 3D case for emitter current crowding.

The spatial dependence of the components (3.4) and (3.5) in Figure 3.2(a) shows that h_{J} and h_{v} are close to 1 in those regions where the hole density matters.

In contrast, the weight function h_{g} shown in Figure 3.2(b) follows mostly the spatial dependence of the bandgap, except for the BC depletion region, where both mobility and hole density are much lower than in the other transistor regions. As a consequence, for the i_{T} formulation of the intrinsic transistor only the impact of h_{g} and, in particular, of the bandgap needs to be taken into account. The influence of h_{J} and h_{v} can be included in the

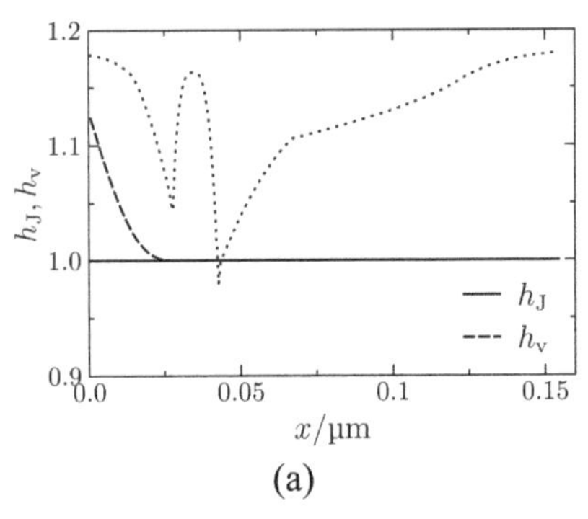

(a)

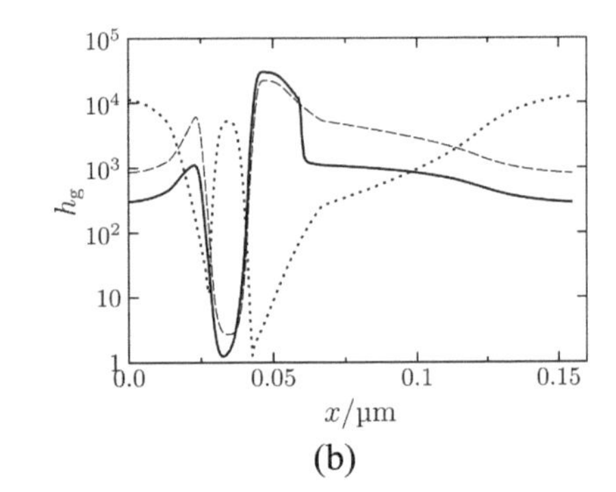

(b)

Figure 3.2 Spatial dependence of the weight functions (a) h_J and h_v for $J_C = 5\text{mA}/\mu\text{m}^2$, and (b) h_g (solid line) for low injection. In both pictures the dotted line shows the 1D doping profile in log-scale. In (b), the dashed line shows the bandgap in linear scale.

3D formulation through the weighted charge formulation (collector current spreading) and the internal base resistance (emitter current crowding).

For compact modeling, it is useful to split Q_{ph} into several components according to the device structure and transistor operation principle. The most suitable split leads to the sum of a zero-bias hole charge Q_{p0} and an excess charge ΔQ_{ph} given by the change Δp of the hole density with non-zero bias:

$$\Delta Q_{ph} = q \int_{x_{E'}}^{x_{C'}} h(x)\Delta p(x)dx = h_{jEi}Q_{jEi} + h_{jCi}Q_{jCi} + Q_{fh}. \quad (3.7)$$

Here, Q_{jEi} and Q_{jEi} are the physical depletion charges of both junctions with the corresponding weight factors h_{jEi} and h_{jCi}, and Q_{fh} is the weighted mobile charge in the transistor. The normalization factor $\overline{\mu_{nB}n_{iB^2}}$ in h_g and (3.1) is chosen such that the weight factor for Q_{p0} becomes 1. The other weight factors in (3.7) are defined as average values in the corresponding transistor region *k*:

$$h_k = \frac{\int_k h(x)\Delta p(x)dx}{\int_k \Delta p(x)dx} = \frac{q\int_k h(x)\Delta p(x)dx}{Q_k}. \quad (3.8)$$

The weighted mobile hole charge in (3.7) is further divided according to the different transistor regions:

$$Q_{fh} = h_{f0}\tau_{f0}I_{Tf} + h_{fE}\Delta Q_{fE} + \Delta Q_{fB} + h_{fC}Q_{fC} + \tau_r I_{Tr}. \quad (3.9)$$

Here, I_{Tf} and I_{Tr} are the forward and reverse transfer currents, respectively, as defined in (3.1), with τ_{f0}, and τ_r being the corresponding (low current) transit times. ΔQ_{fE}, ΔQ_{fB}, and Q_{fC} are the physically stored excess charges in the emitter, base and collector regions at medium and high current densities at forward operation, while h_{f0}, h_{fE}, and h_{fC} are the corresponding weight factors according to (3.8). The weight factor for the base charge is left close to 1 since the value of h_g is close to 1 (cf. Figure 3.2(b)) due to the choice of $\overline{\mu_{\mathrm{nB}} n_{\mathrm{iB}^2}}$. To keep the model simple. Also no dedicated weight factor is used for the reverse charge. Note that although Q_{ph} is related to the actual hole charge, it does not have a physical representation by itself in the transistor. Thus, in contrast to the actual charge, it cannot be measured directly.

In summary, the GICCR in the form of (3.1) represents a physically consistent closed-form description for the transfer current of HBTs, which also provides clear guidance on how additional effects need to be included. The following sections describe the extensions of the GICCR presented above with respect to the advanced SiGeC HBT technology developed in DOTSEVEN and related projects (DOTFIVE, RF2THz).

3.3.2 SiGe HBT Extensions

Within DOTFIVE and DOTSEVEN very different approaches toward SiGe HBT technology development were taken. When investigating the fabricated HBTs, significant differences in the ideality of the transfer current characteristic and in the current density dependence of the corresponding transconductance had been observed. An example of this behavior is shown in Figure 3.3(a). The observed differences were eventually traced to back to different Ge profiles in these technologies, in particular within the BE

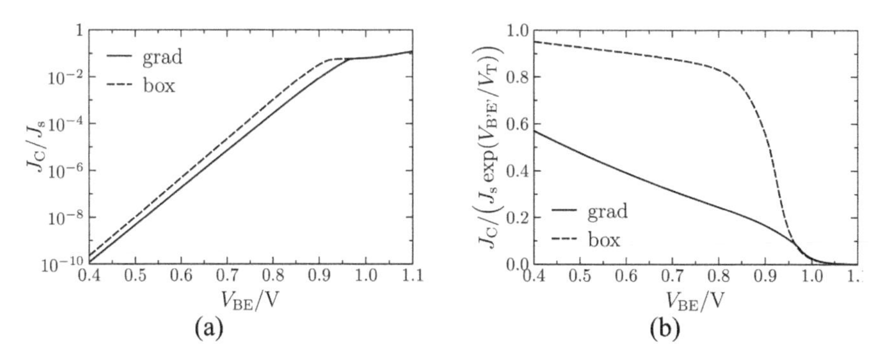

Figure 3.3 (a) Transfer current for transistors with different Ge profiles in the base. (b) Transfer current normalized to their ideal formulation for the same transistors as in (a) at room temperature and $V_{\mathrm{B'C'}} = 0$ V.

space-charge region (SCR), cf. [Paw14a]. While spatially constant Ge profiles show the expected almost ideal behavior, grading the Ge already within the BE SCR led to a significant deterioration in ideality and the transconductance already at medium collector current densities.

Normalizing the transfer current to its ideal form, $I_s \exp(V_{BE}/V_T)$, magnifies the above mentioned non-ideality as shown in Figure 3.3(b). This increase in non-ideality was first noted in [Cra93] and later modeled in [Paa01] by modifying the Gummel number. In this section, the origin of the effect will be briefly reviewed and then its incorporation into the transfer current expression by applying the GICCR will be demonstrated.

A graded Ge profile leads to a spatially dependent bandgap in the base region which in turn leads to a spatially dependent intrinsic carrier density

$$n_i(x) = n_{i0} \exp\left(\frac{\Delta V_g(x)}{2V_T}\right) \tag{3.10}$$

with n_{i0} representing pure Si and $\Delta V_g(> 0)$ as the bandgap reduction due to the Ge mole fraction. Since the bandgap decreased with increasing Ge content, the intrinsic carrier density increases with x.

According to the classical theory for bipolar transistors, the electron density n_e injected at the beginning x_e of the neutral base (cf. Figure 3.4) is given by:

$$n_e = \frac{n_i^2(x_e)}{N_B} \exp\left(\frac{V_{B'E'}}{V_T}\right). \tag{3.11}$$

With increasing forward voltage $V_{B'E'}$ across the BE SCR, x_e moves towards the junction and thus smaller values of n_i. This in turn reduces the injected electron density compared to the ideal case of a spatially independent n_i and leads to the non-ideality observed in Figure 3.4.[2]

The varying n_i enters the GICCR via the weight function (3.4), which needs to be inserted into (3.7) in order to calculate the weight factor h_{jEi} for Q_{jEi}. In the following derivation, a spatially constant base doping profile is assumed for $x_{jE} < x \leq x_{e0}$. Also, for simplifying the expressions, a coordinate transformation is applied such that x_{jE} is chosen as new reference:

$$x' = x - x_{jE}. \tag{3.12}$$

[2]Note that from Figure 3.4 the depletion charge follows as $Q_{jEi} = qN_B(x_{e0} - x_e)$.

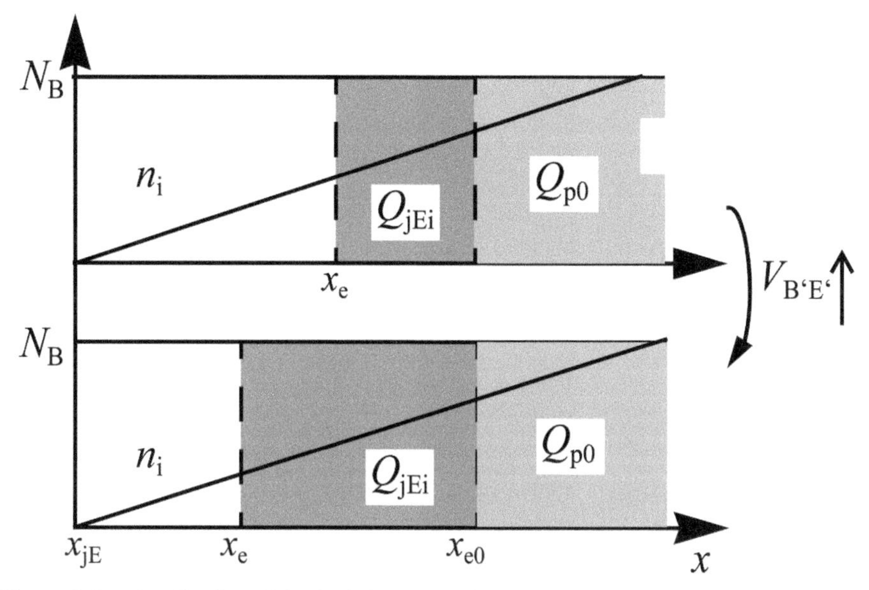

Figure 3.4 Visualization of the depletion charge in the base–emitter space charge region for two bias points with increasing voltage from the top picture to the bottom. The intrinsic carrier density is given in log-scale. x_{jE} relates to the metallurgic BE junction, while x_e and x_{e0}, respectively, are the boundaries of the space charge region for $V_{B'E'} > 0\text{V}$ and $V_{B'E'} = 0\text{V}$, respectively.

Assuming a fully depleted SCR, i.e., $p = 0$ for $x_{jE} \leq x < x_e$, and $p = N_B$ in $x_e \leq x < x_{e0}$, the weight factor is given by:

$$h_{jEi} = \frac{\int_{x'_{e0}}^{x'_e} h(x')\Delta p(x')dx'}{\int_{x'_{e0}}^{x'_e} \Delta p(x')dx'} = \frac{N_B \int_{x'_{e0}}^{x'_e} h(x')dx'}{N_B(x'_e - x'_{e0})} = \frac{\int_{x'_{e0}}^{x'_e} h(x')dx'}{x'_e - x'_{e0}}. \quad (3.13)$$

Inserting (3.4) with (3.10) and assuming a spatially independent electron mobility yields:

$$h_{jEi} = \frac{\overline{\frac{\mu_{nB} n^2{}_{iB}}{\mu_n n_{i,jE}}} \int_{x_{e0}}^{x_e} \exp\left(-\frac{\Delta V_g(x)}{V_T}\right) d_x}{x'_e - x'_{e0}} = \frac{c_{hBE} \int_{x'_{e0}}^{x'_e} \exp\left(-\frac{\Delta V_g(x)}{V_T}\right) dx}{x'_e - x'_{e0}} \quad (3.14)$$

with $n_{i,jE} = n_i(x_{jE})$. The spatially linear bandgap variation is given by:

$$\Delta V_g(x) = \frac{\Delta V_{g,max} x'}{V_T x'_{Vg,max}} = \frac{x'}{a_{ni}}, \quad (3.15)$$

with the maximum bandgap change $\Delta V_{g,max}$ and the location of this maximum, $x'_{Vg,max}$. Inserting this into (3.14) leads to

$$h_{jEi} = -a_{ni} c_{hBE} \frac{\exp\left(-\frac{x'_e}{a_{ni}}\right) - \exp\left(-\frac{x'_{e0}}{a_{ni}}\right)}{x'_e - x'_{e0}}. \tag{3.16}$$

Following classical theory, the width of the SCR is

$$x'_e = x'_{e0}\sqrt{1 - \frac{V_{B'E'}}{V_{DEi}}}, \tag{3.17}$$

which finally leads to

$$h_{jEi} = c_{hBE} \exp\left(-\frac{x'_{e0}}{a_{ni}}\right) \frac{\exp\left(\frac{x'_{e0}}{a_{ni}}\left(1 - \sqrt{1 - \frac{V_{B'E'}}{V_{DEi}}}\right)\right) - 1}{\frac{x'_{e0}}{a_{ni}}\left(1 - \sqrt{1 - \frac{V_{B'E'}}{V_{DEi}}}\right)}. \tag{3.18}$$

A similar expression can also be obtained for an exponential dependence of the base doping profile within $x_{jE} \leq x < x_{e0}$.

Replacing the square root by the grading coefficient z_{Ei}, which allows to capture realistic base doping profile shapes, yields the final model equation

$$h_{jEi} = h_{jEi0} \frac{\exp\left(a_{hjEi}\left(1 - \left(-1\frac{V_{B'E'}}{V_{DEi}}\right)^{z_{Ei}}\right)\right) - 1}{a_{hjEi}\left(1 - \left(-1\frac{V_{B'E'}}{V_{DEi}}\right)\right)^{z_{Ei}}} \tag{3.19}$$

where various quantities in (3.18) have been merged into the model parameters $a_{hjEi} = x'_{e0}/a_{ni}$ and $h_{jEi0} = c_{hBE}\exp(-a_{hjEi})$.

The application of the model is shown in Figure 3.5 for transistors with different Ge shapes. A weaker grading means a smaller value of $\Delta V_{g,max}$ in (3.15) leading to a smaller increase of the weight factor. For all cases, the model equation yields very accurate results. For the box profile, no grading is present, leading to $a_{hjEi} = 0$. In the model, a series expansion of (3.19) is used for small a_{hjEi}, which leads to $h_{jEi} = h_{jEi0}$ for $a_{hjEi} = 0$. Furthermore, the singularities at $V_{B'E'} = 0\,\mathrm{V}$ and at $V_{B'E'} = V_{DEi}$ are avoided in the model formulation.

For a Ge profile increase throughout the entire base region, i.e., including the base portion of the BC SCR, also the weight factor h_{jCi} becomes bias dependent but decreases with increasing voltage $V_{B'C'}$. The behavior of h_{jCi}

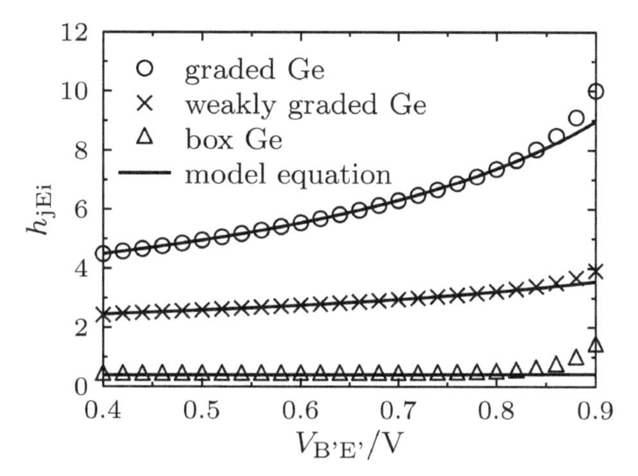

Figure 3.5 Application of the model Equation (3.19) for the weight factor obtained from transistors with different shapes of the Ge profile [Paw15a].

can be described by a similar function as (3.19) and its own set of parameters. The value of h_{jCi} decreases with the slope of the Ge grading, which results in a reduction of the Early effect. A simple physics-based explanation for this is that the Ge grading causes a strong aiding drift field E_{nx} across the base region. Once the injected electron current density J_{n} is dominated by drift, the position of the BC SCR boundary x_{c} at the end of the neutral base has only a weak impact on the value of J_{n}. In other words, in a box Ge profile (i.e., $E_{\mathrm{nx}} \approx 0$) J_{n} is driven the diffusion gradient of the injected carriers. This is visualized in Figure 3.6 by the curve $\zeta \approx 0$, where

$$\zeta = -\frac{E_{\mathrm{nx}}}{V_{\mathrm{T}}/w_{\mathrm{B}}} \tag{3.20}$$

represents the field factor and the normalization factor $V_{\mathrm{T}}/w_{\mathrm{B}}$ corresponds to the equivalent field of a pure diffusion current. For a high field, the carrier gradient disappears as shown in Figure 3.6 for different values of $\zeta > 0$. Hence, moving the location of x_{c} does not change J_{n} anymore, thus resulting in the observed larger Early voltage for graded base HBTs. Since the bias dependence of the corresponding weight factor h_{jCi} is very small, no dedicated model equation for $h_{\mathrm{jCi}}(V_{\mathrm{B'C'}})$ is included in HICUM/L2.

Except for the non-idealities in the transfer current at low injection that were explained before, also the increasing non-linearity of SiGe HBTs with graded Ge compared to transistors with a box Ge profile can be seen even before the onset of high-current effects. An explanation for this phenomenon

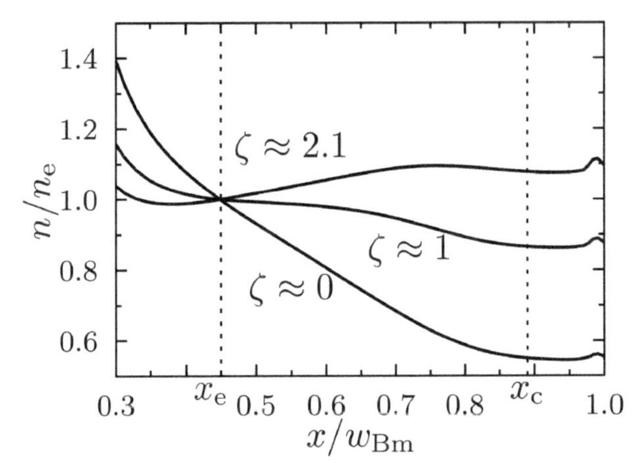

Figure 3.6 Spatial dependence of the electron density normalized to n_e from (3.11) in the neutral base, marked by the vertical dashed lines, for different values of the field factor ζ. The x-axis is normalized to the metallurgical base width w_{Bm} with the BE junction located at $x = 0$.

can be found by a closer look at the different parts of the mobile charges and the relation to the corresponding weight factors according to the GICCR.

The reason for the increased non-ideality is also indicated by the voltage drop from the internal emitter contact E′ at $x = 0$ to $x = x_e$. The voltage drop is given in Figure 3.7 for transistors with different Ge profiles. The corresponding hole densities are given in Figure 3.8(a). It can be seen that although the voltage drop in the emitter is the same for all transistors up to medium current densities, the voltage drop in the BE SCR for a given current density is significantly larger for the transistor with the graded Germanium profile. This is caused by the neutral charge in the SCR. As shown in Figure 3.8(b), for a given collector current density this charge is visibly smaller for the graded Ge compared to the box Ge. Since however the current density is defined by the electron density and the gradient of the quasi-Fermi potential, the reduced carrier density can only be compensated for by an increased voltage drop for maintaining the current density.

For modeling though, only the potential at the E′ is known but the voltage drop not calculated explicitly. However, it was shown in, e.g., [Paw15a, Fri02] that a voltage drop in the internal transistor can be directly expressed by the weight function in the GICCR. The resulting weight factors for the mobile charge in the emitter,

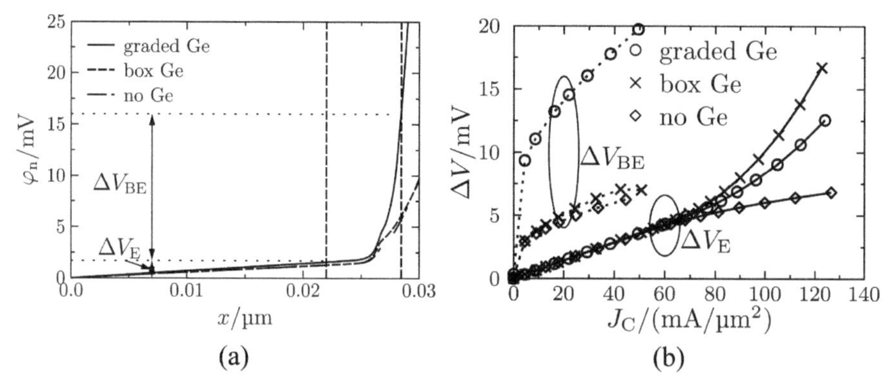

Figure 3.7 (a) Determination of the voltage drop in the neutral emitter (ΔV_E) and in the BE SCR (ΔV_{BE}) from the quasi-fermi potential of the electrons for transistors with different Ge profiles in the base. The dashed lines show the begin and end of the BE SCR. (b) Current dependence of the voltage drops; ΔV_{BE} is shown for bias points only up to the beginning of high-current effects [Paw15a].

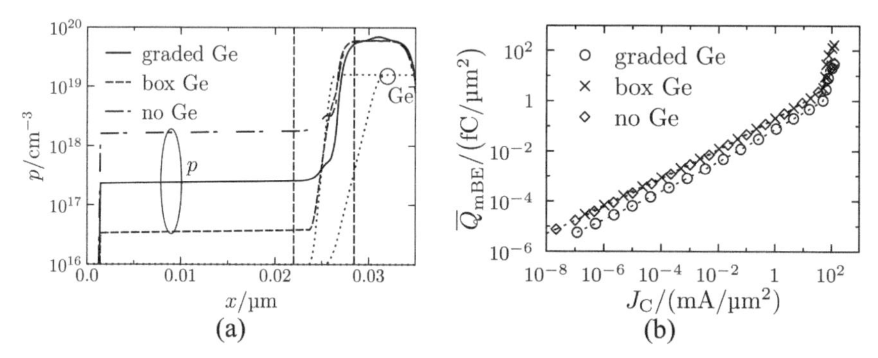

Figure 3.8 (a) Spatial dependence of the hole density for transistors with different Ge profiles. The Ge content is given by the dotted lines for the box (left) and the graded (right) profile. The vertical dashed lines are the same as in Figure 3.7(b) Minority charge as a function of collector current density in the BE SCR for transistors with different Ge profile [Paw15a].

$$h_{fE} = \frac{\overline{\mu_{nB} n^2{}_{iB}}}{\mu_{nE} n^2{}_{iE}}, \tag{3.21}$$

and in the BE SCR,

$$h_{mBE} = \frac{\overline{\mu_{nB} n^2{}_{iB}}}{\mu_{nBE} n^2{}_{ijE}}, \tag{3.22}$$

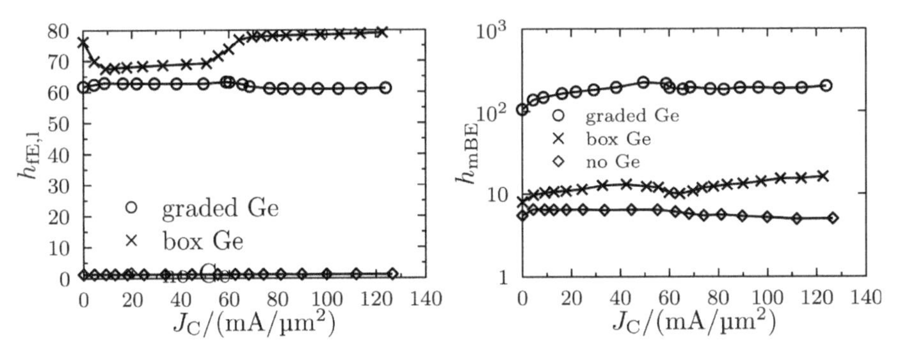

Figure 3.9 Current dependence of the weight factors for the charge stored in the neutral emitter and BE SCR for the transistors with different Ge profiles [Paw15a].

respectively, show just a weak current dependence (see Figure 3.9). While $h_{\mathrm{fE,1}}$ equals 1 for BJTs, it is much lager than 1 for HBTs due to the reduced band-gap in the base of the latter compared to that in the Si emitter, i.e., $n_{\mathrm{iB,SiGe}} \gg n_{\mathrm{iB,Si}}$, but $n_{\mathrm{iE,SiGe}} = n_{\mathrm{iE,Si}}$.

The weight factor h_{mBE} is larger than 1 even for the BJT and box Ge profile due to the bandgap variation across the BE SCR caused by the doping profile (i.e., the bandgap narrowing effect). A Ge grading across the BE SCR can increase the value of h_{mBE} significantly as shown in Figure 3.9.

The low-current component of the mobile charge, $Q_{\mathrm{fl}} = \tau_{\mathrm{f0}} I_{\mathrm{Tf}}$, is experimentally characterized by a low-current transit time τ_{f0}, which is difficult to partition into its components from the different spatial transistor regions. Therefore, in HICUM/L2 the mobile charges in emitter, base, and BC SCR are merged into the low-current minority charge Q_{f1} [cf. (3.9)]. For describing the transfer current, a weighted charge $h_{\mathrm{f0}}\, Q_{\mathrm{fl}}$ has to be inserted according to (3.9). For advanced SiGe HBTs, a weight factor $h_{\mathrm{f0}} > 1$ turned out to be necessary for modeling the transfer characteristics at medium injection levels while maintaining a physics-based value of Q_{p0}.

Two components of Q_{fl}, the one related to the neutral base and to the BC SCR exhibit a dependence on $V_{\mathrm{B'C'}}$. In addition, the Ge grading also causes the weight factor of the neutral base charge to be $V_{\mathrm{B'C'}}$ dependent. Since the $V_{\mathrm{B'C'}}$ dependence is a strong function of the actual transistor design, i.e., the doping concentrations in the base and collector, only a very general model equation according to

$$h_{\mathrm{f0}} = h_{\mathrm{f00}}(1 + a_{\mathrm{hf0,c}} V_{\mathrm{B'C'}}) \tag{3.23}$$

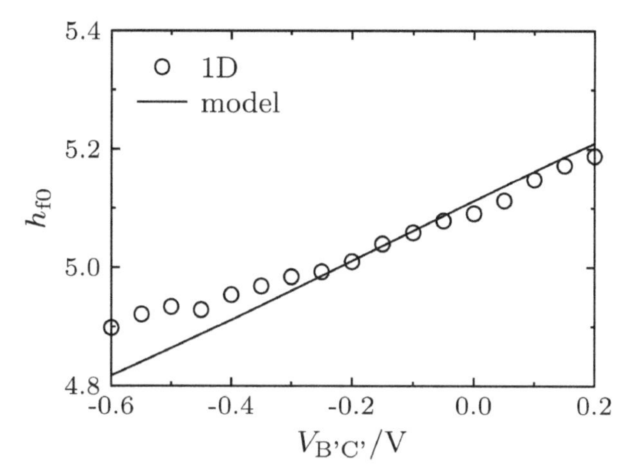

Figure 3.10 Bias dependence of h_{f0} extracted from 1D device simulations and application of the model equation (3.23).

is employed. The bias dependence of h_{f0} based on 1D device simulations is given in Figure 3.10. The extracted values from measured transistors are shown later in Figure 3.28.

The weight factors h_{fE} and h_{fC} for the high-current components ΔQ_{fE} and Q_{fC} in (3.9) were already available in HICUM/L2 prior to version 2.3. For the emitter, (3.21) is applied with a bias-independent value according to Figure 3.9. The weight factor for the collector,

$$h_{fC} = \frac{\overline{\mu_{nB} n^2{}_{iB}}}{\mu_{nC} n^2{}_{iC}} \tag{3.24}$$

is only relevant at high injection since under forward operation the hole carrier density and related charge in the collector are negligible at low and medium injection.

3.3.3 Temperature Dependence

As outlined in the introduction of this chapter, the transfer current exhibits unique temperature effects for graded Germanium transistors compared to box transistors (or BJTs). In addition to the temperature dependence of the charge components, the weight factors play a significant role for correctly describing the temperature dependence of the transfer current. The temperature dependence of the weight factors can be derived systematically from (3.4) and is mostly related to the ratio of the intrinsic carrier densities in the particular transistor region *k* via the region-specific bandgap. The general equation for the weight function temperature dependence therefore reads

$$\frac{h_k(T)}{h_k(T_0)} = \frac{\overline{\mu_{nB}(T)n_{iB}^2(T)}}{\overline{\mu_{nk}(T)n_{ik}^2(T)}} \frac{\overline{\mu_{nk}(T_0)n_{ik}^2(T_0)}}{\overline{\mu_{nB}(T_0)n_{iB}^2(T_0)}} \approx \frac{\overline{n_{iB}^2(T)}}{\overline{n_{iB}^2(T_0)}} \frac{\overline{n_{ik}^2(T_0)}}{\overline{n_{ik}^2(T)}}, \quad (3.25)$$

neglecting the temperature dependence of the mobility. Assuming a linear temperature dependence of the spatially averaged bandgap voltage $\overline{V_{gk}}$ in a particular transistor region *k* yields for the corresponding ratio

$$\frac{\overline{n_{ik}^2(T)}}{\overline{n_{ik}^2(T_0)}} = \left(\frac{T}{T_0}\right)^3 \exp\left[\frac{\overline{V_{gk0}}}{V_T}\left(\frac{T}{T_0} - 1\right)\right] \cong \exp\left[\frac{\overline{V_{gk0}}}{V_T}\left(\frac{T}{T_0} - 1\right)\right] \quad (3.26)$$

with $\overline{V_{gk0}}$ as the toward $T = 0$ extrapolated bandgap voltage. This gives for the corresponding weight factor

$$\frac{h_k(T)}{h_k(T_0)} = \exp\left(\frac{\overline{V_{gB0}} - \overline{V_{gk0}}}{V_T}\left(\frac{T}{T_0} - 1\right)\right). \quad (3.27)$$

Therefore, if the bandgap of region *k* is larger than that of the base region, the value of h_k will decrease with increasing temperature. Except for h_{jCi}, this is indeed the case for all weight factors in HBTs with a graded Ge profile across the entire base.

Following from (3.27), for h_{jEi} the temperature dependence of h_{jEi0} is given by the term

$$h_{jEi0}(T) = h_{jEi0}(T_0) \exp\left(\frac{\Delta V_{gBE}}{V_T}\left(\frac{T}{T_0} - 1\right)\right) \exp(a_{hjEi}(T_0) - a_{hjEi}(T)), \quad (3.28)$$

which has been changed into the more flexible model expression

$$h_{jEi0}(T) = h_{jEi0}(T_0) \exp\left(\frac{\Delta V_{gBE}}{V_T}\left[\left(\frac{T}{T_0}\right)^{\zeta_{VgBE}} - 1\right]\right), \quad (3.29)$$

with

$$\Delta V_{gBE} = \overline{V_{gB0}} - \overline{V_{gjE0}} < 0 \quad (3.30)$$

and the exponent coefficient ζ_{VgBE} as model parameters. In addition to h_{jEi0}, the parameter a_{hjEi} is also temperature dependent due to $x_{e0}(T)$ and V_T in

(3.18) and (3.15). The former can be directly expressed by the corresponding depletion capacitance $C_{jEi0}(T)$, the latter directly by T, yielding

$$a_{hjEi}(T) = a_{hjEi}(T_0)\frac{T_0}{T}\frac{C_{jEi0}(T_0)}{C_{jEi0}(T)} = a_{hjEi}(T_0)\frac{T}{T_0}^{\zeta_{hjEi}}, \tag{3.31}$$

where introducing the exponent coefficient ζ_{hjEi} as model parameter provides more flexibility.

Since a large portion of h_{f0} is related to h_{mBE} which itself depends on V_{gjE0} as well, a similar temperature dependence for h_{f0} is derived reading

$$h_{f0}(T) = h_{f0}(T_0)\exp\left(\frac{\Delta V_{gBE}}{V_T}\left(\frac{T}{T_0} - 1\right)\right) \tag{3.32}$$

The aforementioned model equations are compared with device simulation results in Figure 3.11, showing both the expected decrease of the weight factors and sufficiently high accuracy.

The temperature dependence of the high-current weight factors follows directly from (3.23) and (3.21).

$$h_{f(E,C)}(T) = h_{f(E,C)}(T_0)\exp\left[\frac{\overline{V_{gB0}} - \overline{V_{g(E,C)0}}}{V_T}\left(\frac{T}{T_0} - 1\right)\right] \tag{3.33}$$

Figure 3.12 shows the corresponding comparison with device simulation results.

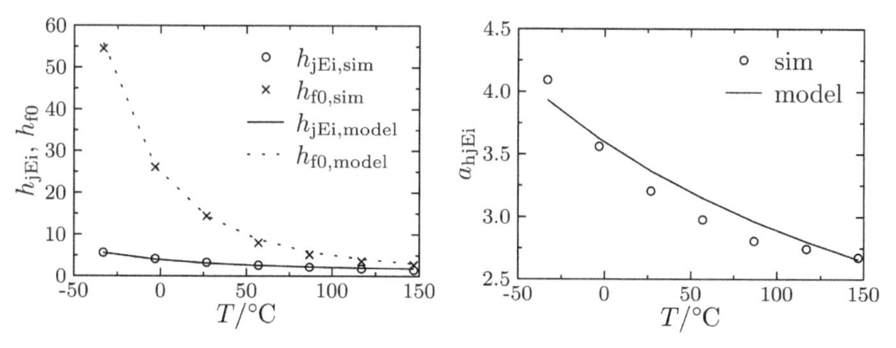

Figure 3.11 Application of the model Equations (3.29) and (3.32) to $h_{jEi0}(T)$ and $h_{f0}(T)$ as well as Equation (3.31) to a_{hjEi} obtained from 1D device simulations [Paw15a].

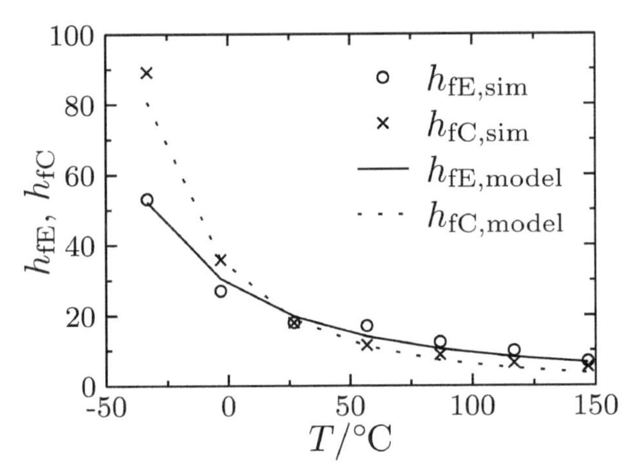

Figure 3.12 Application of (3.33) to the high current weight factors h_{fE} and h_{fC} obtained from 1D device simulations [Paw15a].

3.4 Charge Storage

3.4.1 Critical Current

Based on the Kirk-effect, the critical current I_{CK} characterizes the onset of high-current effects. It is modeled in HICUM/L2 by the equation

$$I_{CK} = \frac{V_{Ci}}{R_{Ci0}\left(1+\left(\frac{V_{Ci}}{V_{lim}}\right)^{\delta_{ck}}\right)^{\frac{1}{\delta_{ck}}}}\left[1+\frac{v+\sqrt{v^2+a_{ICKpt}}}{2}\right] \tag{3.34}$$

with $v = (V_{Ci} - V_{lim})/V_{PT}$ as argument,

$$V_{Ci} = V_{C'E'} - V_{C'E's} \tag{3.35}$$

as effective collector voltage and $V_{C'E's}$ as the internal CE saturation voltage, and V_{lim} corresponding to the electric field separating the ohmic from the saturation region in the velocity versus field relation.

For low values of V_{Ci} the entire epi-collector region becomes neutral at the onset of high current densities. This is represented in (3.34) by the first term, where the V_{Ci}-dependent term in the denominator models the field dependent mobility. The low-field mobility has been absorbed in the model parameter R_{Ci0}, which resembles the ohmic resistance of the entire epi-collector region with an average doping concentration N_{Ci}. The parameter

δ_{ck} was set to 2 in versions of HICUM/L2 prior to 2.3, but has recently been introduced as model parameter in order to allow a more flexible voltage-dependent (i.e., collector field) description of I_{Ck} (e.g., $\delta_{ck} = 1$ for pnp transistors).

The last term within the square brackets in (3.34) represents the high-voltage solution which is characterized by the collector punch-through voltage

$$V_{PT} = \frac{qN_{Ci}}{2\varepsilon} w_{Ci}^2. \tag{3.36}$$

While the collector doping has been continuously increasing for high-speed HBTs, the collector width w_{Ci} has decreased, yielding smaller and smaller values for V_{PT}. Furthermore, a_{ICKpt} was recently introduced as a model parameter (rather than a fixed parameter) for the hyperbolic smoothing function that connects the low- and high-voltage regions in order to provide a highly accurate modeling of the (quasi-)saturation region in the output characteristics and to avoid possible kinks due to non-physical parameter values [Cel14].

The effect of δ_{ck} on the voltage dependence of I_{CK} is shown in Figure 3.13(a). The lower δ_{ck} reduces the slope at lower voltages (i.e., field in the collector), but the asymptotic value for I_{CK} will be the same at very large voltages since the saturation velocity is not changed by the parameter. Figure 3.13(b) exhibits the impact of a_{ICKpt}. The possible kink in I_{CK} for too small values of a_{ICKpt} can be clearly observed.

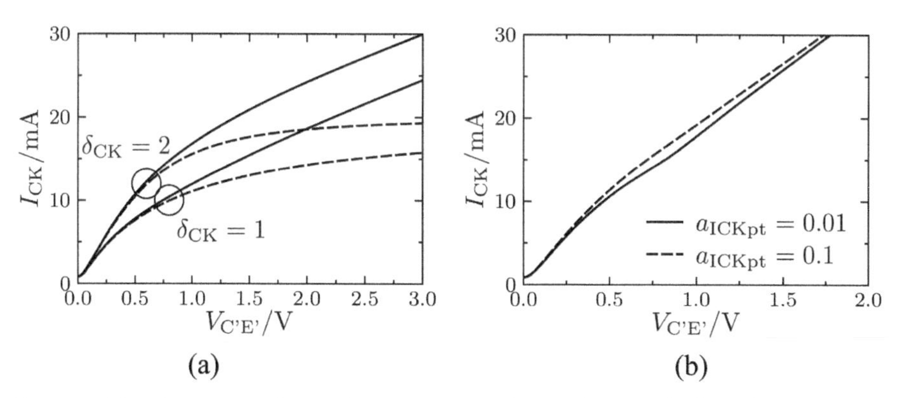

Figure 3.13 (a) Impact of δ_{ck} on the voltage dependence of I_{CK}. Solid lines show the actual I_{CK} while dashed lines show the results for the low-voltage portion of [i.e., the first term of (3.34), neglecting punch-through]. (b) Impact of a_{ICKpt} on I_{CK} for a very small punch-through voltage.

3.4.2 SiGe Heterojunction Barrier

The changing composition from SiGe back to Si within the BC SCR causes a barrier in the valence band. Due to the very small difference in electron affinity between Si and SiGe (for typical Ge concentrations not exceeding 30%), this barrier is almost completely given by the difference in bandgap between the Ge (peak) location and the pure Si collector region. For low current densities and CE voltages beyond strong saturation, the barrier is typically masked by the electric field in the BC SCR. However, at high current densities the electric field in the BC SCR starts collapsing due to the compensation of N_{Ci} by a high mobile carrier concentration. Since the BC barrier initially prevents holes from being injected into the collector (unlike in a BJT) not only the increase in electron current density with $V_{\mathrm{B'E'}}$ becomes limited but also a dipole layer starts to form around the barrier [Sch10]. This leads to a shift of the barrier height from the valence band into the conduction band and hole injection from the base into the collector to enable a further increase in current density via additional diffusion. The formation of the barrier is highlighted in Figure 3.14 where also the barrier height in the conduction band, $\Delta V_{\mathrm{C,bar}}$, is defined for the case of high current densities.

Resulting from observing its current dependence, the barrier height in HICUM/L2 is modeled by a simple empirical expression,

$$\Delta V_{\mathrm{C,bar}} = V_{\mathrm{C,bar}} \exp\left(-\frac{2}{i_{\mathrm{bar}} + \sqrt{i_{\mathrm{bar}}^2 + a_{\mathrm{Cbar}}}} \right), \tag{3.37}$$

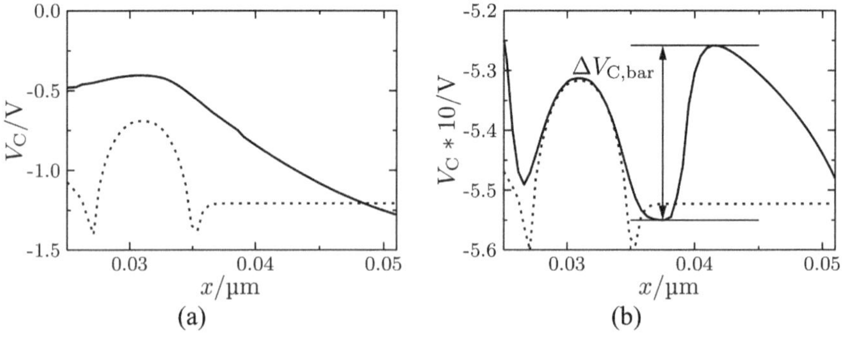

Figure 3.14 Spatial dependence of the conductance band edge for low and high current densities (solid lines), highlighting the presence of the barrier at high current densities. The dashed line shows the doping profile of the transistor just for reference.

with the normalized current

$$i_{\mathrm{bar}} = \frac{I_{\mathrm{Tf}} - I_{\mathrm{CK}}}{I_{\mathrm{C,bar}}} \tag{3.38}$$

The model is based on the observation that, independent of the collector voltage, the barrier shows almost the same current dependence starting from an onset current. For the latter, I_{CK} is used since usually this onset current correlates with the classical Kirk-effect as long as the barrier is located not too far away from the metallurgic junction in the collector. Figure 3.15 shows the current dependence of $\Delta V_{\mathrm{C,bar}}$ obtained from 1D device simulation in comparison with (3.37) for a wide range of (internal) CE voltages.

The impact of the BC barrier on the mobile charge and associated transit time is split into two distinct components [Sch10]. First, the charge storage contribution in the base region caused by the BC barrier only is calculated by

$$\Delta Q_{\mathrm{Bf,b}} = \tau_{\mathrm{Bfvs}} I_{\mathrm{Tf}} \left[\exp\left(\frac{\Delta V_{\mathrm{C,bar}}}{V_{\mathrm{T}}} \right) - 1 \right]. \tag{3.39}$$

Second, the classical high-injection charge in the collector and its associated base component (triggered by the collector related high-current effect) [Sch99] is extended by a barrier term,

$$\Delta Q_{\mathrm{fh,c}} = \tau_{\mathrm{hCs}} I_{\mathrm{Tf}} w^2 \exp\left(\frac{\Delta V_{\mathrm{C,bar}} - V_{\mathrm{C,bar}}}{V_{\mathrm{T}}} \right), \tag{3.40}$$

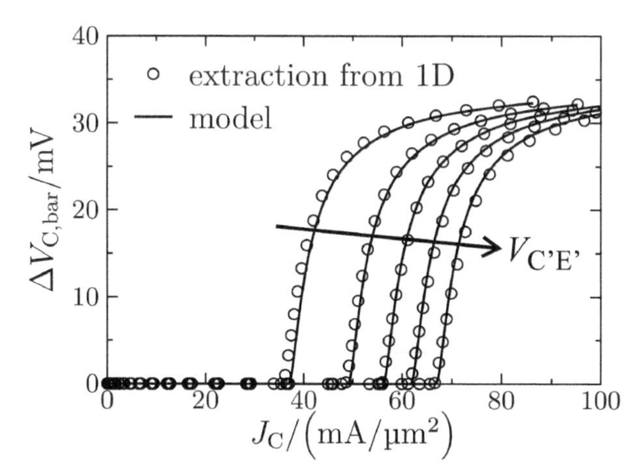

Figure 3.15 Modeling of the current dependence of the heterojunction barrier voltage for different voltages $V_{\mathrm{C'E'}}/V = [0.3; 0.6; 0.9; 1.2; 1.5]$.

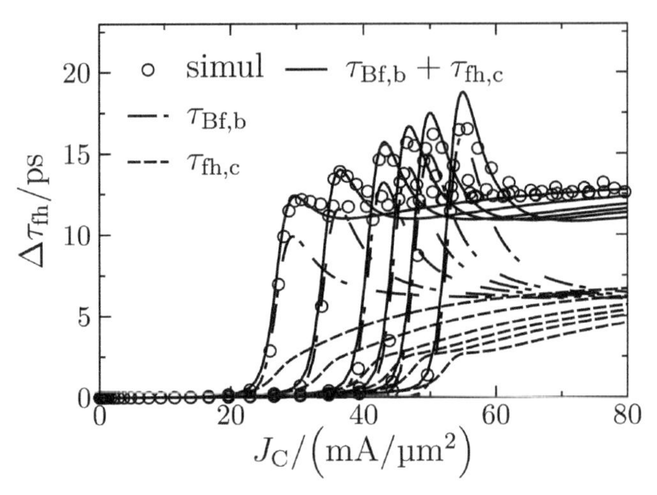

Figure 3.16 Transit times of a SiGe HBT showing the BC barrier effect: comparison of Equations (3.39) and (3.40) with results from 1D device simulation for different voltages $V_{C'E'}/V = [0.3; 0.5; 0.8; 1.0; 1.2; 1.5]$.

which causes a delayed nonlinear increase until the barrier is built up in the conduction band. Figure 3.16 shows the corresponding transit times obtained from the charge formulations above as

$$\tau = \left.\frac{dQ}{dI_{Tf}}\right|_{V_{C'E'}} \tag{3.41}$$

and compared to 1D device simulation results. The different components according to (3.39) and (3.40) are drawn separately as well as their sum which yields the small-signal transit time seen at the terminal of the 1D transistor.

3.5 Intra-Device Substrate Coupling

At high frequencies, the signal coupling between the collector of a transistor and its surrounding substrate can strongly affect the small-signal behavior, especially the HF output impedance. Different signal paths have to be distinguished. Since the collector (i.e., the buried layer) of a Si-based HBT usually forms a pn junction with the substrate, a signal path to the bulk substrate contacts exists across the area component of the CS junction capacitance C_{jSa} in Figure 3.17. In addition, a signal path through the perimeter junction and shallow or deep trench exists which is represented by C_{jSp}, C_{STI}, and C_{DTI}

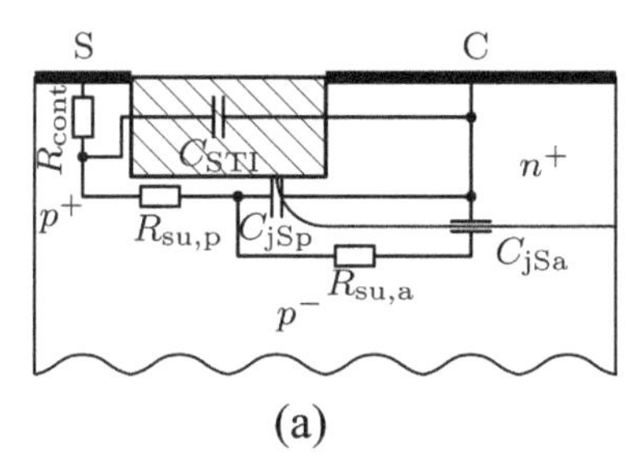

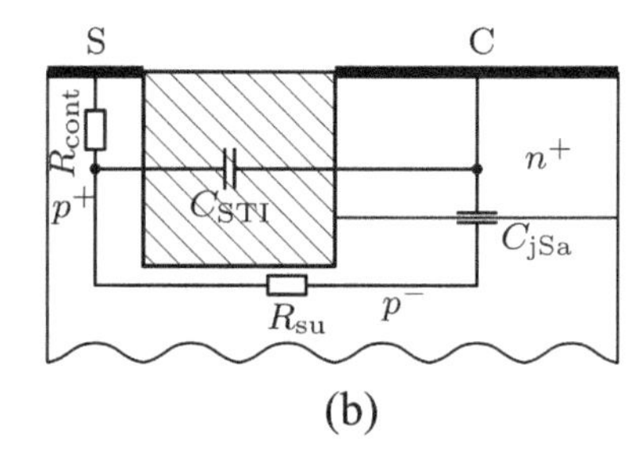

Figure 3.17 Sketch of the cross section for (a) a junction isolated (with partial trench-isolation) and (b) a deep trench isolated collector including all relevant elements of the most simple equivalent circuit for modeling intra-device substrate coupling. Note that for all series resistance a parallel capacitance exists due to the permittivity of the substrate and that due to changes of the substrate-collector SCR all depletion capacitances $C_{\mathrm{jS(a,p)}}$ and series resistances $R_{\mathrm{Su,(a,p)}}$ are bias dependent [Sch10, Paw15a].

in Figure 3.17. This perimeter path is much shorter than the bulk path if the substrate contact surrounds the transistor and is placed as close as possible to the collector, which is the case for device characterization in test structures. This coupling effect is called *intra*-device coupling. However, in circuits the situation is quite different since typically the surrounding substrate contact is omitted and the substrate is contacted somewhere on the chip. In this case, not only the transistor output impedance has a different frequency dependence but there is also signal coupling though the substrate directly between transistors. This effect is called *inter*-device substrate coupling. Both intra- and inter-device coupling are strongly layout dependent.

It is important to understand for both modelers and circuit designers that the CMs delivered in PDKs should be consistent with the p-cells offered to circuit designers. If the p-cells do not contain a surrounding substrate contact (in contrast to the characterization structures), then the PDK model should not include intra-device coupling. In this case, in which the circuit layout is unknown to the modeling engineers, it is the responsibility of the circuit designer to determine the substrate coupling and cross-talk related impedance network for each (critical) transistor!

In this section, the discussion is limited to intra-device coupling with a connected substrate ring. The general impact of the signal coupling on the output conductance $g_{\mathrm{o}} = \mathrm{Re}\{\underline{Y}_{22}\}$ is visualized in Figure 3.18. It can be seen that, for low and medium current densities, substrate coupling leads to a strong increase of g_{o} already at lower frequencies. This increase is proportional to the square of the frequency.

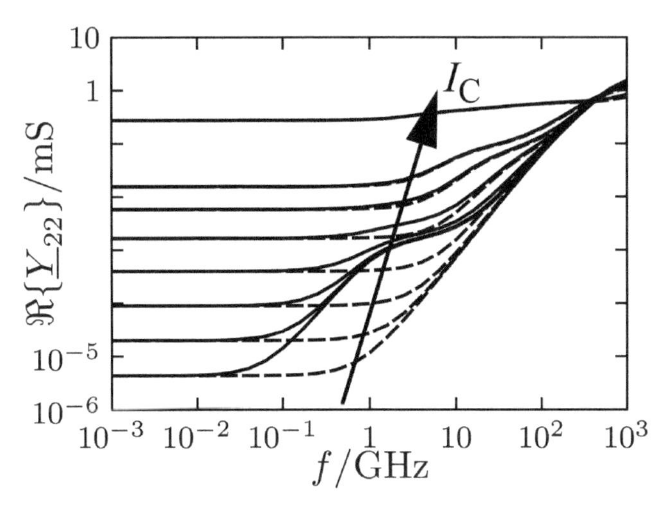

Figure 3.18 Impact of intra-device substrate coupling on the dynamic output-conductance given by the real part of $\underline{Y}_{22}$. Solid lines show the actual values including substrate coupling while the dashed lines show results with an ideally open substrate.

The elements of the equivalent circuit in Figure 3.17 follow directly from the transistor structure. The direct path between the collector and the substrate may be represented either by the CS perimeter depletion capacitance C_{jSp} for the case of pure junction isolation or by a bias-independent (at least to first order) deep trench oxide capacitance C_{DTI} in case of a deep trench isolation. For a combination of a shallow trench and junction isolation, a combination of bias-dependent and -independent capacitances is required. Depending on the spatial distance between the junction and the substrate contact significant series resistances can exist in the various signal paths. For each of the bulk related series resistances a parallel capacitance C_{Su} exists due to the permittivity of the substrate.

All elements given in Figure 3.17 are lumped elements which represent highly distributed effects that depend on the transistor dimensions and the operating frequency. Therefore, even more complicated equivalent circuits than the one in Figure 3.17 may be required to correctly capture the frequency behavior of the output impedance at mm- and sub-mm-wave frequencies. For accurately capturing the impact of intra- and inter-device coupling, the topology of the respective equivalent circuit along with its element values can generally only be obtained from analyzing the actual layout of all components *after* designing the circuit.

In HICUM/L2 the equivalent circuit shown in Figure 3.19 was chosen which is capable of capturing intra-device substrate coupling as it is encountered during device characterization and model parameter extraction as well as during circuit design if p-cells with surrounding substrate contacts are employed. In Figure 3.19, C and S are the terminal collector and substrate contact, R_{Cx} is the external collector resistance, C_{jS} is the bottom component of the SC depletion capacitance, and R_{Su} and C_{Su} represent the connection through the bottom part of substrate. These elements correspond to C_{jSa} and R_{Su} (in Figure 3.17(b)) or $R_{\mathrm{Su,a}}$ (in Figure 3.17(a)), respectively.

The perimeter substrate capacitance C_{SCp} follows from C_{STI} and C_{jSp} where it is important to realize that in case of a combined trench and junction isolation (cf. Figure 3.17(a)) portions of C_{jSp} may be included in C_{jS} depending on the width of the trench oxide and therefore the value of $R_{\mathrm{Su,p}}$.

The additional series resistance R_{cont} along the side-wall of the trench-oxide is not included in the CM in order to reduce the node count for low-frequency operation. As demonstrated in Figure 3.19 it can be connected to the substrate terminal of the CM in a subcircuit. Similar to the discussion for C_{SCp}, the spatial dimensions and relations between the circuit element values define whether $R_{\mathrm{Su,p}}$ (if present) is merged into R_{cont} or R_{Su}.

In contrast to C_{jS} which is modeled depending on the internal SC voltage $V_{\mathrm{S'C'}}$, the bias dependence of C_{SCp} can be activated or deactivated by the user to include the contributions of the perimeter depletion capacitance and the constant trench oxide, respectively.

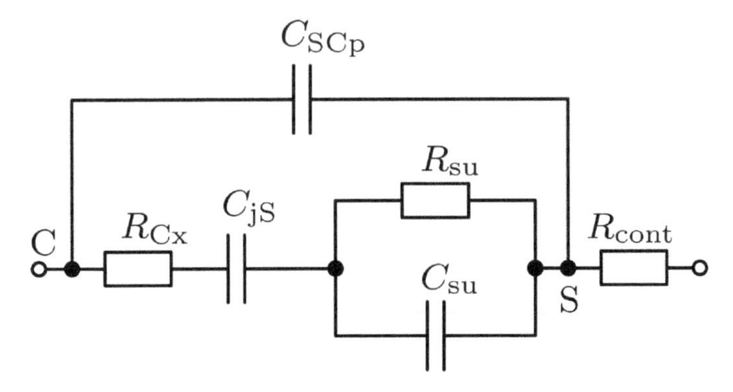

Figure 3.19 Substrate coupling equivalent circuit in HICUM/L2.

3.6 SiGe HBT Parameter Extraction

For various reasons, it should *not* be attempted to determine the parameters of a sophisticated physics-based compact HBT model such as HICUM/L2 from the characteristics of just a single transistor[3]. First, the equivalent circuit represents still to some extent the distributed device structure; for instance, there are internal and perimeter related elements. Also, the transfer current results from a well-defined transformation of a two-transistor behavior into a single transistor representation (cf. the effective emitter area section later) and, e.g., the series resistance values are scaled based on sheet or contact resistivities and device dimensions. The information corresponding to those partitionings and transformations can simply not be obtained from measurements on a single device. Second, a one-hat-fits-all single-device geometry is never being used in analog RF circuit design. In fact, exploiting a process technology (and amortizing the cost for its development as rapidly as possible) requires optimizing the circuits by using suitable and typically different transistor sizes and configuration for the different applications even within the circuits. This requires to cover a certain range of device sizes during parameter extraction. Third, utilizing device size adds another independent dimension and set of data points for determining the unknown parameters and increases the chances for obtaining physics-based parameter values. The number of independent data can be increased by adding special test structures to the (test) transistors. Fourth, only a physics-based set of parameters maximizes the use of a physics-based CM by enabling statistical and predictive circuit design and modeling.

An extraction follows a certain procedure that is preferably designed such that as many as possible model parameters are determined independently. The description below will reflect that sequence, starting with an overview of series resistance determination. Then, the methods for those parameters are covered that have been introduced in the extended equations. Finally, an extended concept for geometry-scalable parameter extraction is described.

Compact models are supposed to represent the typical characteristics of a process. For their first delivery along with the process qualification only limited statistics are available which need to be utilized though for selecting a proper die for parameter extraction. Since it is unlikely that all process control monitor values of an available die match all their nominal values from

[3]Reasonable results have been obtained though for at least all parameters related to the internal transistor after careful deembedding of all external elements and heavy utilization of optimization [Ros13].

wafer tests, the extracted model parameters need to be "shifted" properly (e.g., [Sch05]) to the nominal values. This is possible only with a physics-based model or, more precisely, with the associated model parameters having a clear physics-based relation to process parameters.

3.6.1 Extraction of Series Resistances

Preferably, a series resistance can be determined from its components as shown in Figure 3.20 for the example of the base region. Each different structural region in the cross section is represented by a resistor element. The value of the latter can always be calculated from a specific resistivity, the length of the region in direction of the current flow (b), and its cross-sectional area. Except for contacts, semiconductor layers are typically characterized by the sheet resistance r_S, which eliminates the need for knowing the spatially dependent vertical doping profile and reduces the cross-sectional area to the lateral layout dimension (l) perpendicular to the current flow. Contact resistances are calculated either from an area-specific resistivity (in $\Omega\mu\mathrm{m}^2$) if the current crosses the contact cross-sectional area vertically or from a length-specific resistivity (in $\Omega\mu\mathrm{m}$) if the current flows to the side. The contact in Figure 3.20. belongs to the latter category. Thus, the overall external base series resistance of the structure example in Figure 3.20 reads:

$$R_{\mathrm{Bx}} = \underbrace{\frac{\rho_{\mathrm{B,c}}}{l_{\mathrm{Bc}}}}_{\text{Contact}} + \underbrace{r_{\mathrm{BS,po}}\frac{b_{\mathrm{po}}}{l_{\mathrm{B,po}}}}_{\text{poly-Si on oxide}} + \underbrace{r_{\mathrm{BS,pm}}\frac{b_{\mathrm{pm}}}{l_{\mathrm{B,pm}}}}_{\text{mono-Si on poly-Si}} + \underbrace{r_{\mathrm{BS,l}}\frac{b_{\mathrm{l}}}{l_{\mathrm{E,l}}}}_{\text{Link (spacer)}} \tag{3.42}$$

where $\rho_{\mathrm{B,c}}$ is the base contact resistivity, $r_{\mathrm{BS,ab}}$ is the sheet resistance of the region "ab," and the b_{ab} and l_{ab}, respectively, are the widths and (effective) lengths of each region.

The form (3.42) allows to calculate the base series resistance for all sizes and even for changes in the dimensions and doping profiles as they occur during fabrication (resulting in process tolerances) and process development. A general and accurate formulation of (3.42) can be found in [Sch08a, Sch10], which is applicable to all common contact configurations and SiGe HBT architectures. Obviously, to employ (3.42) for generating R_{Bx} for a given HBT device size, the parameters of each component need to be known. The determination of the actual (i.e., not drawn) dimensions can be obtained from TEM and SEM measurements. Once the process is qualified, the relation between each actual and drawn dimension can be established, so that for

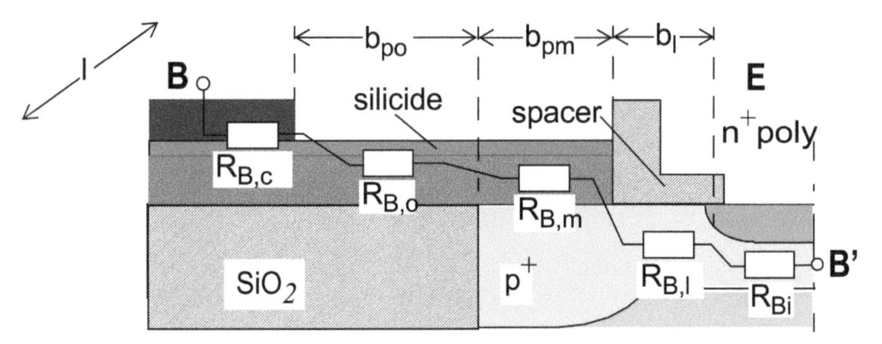

Figure 3.20 Illustration of the base series resistance components and their relation to the HBT cross section (schematic).

model generation the (drawn) layout dimension can be used. The specific contact and sheet resistance can be determined based on very simple (DC) test structures [Sch88, Sch08b] as shown in Figure 3.21 for the example of the contact and poly-on-mono region. Forcing a current from B1 and B3 and measuring the voltage drops between (a) B1 and B3 and (b) B1 and B2 using a Kelvin contact configuration gives two resistance values for the two unknowns $\rho_{B,c}$ and $R_{BS,po}$. Extending the structure to include more base

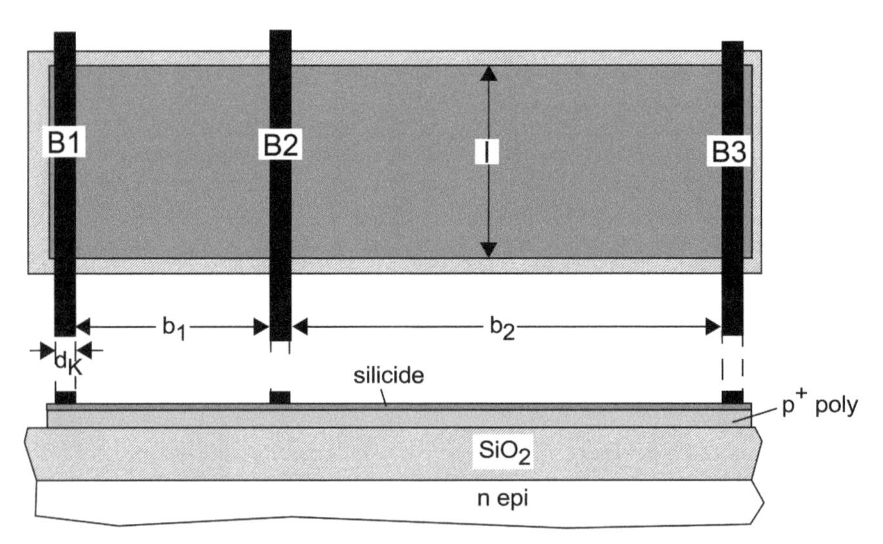

Figure 3.21 Typical test structure (a.k.a. contact chain) used for determining the specific electrical resistivities of the external base resistance components. B1, B2, B3 designates the contacts.

layers (as in Figure 3.20) allows the successive determination of all other sheet resistances. The principle described above can be applied also to the determination of the components of the external collector resistance.

From transistor theory (e.g., [Pri67, Sch10]), the bias-dependent internal DC base resistance is generally given by

$$R_{\mathrm{Bi}} = r_{\mathrm{SBi}} \frac{w_{\mathrm{E}}}{l_{\mathrm{E}}} g_{\mathrm{i}}(b_{\mathrm{E}}, l_{\mathrm{E}}) \psi_{\mathrm{dc}}(I_{\mathrm{Bi}}, r_{\mathrm{SBi}}, b_{\mathrm{E}}, l_{\mathrm{E}}), \tag{3.43}$$

where r_{SBi} is the sheet resistance of the internal base region (i.e., under the emitter), $b_{\mathrm{E}}(l_{\mathrm{E}})$ is the (effective) emitter width (length), I_{Bi} is the internal base current, g_{i} is a geometry factor that takes into account all common emitter shapes [Sch91, Sch92, Sch10], and Ψ_{dc} is the emitter current crowding function. The latter can be neglected (i.e., $\Psi_{\mathrm{dc}} = 1$) for all modern SiGe HBT process technologies. For a given HBT size, only the zero-bias sheet resistance

$$R_{\mathrm{Bi0}} = r_{\mathrm{SBi0}} \frac{w_{\mathrm{E}}}{l_{\mathrm{E}}} g_{\mathrm{i}}(w_{\mathrm{E}}, l_{\mathrm{E}}), \tag{3.44}$$

is required as parameter in HICUM. Hence, r_{SBi0} needs to be extracted from measured data. The test structure of choice here is the transistor tetrode [Sch88, Rei91, Sch07]. The principle method for extracting r_{SBi} over a wide reverse and forward bias range *under simultaneous transistor operation* is described in detail in [Rei91]. There have been other proposals on how to utilize the tetrode for determining r_{SBi} or R_{B}. Most recently [Sch17], the existing methods have been applied to various advanced process technologies. The corresponding comparison clearly indicates that the method in [Rei91] is the most accurate over the widest bias range (including even high the high-current region).

Furthermore, the extracted bias-dependent sheet resistance of the internal base, r_{SBi}, is utilized to determine the zero-bias hole charge of the transistor Q_{p0}, which is required for the accurate calculation of the transfer current. Due to the Early-effect and injected minority charge, the base sheet resistance

$$r_{\mathrm{SBi}} \cong r_{\mathrm{SBi0}} \frac{Q_{\mathrm{p0}}}{Q_{\mathrm{p0}} + Q_{\mathrm{jEi}} + Q_{\mathrm{jCi}} + Q_{\mathrm{f}}}, \tag{3.45}$$

is modulated by the voltages of both SCRs through the depletion charges Q_{jEi} and Q_{jCi} and the diffusion charge Q_{f}. Note that above equation is slightly simplified by neglecting the bias dependence of the hole mobility. Figure 3.22. shows the extracted r_{SBi} curves in the low bias range for two different technologies. In each case, the different bias conditions lead to a

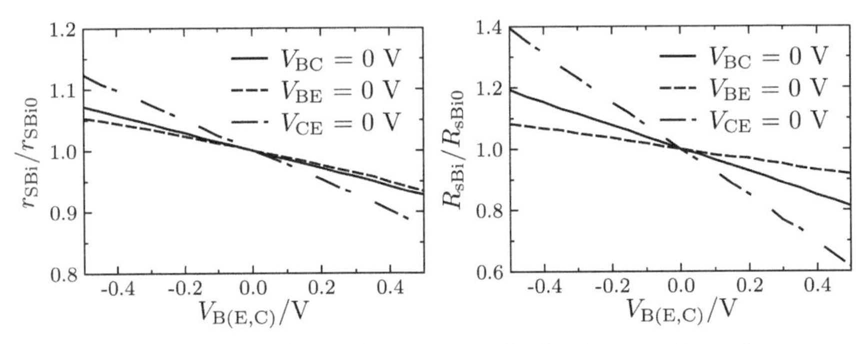

Figure 3.22 Internal base sheet resistance, normalized to its zero-bias value r_{SBi0}, as extracted from tetrodes for different bias conditions and technologies.

different contribution of the depletion charges in (3.45), while mobile charge is negligible in al cases. The stronger impact of the BE junction charge compared to the BC junction charge is clearly visible by the larger slope for $V_{BC} = 0$ V. Also, for the technology on the r.h.s. a larger impact of the space charges on Q_{p0}, can be observed, indicating a lower base doping at a similar width.

While it is possible for BJTs (i.e., bipolar technologies until the early 1990s) to determine the emitter resistance from dedicated test structures, this has become impossible for modern SiGe HBTs. The reason for this is that the by far largest contribution to the emitter resistance still comes from the interface between the poly- and mono-silicon, but that its formation is intrinsically coupled to the emitter (poly-)layer formation; i.e., there is no separate emitter implant anymore. Therefore, the emitter resistance has to be measured directly on a HBT structure (see below). But this should be done on various geometries in order to obtain the simple and usually sufficient scaling equation,

$$R_E = \rho_{E,C}/A_{E0} \tag{3.46}$$

for a single emitter window. Here, $\rho_{E,c}$ is the poly-to-mono-Si contact resistivity and A_{E0} is area of this interface, which corresponds to the area of the actual emitter window opening.

Some modelers still question the accuracy of transistor theory and prefer to determine series resistances directly from a given HBT structure. Over the past >60 years of bipolar transistor technology development, many different methods were proposed. There often quite different results can become confusing especially for young modeling engineers and the question arises which of these methods work at all and which ones are actually applicable to advanced SiGe HBT technologies. To answer these questions, recently several studies on the extraction methods for base, collector and emitter resistance have been completed, in which the various methods were applied to SiGe HBTs with widely varying emitter sizes from six different process technologies, ranging from established production to the most advanced prototyping processes. In all cases, the (absolute) accuracy of each method was assessed by applying it to a complete HICUM/L2 model for the particular process and comparing the obtained resistance with the known one of the model. This approach also allows the investigation and identification of the causes of observed failures. The results are briefly summarized below; for detailed results, the reader is referred to the corresponding references.

In [Kra15], a comprehensive and detailed study of nine widely used methods (and their variants) for extracting the emitter resistance R_E is presented. Using high-performance and high-voltage devices with a wide range of up to 12 emitter sizes, the results of this study are believed to be representative for the actual accuracy and applicability of the various R_E extraction methods. It was found that *none* of the existing methods works reliably across *all* process technologies. The most important causes of deviations are the strongly simplified equivalent circuit and the neglect of important physical effects (such as high-injection, CB barrier effect, self-heating) in the derivation of the methods. The two methods working mostly are the ones in [Paw14b, Hue04], but the one in [Paw14b] is based on the occurrence of self-heating. Methods based on fly-back/open-collector and impact-ionization, which increase the risk of device destruction, are not reliable in practice. This also applies to HF small-signal methods, where the impact of R_E is masked by the much higher base resistance. The Ning-Tang method is the least reliable of all methods.

A detailed quantitative analysis of the most widely used methods for determining the base resistance directly from a transistor was performed in [Paw16] for a wide range of emitter geometries. The CM-based assessment clearly revealed that all methods only enable the extraction of the external base resistance, while the determination of the internal (bias dependent) base resistance is either impossible or, at best, limited to just a very narrow bias range, typically at very high injection, and are not very accurate. Small-signal HF methods, when operated at very low V_{CE} values, yield the best results, although still not with reliable accuracy across all technologies. A major cause of the failure or inaccuracy of the DC-based methods is self-heating. Thus, the use of DC operation-based methods cannot be recommended since self-heating will rather increase in future technologies. The application of the extraction methods to experimental data confirmed the large spread in the

R_{Bx} results (for the same transistor structure) that was already observed from the CM. From both experimental and model based data, it was found that although some methods work reasonably well for some process technologies no method yields reliable accuracy for *all* technologies. Overall, it can be concluded that an accurate determination of both the *total* base resistance R_{B} and the external base resistance R_{Bx} from widely used single-transistor-based methods is highly unreliable even if small-signal methods are employed.

Finally, a comprehensive and detailed study of eight widely used methods (and their variants) for extracting the DC collector series resistance was presented in [Paw18]. Again, no method yielded accurate and reliable results across all technologies. But RF methods that rely on just a simple equivalent circuit, some of the substrate transistor-based methods, and the open-emitter method, yield overall the best results. The most important causes of deviations are the neglect or too strong simplification of the description of important physical effects (such as self-heating, high-injection). Note, the none of the methods yields any reasonable result for the bias-dependent internal collector resistance.

The recommendation for all single HBT-based series resistance extraction methods is to apply them to the utilized CM again in order to verify whether the same result is obtained. If not, then the method is unsuitable for the chosen model (equivalent circuit).

3.6.2 Extraction of the Transfer Current Parameters

The extraction of the parameters for the transfer current of HICUM/L2 can be applied based on the methods described in [Ber02, Paw11]. The usual method is to consider different operating ranges, where only a single or very few unknown parameters exist. The most convenient way is to start with the low current weight factors h_{jEi} and h_{jCi} for the depletion charges. The extraction of the latter is not discussed here further.

For $V_{\mathrm{BC}} = 0$ V and low V_{BE}, where the voltage drop across the series resistances is negligible, the DC collector current is expressed by

$$I_{\mathrm{C}} = I_{\mathrm{T}} = \frac{c_{10}}{Q_{\mathrm{p0}} + h_{\mathrm{jEi}} Q_{\mathrm{jEi}}} \exp\left(\frac{V_{\mathrm{BE}}}{V_{\mathrm{T}}}\right). \tag{3.47}$$

The method described in [Ber02] rewrites the above equation as

$$\frac{\exp(V_{\mathrm{BE}}/V_T)}{I_{\mathrm{T}}} = \frac{Q_{\mathrm{p0}}}{c_{10}} + \frac{h_{\mathrm{jEi}}}{c_{10}} Q_{\mathrm{jEi}} = f(Q_{\mathrm{jEi}}). \tag{3.48}$$

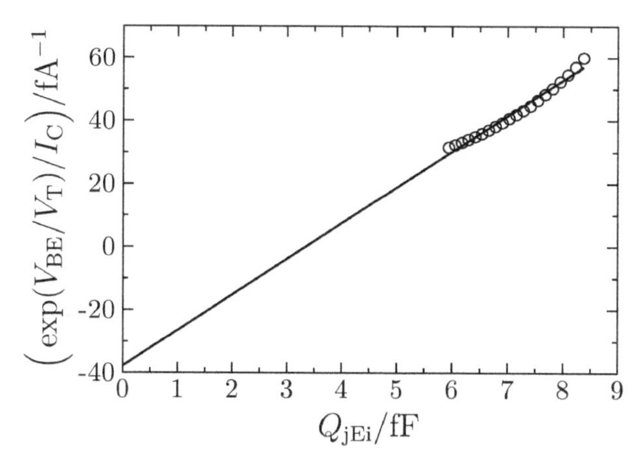

Figure 3.23 Extraction of h_{jEi} based on (3.48).

Using the physics-based value for Q_{p0} extracted from tetrodes, the parameters c_{10} and h_{jEi} can be extracted from the slope and intercept of the above function $f(Q_{jEi})$. Results for applying this method to transistors with a graded Ge profile in the base are visualized in Figure 3.23. As can be seen, the data obtained from measurements do no yield a straight line as expected from (3.48) but show a curvature. Depending on this curvature, the extraction may yield a non-physical negative value for the intercept with the y-axis, given by Q_{p0}/c_{10}. The curvature is caused by the non-constant h_{jEi} present for transistors with a graded Ge profile. Hence, according to (3.19), the parameter a_{hjEi} enters into (3.48) as an additional unknown. While the method in [Paw11] works reliably for numerical device simulations due to the optimal accuracy of the results (i.e., no impact of measurement noise), results may become unreliable due to small noise in the measurements. The method in [Ste12] is based on a known saturation current and a normalized h_{jEi}, which are not so simple to extract without known a_{hjEi}.

In this section, an alternative method is presented which is based on [Ste12] without involving the additional unknown. Rather than performing a linear fit, taking the derivative of (3.48) leads to the differential equation

$$C = f(V_{BE}) + \frac{df(V_{BE})}{dV_{BE}} \frac{Q_{jEi}}{C_{jEi}} \tag{3.49}$$

with

$$f(V_{BE}) = \frac{h_{jEi}(V_{BE})}{c_{10}} \quad \text{and} \quad C = \frac{d\left[\frac{\exp(V_{BE}/V_T)}{I_T}\right]}{dQ_{jEi}}. \tag{3.50}$$

This differential equation can be solved numerically, requiring an initial value $f_1 = f(V_{\rm BE1})$ with $V_{\rm BE1}$ being the minimum $V_{\rm BE}$ with reliable current values. For calculating $a_{\rm hjEi}$, (3.19) is altered into

$$f = f_1 \frac{\exp\left(a_{\rm hjEi}\left(\left(1-\frac{V_{\rm BE1}}{V_{\rm DEi}}\right)^{z_{\rm Ei}} - \left(1-\frac{V_{\rm B'E'}}{V_{\rm DEi}}\right)^{z_{\rm Ei}}\right)\right)-1}{a_{\rm hjEi}\left(\left(1-\frac{V_{\rm BE1}}{V_{\rm DEi}}\right)^{z_{\rm Ei}} - \left(1-\frac{V_{\rm B'E'}}{V_{\rm DEi}}\right)^{z_{\rm Ei}}\right)}$$

$$= f_1 \frac{\exp(u)-1}{u}, \tag{3.51}$$

i.e., shifting the reference from $V_{\rm B'E'} = 0$, as it is the case in (3.19), to $V_{\rm B'E'} = V_{\rm BE1}$. Note that c_{10} is a constant value and, therefore, does not change the shape of the resulting curve. Defining $w = f/f_1$ according to [Stel2] allows the calculation of u by

$$u = -\frac{1}{w} - W_{-1}\left\{-\frac{1}{w}\exp\left(-\frac{1}{w}\right)\right\}, \tag{3.52}$$

with W_{-1} being the negative branch on the Lambert-W function, and finally $a_{\rm hjEi}$ by

$$a_{\rm hjEi} = \frac{u}{\left(1-\frac{V_{\rm BE1}}{V_{\rm DEi}}\right)^{z_{\rm Ei}} - \left(1-\frac{V_{\rm B'E'}}{V_{\rm DEi}}\right)^{z_{\rm Ei}}}, \tag{3.53}$$

The form of the extracted curve for f from (3.49) strongly depends on the chosen initial value f_1. Therefore, if the shape does not agree with that of the model equation, a non-constant $a_{\rm hjEi}$ is extracted.

The actual extraction therefore is based on an iterative change of f_1 until the relative standard deviation of $a_{\rm hjEi}$ is minimized, i.e., the most constant $a_{\rm hjEi}$ is obtained. A bi-section method is a reliable choice for the algorithm here. The application is visualized in Figure 3.24. Usually, for too small values of f_1, a decreasing curve of $a_{\rm hjEi}$ is obtained while too large values of f_1 lead to increasing values. The center curve in the plot shows the actual solution of the iteration.

The remaining steps for extracting $h_{\rm jEi0}$ and c_{10} follow [Paw11]. Utilizing the extracted $a_{\rm hjEi}$, (3.48) is altered to

$$\frac{\exp(V_{\rm BE}/V_{\rm T})}{I_{\rm T}} = f(hQ_{\rm jEi}) \tag{3.54}$$

with $h = h_{\rm jEi}/h_{\rm jEi0}$ from (3.19), where $a_{\rm hjEi}$ is the only parameter entering, which resolves the bending of the extraction curve and leads to the correct signs of the extracted parameters as demonstrated in Figure 3.25.

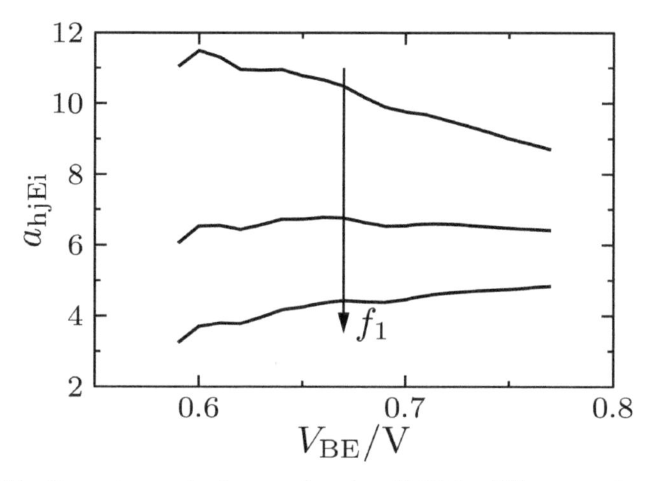

Figure 3.24 Extraction results for a_{hjEi} based on (3.53) for different starting values f_1.

A linear extrapolation finally allows extracting c_{10} and h_{jEi0} from

$$c_{10} = \frac{Q_{\text{p0}}}{y_0} \text{ and } h_{\text{jEi0}} = \frac{m}{y_0} Q_{\text{p0}} \tag{3.55}$$

with m being the slope of the straight line and y_0 the intercept with the y-axis. The extraction is performed at low injection, where series resistances and self-heating have only negligible impact during extraction of h_{jEi}

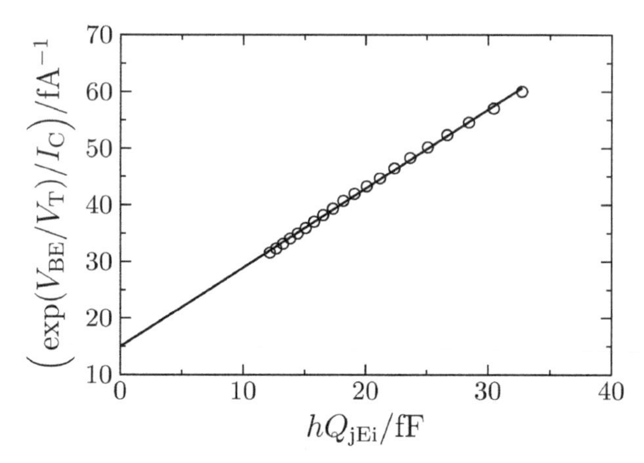

Figure 3.25 Extraction of h_{jEi} based on (3.54) including the correction based on the extracted a_{hjEi}.

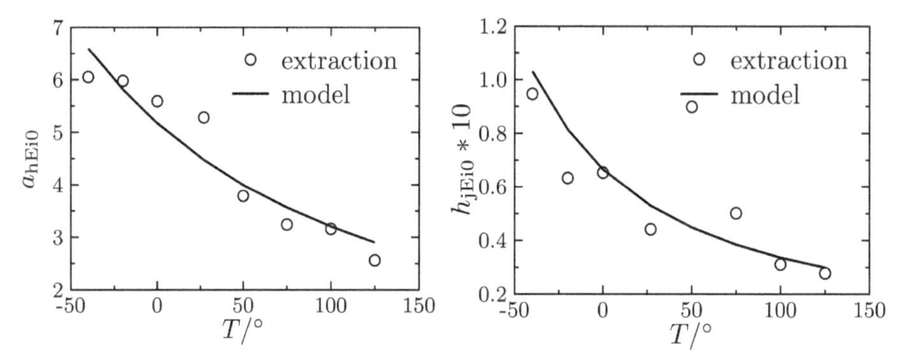

Figure 3.26 Extraction results and application of the temperature model for a_{hjEi} (3.31) and h_{jEi0} (3.29).

and c_{10}. Therefore, the values for a_{hjEi}, h_{jEi0}, and c_{10} are extracted for the given ambient temperature. Results are shown in Figure 3.26.

The extraction of the high-current weight factors is performed by stepwise inclusion of the related diffusion charges from (3.9). For obtaining correct results, the voltage drops across the series resistances as well as the actual device temperature due to self-heating have to be calculated. Starting with the low-current minority charge $\tau_{f0} I_{\text{Tf}}$, the weight factor is calculated from

$$h_{\text{f0}} = \frac{c_{10}}{I_{\text{T}}} \exp\left(\frac{V_{\text{B'E'}}}{V_{\text{T}}}\right) - (Q_{\text{p0}} + h_{\text{jEi}} Q_{\text{jEi}} + h_{\text{jCi}} Q_{\text{jCi}}). \tag{3.56}$$

The application of above equation to data in Figure 3.27 shows a bias dependence of h_{f0}, which has to be obtained at low current densities though, yielding the results shown in Figure 3.27(b). The increase at high current densities in Figure 3.27(a) is caused by the so-far-neglected charge components and does not yield the correct h_{f0}.

The temperature dependence of h_{f0} in Figure 3.27(b) exhibits visible steps between the values of different temperature ranges. They are caused by the V_{BC} dependence of h_{f0}. Extracting h_{f0} for a given ambient temperature yields the results displayed as crosses in Figure 3.28. Since these values are still affected by the self-heating-related temperature increase inside of the device, applying the temperature dependence given in Figure 3.27(b) and correcting only the changes due to self-heating effects yields the results given by circles in Figure 3.28 which still displays a weak bias dependence.

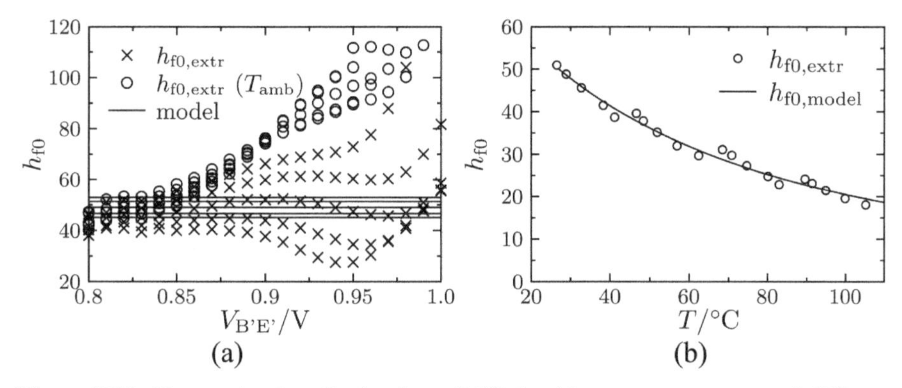

Figure 3.27 Extracted values for h_{f0} from (3.56) for (a) room temperature and different V_{BC}. (b) Extracted values chosen at low current densities for different temperatures and V_{BC} [Paw15a].

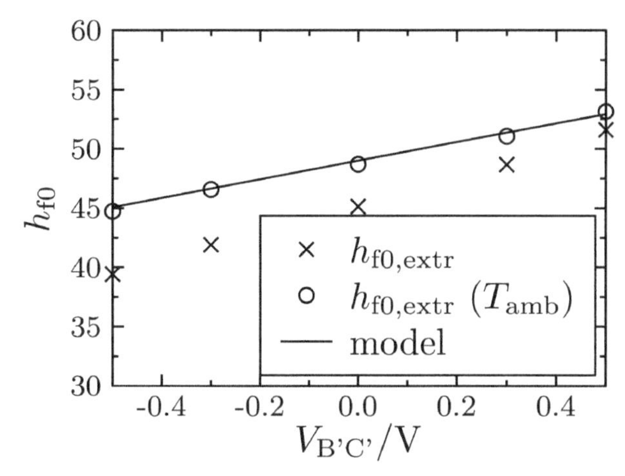

Figure 3.28 Extraction results for the bias-dependent h_{f0} at room temperature for the results given in Figure 3.27(b) [Paw15a].

The extraction of h_{fE} follows the same steps by further including $h_{f0}\tau_{f0}I_{Tf}$ into (3.56), thus calculating h_{fE} from

$$h_{fE} = \frac{c_{10}}{I_T} \exp\left(\frac{V_{B'E'}}{V_T}\right) - (Q_{p0} + h_{jEi}Q_{jEi} + h_{jCi}Q_{jCi} + h_{f0}\tau_{f0}I_{Tf}). \quad (3.57)$$

The data in Figure 3.29(a) initially show a strong dependence on I_T and V_{BC}, which is caused by self-heating and the temperature dependence of h_{fE}. After taking into account the temperature dependence according to

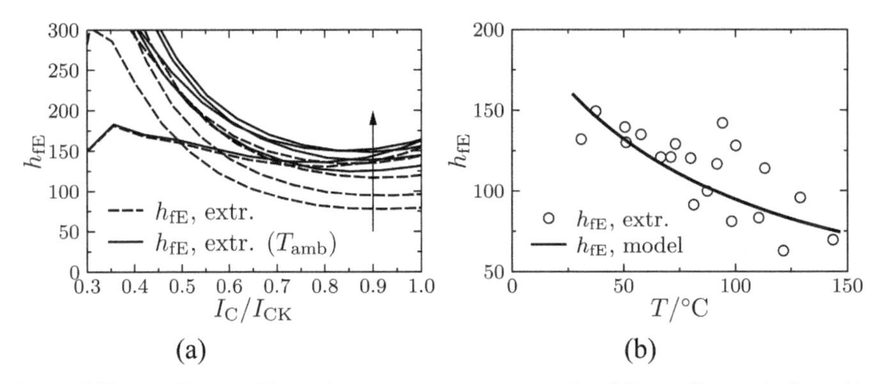

Figure 3.29 (a) Extracted h_{fE} values at room temperature for different V_{BC} as indicated by the arrow. (b) Extracted temperature dependence [Paw15a].

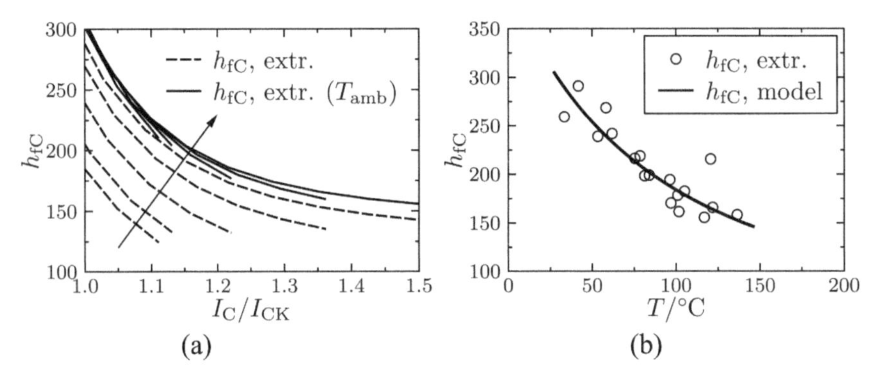

Figure 3.30 (a) Extracted values for h_{fC} at room temperature for different V_{BC} as indicated by the arrow. (b) Extracted temperature dependence [Paw15a].

Figure 3.29(b), the bias dependence of h_{fE} almost vanishes (see solid curves in Figure 3.29(a)).

As the final step, h_{fC} is extracted analogously by also including $h_{fE}\Delta Q_{Ef}$ and ΔQ_{Bf} into the calculations. Note that ΔQ_{Bf} is weighted by 1. Also for h_{fC}, a strong bias dependence is observed initially, which disappears though after correctly including the effect of self-heating.

3.6.3 Physics-Based Parameter Scaling

HICUM/L2 has emerged as one of the industry standard bipolar transistor models and was therefore selected as the backbone of the compact transistor modeling strategy in DOTSEVEN. From its first development on in the

1980s, the model has been formulated with geometry scaling capability in mind since this feature has been crucial to achieve the optimum circuit performance for a given process technology. Geometry scaling is fundamentally based on a physics-based model formulation and parameter extraction strategy. Unlike compact MOSFET models, bipolar transistor models do not include scaling equations in their simulator code for several reasons. First, the large variety of possible contact arrangements and structures makes such equations complicated. Second, the accurate calculation of the impact of some effects, such as the geometry dependence of the substrate and thermal coupling, requires the solution of implicit or even differential equations which are difficult to program in simulator-supported description languages including Verilog-A.

Therefore, the development and implementation of HBT geometry scaling formulations is typically left to the foundry or model user. Due to the physics-based formulation of HICUM/L2, its important model parameters can be expressed readily in terms of transistor geometry. Within DOTSEVEN, significant effort was spent on further improving the model's geometry-scaling capabilities for most advanced SiGe HBT structures and for developing reliable geometry-scalable parameter extraction methodologies. Besides enabling the selection of the optimal transistor size for any given circuit application, another benefit is the reduction of the so-called "parameter extraction noise." The latter is a well-known and undesired effect that results in erratic and unpredictable parameter variations with respect to transistor geometry. It occurs when correlated model parameters are extracted on a set of individual transistors by numerical optimization. Geometry-scalable parameter extraction methods have the additional advantage of smoothing random measurement errors and allowing the detection of systematic measurement errors due to, e.g., measurement (equipment) limitations as well as test structure layout and process issues. Finally, scalable models ensure that important model parameters (like the base resistance) behave properly with respect to geometry.

3.6.3.1 Standard geometry scaling equation

The most important concept regarding the scaling equations used in conjunction with HICUM/L2 is probably the concept of an *effective* (electrical) emitter area. At low current densities, (i.e., for negligible voltage drops across series resistances), the total collector current can be split into an internal and perimeter portion, each related to a specific emitter region (see Figure 3.31).

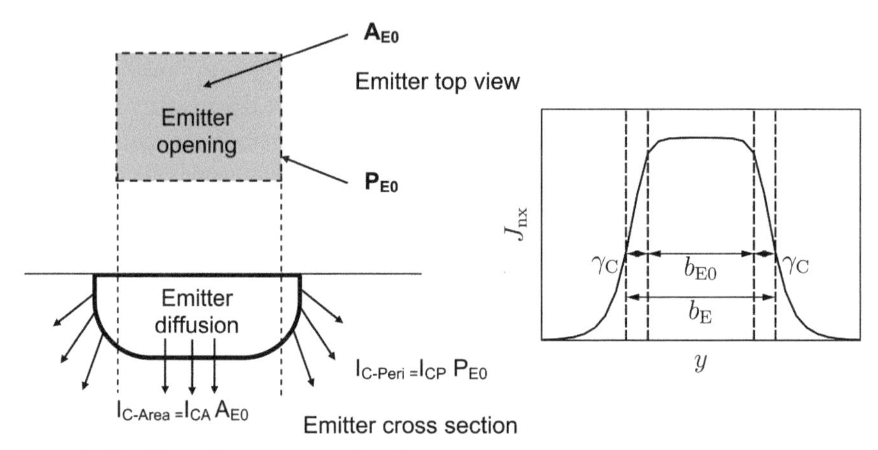

Figure 3.31 The collector current flow in the emitter can be split into an intrinsic portion related to the emitter area and a portion related to the perimeter only. The right picture shows the spatial distribution of the vertical electron current density. Marked are the actual emitter width b_{E0} as well as the effective emitter width b_{E}, calculated from b_{E0} and γ_{C} (cf. 3.58, 3.60).

Assuming no impact of collector avalanche or tunneling effects, this graphical concept can be expressed mathematically by

$$I_{\mathrm{C}} = I_{\mathrm{CA}} A_{\mathrm{E0}} + I_{\mathrm{CP}} P_{\mathrm{E0}}, \tag{3.58}$$

where $A_{\mathrm{E0}} = b_{\mathrm{E0}} l_{\mathrm{E0}}$ is the actual emitter window area and $P_{\mathrm{E0}} = 2b_{\mathrm{E0}} + 2l_{\mathrm{E0}}$ is the actual emitter window perimeter[4], both given by the window opening and interface area of the emitter poly-silicon with the mono-silicon region. Furthermore, $I_{\mathrm{CA}} A_{\mathrm{E0}}$ and $I_{\mathrm{CP}} P_{\mathrm{E0}}$, respectively, are the collector components resulting from carrier injection across the emitter window area and window perimeter, respectively. Equation (3.58) can be conveniently reformulated as

$$I_{\mathrm{C}} = I_{\mathrm{CA}} A_{\mathrm{E0}} \left(1 + \frac{I_{\mathrm{CP}} P_{\mathrm{E0}}}{I_{\mathrm{CA}} A_{\mathrm{E0}}}\right). \tag{3.59}$$

By introducing the process-specific parameter γ_{C}, defined as the ratio of perimeter-specific to area-specific collector current, (3.59) defines the *effective electrical* emitter area

$$A_{\mathrm{E}} = A_{\mathrm{E0}} \left(1 + \gamma_{\mathrm{C}} \frac{P_{\mathrm{E0}}}{A_{\mathrm{E0}}}\right), \tag{3.60}$$

[4]Simplified equation neglecting corner contributions and possible corner rounding.

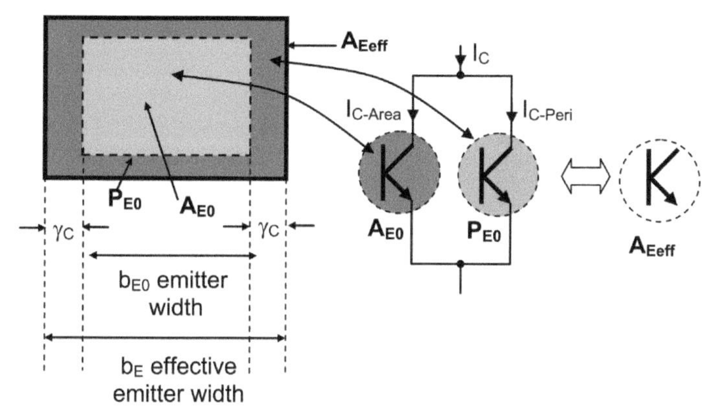

Figure 3.32 Schematic illustration of the effective emitter area concept. The area and perimeter currents are gathered in a single effective contribution.

which reduces the two current components of (3.58) to a single component with a clearly defined geometry dependence. Therefore, as can be seen from Figure 3.32, this approach allows to represent the internal and perimeter transistor by a single-transistor model having an effective emitter area A_{E}. This is obviously advantageous over a two-transistor model approach in terms of computational efficiency and parameter extraction effort. This concept can be easily extended to the modeling of other current components as well as to the charges (and capacitances) of a transistor structure.

According to (3.59), plotting $I_{\mathrm{C}}/A_{\mathrm{E0}}$ versus $P_{\mathrm{E0}}/A_{\mathrm{E0}}$ allows to extract the geometry scaling-related parameters I_{CA} and γ_{C} from the y intercept and the slope of the curves. This procedure is also known as the P/A (perimeter over area) approach. The application of this concept to experimental data of a DOTSEVEN process in Figure 3.33 displayed the excellent scalability of the collector current. Notice that the use of data from several structures helps averaging out local process variations as compared to performing the extraction from just a single device.

3.6.3.2 Generalized scaling equations

During the various projects (DOTSEVEN, DOTFIVE, RF2THz) deviations from the P/A scaling were observed for some process versions, [Paw13, Paw15b]. It turned out though that such non-standard scaling behavior could be captured by an extension of the standard P/A approach. Figure 3.34 depicts schematically this extension. Again, the goal is to combine the models associated with each lateral region of the transistor into a single model representing a transistor with an effective emitter area A_{E}. The extension relies

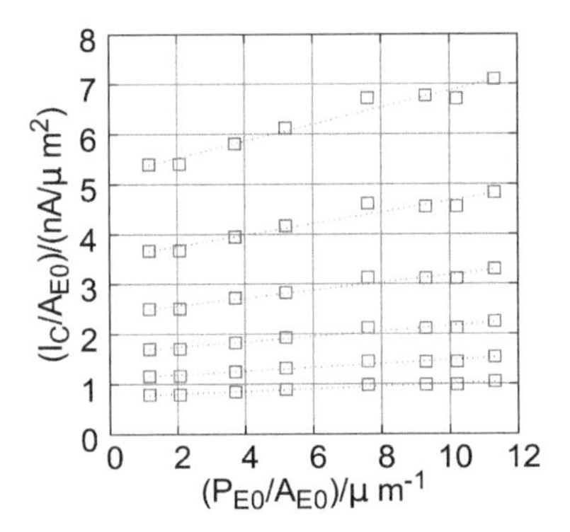

Figure 3.33 Experimental data of I_C/A_{E0} versus P_{E0}/A_{E0} for $V_{BE} = 0.45$ 0.5 V in steps of 10 mV. A_{E0} and P_{E0}, respectively, are the actual emitter window area and perimeter, respectively. The drawn emitter dimensions are $(0.31, 0.35, 0.4, 0.53, 0.7, 1.2, 2.2) \times 10\,\mu m^2$. Symbols represent measured data and dashed lines results from linear regression.

on considering the four different components, defined by an injection across the window area, width and length related perimeter junctions, and corner junctions, separately. Obviously, the simple P/A approach in Figure 3.32 is just a special case of this extended generalized linear scaling approach and is obtained when the specific currents related to the width, length, and corner are merged into a single perimeter related specific current.

In order to properly extract the scalable model parameters for the generalized scaling approach, a matrix of test structure is required with the same set of emitter widths for at least two emitter lengths (see Figure 3.35). From the

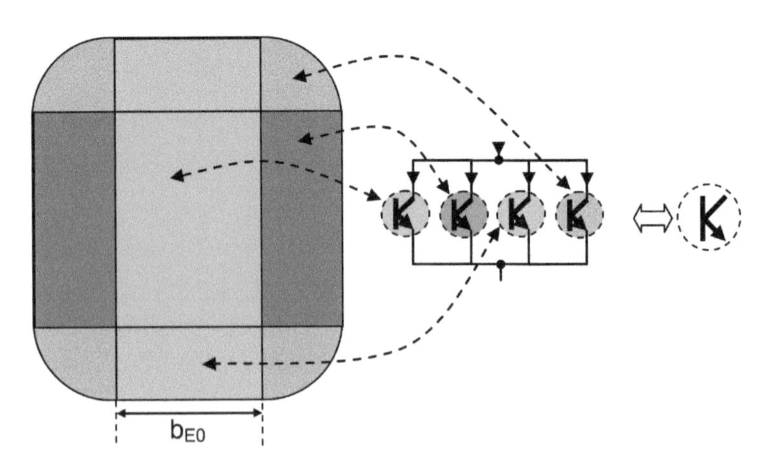

Figure 3.34 Schematic illustration of the generalized effective emitter area concept.

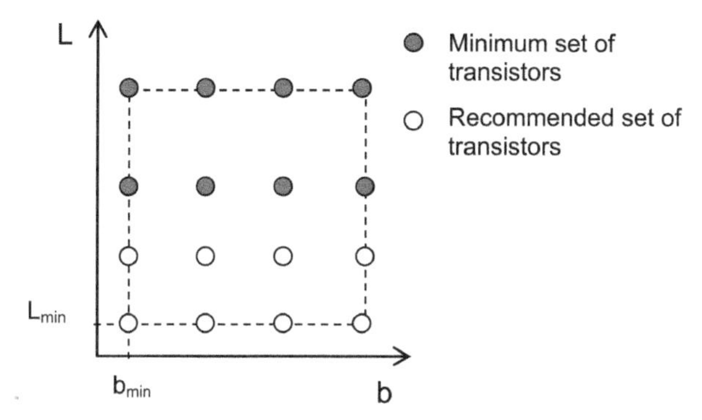

Figure 3.35 Required matrix of test structures for scalable parameter extraction in the case of the generalized scaling laws.

corresponding measurements, one can obtain a complete set of parameters for each of the different transistors (and associated lateral regions) depicted in Figure 3.34. These parameters then need to be transformed into a set for a single-transistor model. This has been implemented with the help of a Taylor series expansions in the geometry scaling equations for HICUM/L2.

3.7 Compact Model Application to Experimental Data

Within DOTSEVEN and related research projects (such as DOTFIVE, RF2THz), HICUM/L2 model parameters were determined from the electrical characteristics measured for many process runs and technology versions. Since the vast amount of data and comparisons cannot be displayed here due to the lack of space, just an overview is given that is based on selected publications[5] and the complete version of HICUM/L2 with all the previously described extensions.

An overview on the overall parameter extraction approach and employed procedures is given in [Paw11], while [Paw14a] highlights the improvements in modeling the transfer current of SiGe HBTs using the extensions described in Section 3.3 and providing a guideline for extracting the new parameters. Very detailed information on both parameter extraction and model comparisons for different process technologies have been given in [Paw15a][Kra15b][Ros17]; these include a large number of DC, AC and

[5]Note that – in contrast to circuit design – the (successful) results of compact modeling can typically not be published for subsequent (and intermediate) process technology versions with improving electrical performance.

large-signal results for a large variety of transistor structures of the technologies developed in different research projects.

A general overview on the modeling results of DOTSEVEN was given in [Paw15b] focussing on a veriety of characteristics and operating conditions. The application of the compact model with a focus on decomposing the impact of different physical effects for guiding process technology development has been given in [Paw13], [Kor15], [Paw17a] and [Paw17b].

High-frequency noise, including the correlation between collector current and stored base charge and its generic implementation in circuit simulators, was discussed in [Her12]. There, the applicability of the noise correlation formulation in HICUM/L2 was verified based on the results of the Boltzmann transport equation for frequencies at least up to 500 GHz. Noise measurements at such frequencies are presently impossible so that experimental verifications have been restricted to 50 GHz so far [Her12], [Sak15b], [Sak15]. Here, the accuracy of the model has allowed the decomposition of the various physical noise mechanisms within the transistor, yielding valuable insights into their magnitude and relative importance.

Applications of the compact model with a focus also on circuit results have been demonstrated in [Ard15][Sch16c][Lia17][Sch18b].

References

[Ard15] Ardouin, B. et al. (2015). “Compact Model Validation Strategies Based on Dedicated and Benchmark circuit blocks for the mm-Wave Frequency Range”, (inv.) Proc. CSICS, New Orleans, 4p.

[Ber02] Berger, D., Celi, M., Schröter, M., Malorny, M., Zimmer, T., and Ardouin, B. (2002). “HICUM parameter extraction methodology for a single transistor geometry,” in *Proceedings of the Bipolar/BiCMOS Circuits and Technology Meeting (BCTM)*, Atlanta, 116–119.

[Cel14] Céli, D. (2014). “Investigation on bias dependence of critical current ICK in HICUM models,” in *Proceedings of the 27th BAK*, Crolles.

[Cra90] Crabbe, E. F., Patton, G. L., Stork, J. M. C., Comfort, J. H., Meyerson, B. S., and Sun, J. Y. C. (1990). “Low temperature operation of Si and SiGe bipolar transistors,” in *Proceedings of the Technical Digest of International Electron Devices Meeting (IEDM)* (Montgomery, AL: IEDM), 17–20.

[Cra93] Crabbe, E. F., Cressler, J. D., Patton, G. L., Stork, J. M. C., Comfort, J. H., Sun, J. Y. C. (1993). Current gain rolloff in graded-base SiGe heterojunction bipolar transistors. *IEEE Electron Device Lett.* 14, 193–195.

[Fri02] Friedrich, M. (2002). *An Analytical Mode for Simulating Silicon-Germanium Heterojunction Bipolar Transistors in Integrated Circuits.* Ph.D. thesis Ruhr-Universität Bochum, Bochum.

[Hei16] Heinemann, B., Rücker, H., Barth, R., Bärwolf, F., Drews, J., Fischer, G., et al. (2016). "SiGe HBT with fT/fmax of 505 GHz/ 720 GHz," in *Proceedings of the IEEE IEDM Technical Digest* (IEEE: San Francisco, CA).

[Her12] Herricht, J., Sakalas, P., Ramonas, M., Schröter, M., Jungemann, C., Mukherjee, A., et al. (2012). Systematic method for usage of correlated noise sources in compact models for high frequency transistors. *IEEE Trans. Microw. Theory Technol.* 60, 3403–3412.

[Hue04] Huerta, A., Vanhoucke, T., and van Noort, W. D. (2004). Electrical characterization of high-performance SiGe:C HBT's focus on emitter-base junction. *Philips Res.* 8, 200–221.

[Kor15] Korn, J., Ruecker, H., Heinemann, B., Pawlak, A., Wedel, G., Schröter, M. (2015). "Experimental and theoretical study of fT for SiGe HBTs with a scaled vertical doping profile," in *Proceedings of the IEEE BCTM*, Boston, 117–120.

[Kra15a] Krause, J., and Schröter, M. (2015). Methods for determining the emitter resistance in SiGe HBTs: A review and evaluation across different technologies. *IEEE Trans. Electron Dev.* 62, 1363–1374.

[Kra15b] Krause, J. (2015). "Model parameter extraction for very advanced heterojunction bipolar transistors," Dissertation, Chair for Electron Devices and Integrated Circuits, TU Dresden.

[Lac14] Lachner, R. (2014). Towards 0.7 terahertz silicon germanium heterojunction bipolar technology – The DOTSEVEN project. *ECS Trans.* 64, 21–37.

[Leh14] Lehmann, S., Zimmermann, Y., Pawlak, A., and Schröter, M. (2014). Characterization of the static thermal coupling between emitter fingers of bipolar transistors. *IEEE Trans. Electron Dev.* 61, 3676–3683.

[Lia17] Liang, W., Pawlak, A., Sakalas, P. and Schröter, M. (2017). "'96 GHz 4.7 mW low-power frequency tripler with 0.5 V supply voltage", *Electron. Lett.*, 53(19), 1308–1310.

[Muk16] Mukherjee, A., Pawlak, A., Schröter, M., Celi, D., and Huszka, Z. (2016). "Implementation and quality testing for compact models implemented in Verilog-A," in *Proceedings of the DATE*, Dresden, 403–408.

[Paa01] Paasschens, J., Kloosterman, W. J., and Havens, R. J. (2001). "Modelling two SiGe HBT specific features for circuit simulation," in

Proceedings of the Bipolar/BiCMOS Circuits and Technology Meeting (BCTM), Boston, 38–41.

[Paw11] Pawlak, A., Schröter, M., Krause, J., Celi, D., and Derrier, N. (2011). "HICUM/2 v2.3 parameter extraction for advanced SiGe-heterojunction bipolar transistors," in *Proceedings of the IEEE Bipolar/ BiCMOS Circuits and Technology Meeting (BCTM),* Boston, 195–198.

[Paw13] Pawlak, A., Schröter, M., and Fox, A. (2013). Geometry scalable model parameter extraction for mm-wave SiGe-heterojunction transistors," in *Proceedings of the IEEE BCTM*, Boston, 127–130.

[Paw14a] Pawlak, A., and Schröter, M. (2014). An improved transfer current model for RF and mm-wave SiGe(C) heterojunction bipolar transistors. *IEEE Trans. Electron Dev.* 61, 2612–2618.

[Paw14b] Pawlak, A., Lehmann, S., and Schröter, M. (2014). "A Simple and Accurate Method for Extracting the Emitter and Thermal Resistance of BJTs and HBTs," in *Proceedings of the IEEE BCTM*, Boston.

[Paw15a] Pawlak, M. (2015). *Advanced Modeling of Silicon-Germanium Heterojunction Bipolar Transistors*. Ph.D. thesis, Dissertation, Chair for Electron Devices and Integrated Circuits, TU Dresden.

[Paw15b] Pawlak, A., Lehmann, S., Sakalas, P., and Schröter, M. (2015). "SiGe HBT modeling for mm-wave circuit design," in *Proceedings of the IEEE BCTM*, Boston, 149–156.

[Paw16] Pawlak, A., Krause, J., Wittkopf, H., and Schröter, M. (2016). Single transistor based methods for determining the base resistance in SiGe HBTs: Review and evaluation across different technologies. *IEEE Trans. Electron Dev*. 63, 4591–4602.

[Paw17a] Pawlak, A., and Schröter, M. (2017). "Modeling of SiGe HBTs with (fT, fmax) of (340, 560) GHz based on physics-based scalable model parameter extraction," *17th IEEE Topical Meeting on Silicon Monolithic Integrated Circuits in RF Systems* (SiRF), Phoenix, AZ, 100–102.

[Paw17b] Pawlak, A., Heinemann, B., and Schröter, M. (2017). "Physics-based modeling of sige HBTs with fj of 450 GHz with HICUM *Level 2," IEEE Bipolar/BiCMOS Circuits and Technology Meeting (BCTM)*, Miami, FL, 134–137.

[Paw18] Pawlak, A., Krause, J., and Schröter, M. (2018). Methods for determining the collector series resistance in SiGe HBTs: A review and evaluation across different technologies.

[Pri67] Pritchard, R. L. (1967). *Electrical Characteristics of Transistors*. New York, NY: McGraw-Hill.

[Rei91] Rein, H.-M., and Schröter, M. (1991). Experimental determination of the internal base sheet resistance of bipolar transistors under forward-bias conditions. *Solid-State Electron.* 34, 301–308.

[Ros13] Rosenbaum, T., Schröter, M., Pawlak, A., and Lehmann, S. (2013). "Automated transit time and transfer current extraction for single transistor geometries," in *Proceedings of the IEEE BCTM* (Bordeaux: IEEE), 25–28.

[Ros17] Rosenbaum, T. (2017). "Performance prediction of a future silicon-germanium heterojunction bipolar transistor technology using a heterogeneous set of simulation tools and approaches", Dissertation, Chair for Electron Devices and Integrated Circuits, TU Dresden.

[Rue12] Rücker, H., Heinemann, B., and Fox, A. (2012). "Half-terahertz SiGe BiCMOS technology," in *Proceedings of the IEEE 11th Topical Meeting on Silicon Monolithic Integrated Circuits in RF Systems (SiRF)*, Santa Barbara, CA, 133–136.

[Sak15a] Sakalas, P., Schröter, M., and Zirath, H. (2015). mm-Wave noise modeling in advanced SiGe and InP HBTs. *J. Comput. Electronics*,14, 62–71.

[Sak15b] Sakalas, P., and Schröter, M. (2015). "Noise in advanced bipolar transistors at mm-wave frequencies", (inv.), *23rd Int'l Conf. on Noise and Fluctuations (ICNF)*, p. 6.

[Sch88] Schröter, M. (1988). *A Compact Physical Large-Signal Model for High-Speed Bipolar Transistors with Special Regard to High Current Densities and Two-Dimensional Effects.* Ph.D. thesis, Ruhr-University Bochum, Bochum.

[Sch91] Schröter, M. (1991). Simulation and modeling of the low-frequency base resistance of bipolar transistors in dependence on current and geometry. *IEEE Trans. Electron Dev.* 38, 538–544.

[Sch92] Schröter, M. (1992). Modeling of the low-frequency base resistance of single base contact bipolar transistors. *IEEE Trans. Electron Dev.* 39, 1966–1968.

[Sch99] Schröter, M., and Lee, T.-Y. (1999). A physics-based minority charge and transit time model for bipolar transistors. *IEEE Trans. Electron Dev.* 46, 288–300.

[Sch05] Schröter, M, Wittkopf, H., Kraus, W. (2005). "Statistical modeling of bipolar transistors," in *Proceedings of the Bipolar Circuits and Technology Meeting (BCTM)*, Santa Barbara, CA, 54–61.

[Sch07] Schröter, M., and Lehmann, S. (2007). "The rectangular bipolar transistor tetrode structure and its application," in *Proceedings of the ICMTS*, Tokyo, 206–209.

[Sch08a] Schröter, M., Krause, J., Lehmann, S., and Celi, D. (2008). Compact layout and bias dependent base resistance modeling for advanced SiGe HBTs. *IEEE Trans. Electron Dev*. 55, 1693–1701.

[Sch08b] Schröter, M., Krause, J., Aufinger, K., and Ardouin, B. (2008). *Test Structures for SiGe HBT Compact Modeling, Parameter Extraction and Circuit Design*. DOTFIVE Report, No. D4.3.1.

[Sch10] Schröter, M., and Chakravorty, A. (2010). *Compact Hierarchical Modeling of Bipolar Transistors with HICUM*. Singapore: World Scientific.

[Sch11a] Schröter, M., Wedel, G., Heinemann, B., Jungemann, C., Krause, J., Chevalier, P., et al. (2011). Physical and electrical performance limits of high speed SiGeC HBTs – Part I: vertical scaling. *IEEE Trans. Electron Dev.* 58, 3687–3696.

[Sch11b] Schröter, M., Krause, J., Rinaldi, N., Wedel, G., Heinemann, B., Chevalier, P., et al. (2011). Physical and electrical performance limits of high-speed SiGeC HBTs – Part II: lateral scaling. *IEEE Trans. Electron Dev.* 58, 3696–3706.

[Sch16a] Schröter, M., Rosenbaum, T., Chevalier, P., Heinemann, B., Voinigescu, S., Preisler, E., et al. (2017). SiGe HBT technology: future trends and TCAD based roadmap. *Proc. IEEE*, 105, 1068–1086.

[Sch16b] Schröter, M., Lehmann, S., and Pawlak, A. (2016). "Why is there no internal collector resistance in HICUM?" in *Proceedings of the IEEE Bipolar/BiCMOS Circuits and Technology Meeting (BCTM)*, New Brunswick, NJ, 142–145.

[Sch16c] Schröter, M., Boeck, J., d'Alessandro, V., Fregonese, S., Heinemann, B., Jungemann, C., et al. (2016). "The EU DOTSEVEN project: overview and results," in *Proceedings of the IEEE CSICS*, Austin, TX, 4.

[Sch18a] Schröter, M., and Pawlak, A. (2018). The bipolar transistor tetrode for determining the base resistance: review and qantitative assessment.

[Sch18b] Schröter, M. and Pawlak, A. (2018). "SiGe heterojunction bipolar transistor technology for sub-mm-wave electronics - State-of-the-art and future prospects", *18th IEEE Topical Meeting on Silicon Monolithic Integrated Circuits in RF Systems (SiRF)*, Anaheim, CA, 60–63.

[Ste12] Stein, F., Huszka, Z., Derrier, N., Maneux, C., and Celi, D. (2012). "Extraction of the emitter related space charge weighting factor parameters of HICUM L2.30 using the Lambert W function," in *Proceedings of the IEEE Bipolar/BiCMOS Circuits and Technology Meeting (BCTM)*, Portland, OR, 1–4.

4

(Sub)mm-wave Calibration

M. Spirito[1] and L. Galatro[1,2]

[1]Electronic Research Laboratory, Delft University of Technology, The Netherlands
[2]Vertigo Technologies B.V., The Netherlands

4.1 Introduction

High-frequency characterization of active and passive devices is carried out by extracting the scattering parameters of the component (often in a two-port configuration) employing a vector network analyzer (VNA). This class of instruments allows to characterize the response of the device under test (DUT) over a broad frequency range (exceeding 1 THz [Dio17]) at a user-defined reference plane. In order to define such reference planes and remove all the imperfections of the measurement setup (i.e., cable and receiver conversion losses, amplitude and phase tracking errors, and other statistical errors), a calibration procedure [Ryt01] needs to be carried out prior to the measurement. The calibration procedure employs the knowledge of the devices used (i.e., standards) to solve the unknowns representing the measurement setup response (often referred to as error terms). The derived error terms allow then to remove the imperfections of the setup, during the measurement procedure. The accuracy of the calibration is then directly dependent on the accuracy with which the standards are known [Stu09]. In the literature, different calibration techniques have been presented, often trading off (more) knowledge on the response of the standard device for (lower) space occupancy (i.e., when considering SOLR/LRM [Fer92; Dav90] calibrations versus TRL type ones [Eng79]). Traditionally, calibration techniques requiring little standards knowledge (e.g., TRL, LRL) have been considered the most accurate, with TRL reaching metrology institute precision, by only requiring the information of the characteristic impedance of the line [Eng79]. In this chapter the focus will be placed only on TRL calibration techniques due to their best compatibility with millimeter- and sub-millimeter-wave characterization. For a more extensive discussion on the various possible calibration techniques the reader is referred to [Tep13]. Calibration techniques for on-wafer measurements typically consist of a probe-level calibration (first-tier) performed on a low-loss substrate (i.e., alumina or fused silica) [Eng79; Eul88; Dav90; Mar91a]. This probe-level calibration is then transferred to the environment where the DUT is embedded in and often, to increase the measurement accuracy, this calibration is augmented with a second-tier on-wafer calibration/de-embedding step. This allows moving the reference plane as close as possible to the DUT, by removing the parasitics associated with the contact pads, the device-access lines, and the vias [Tie05]. In this chapter we will first review the challenges and potential solutions associated with first-tier calibrations performed on low-loss substrates, then the approach to design calibration kits integrated in the back-end-of-line of silicon based technology will be presented, and finally a direct de-embedding/calibration strategy, capable of setting the reference plane at the lower metal layer of a technology stack, will be described.

4.2 Multi-mode Propagation and Calibration Transfer at mm-wave

The different propagating modes supported by a coplanar wave guide (CPW) are qualitatively sketched in Figure 4.1. The CPW mode, characterized by opposite direction of the fields across the slots, represents the intended propagation mode and is often referred to as *CPW differential mode.* The CPW mode characterized by in-phase direction of the fields across the slots (W_{GAP}) represents an unwanted radiating mode and is often referred to as *CPW common mode.* The TM_{n} and TE_{m} modes are surface waves propagating along the grounded dielectric slab. Their cut-off frequencies

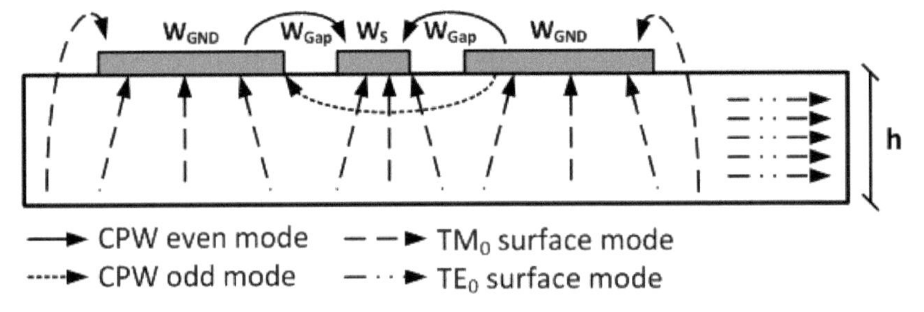

Figure 4.1 Cross section of a coplanar wave guide (CPW) with finite ground planes, and sketches of the E field distributions of the first propagating modes supported.

($n > 0$ and $m \geq 1$) are functions of the height and dielectric constant of the substrate [Poz04]. The overall effect of the unwanted modes described above is an increase of the transmission line losses (i.e., $|S_{21}|$) and the generation of ripples on the transmission parameter (i.e., S_{21}) of the CPW. The ripples are the results of interference (constructive or destructive depending on the frequency) between the unwanted modes, reflected by discontinuities (i.e., dielectric constant changes), and the intended CPW mode. The lines conventionally employed for probe-level TRL calibrations, are:

- *The thru standard:* A CPW line with a physical length in the order of 200–250 μm,
- *The line standard:* A CPW line providing an insertion phase of 90° at the center of the calibration band.

The analysis presented in this section is based on numerical 3D EM simulations, i.e., using Keysight EM Pro.

4.2.1 Parallel Plate Waveguide Mode

During the calibration procedure the substrate is placed on a metallic wafer chuck, creating effectively a grounded coplanar waveguide (GCPW) structure, as shown in Figure 4.2(a–c). This structure supports, in addition to the modes shown in Figure 4.1, also a parallel plate waveguide (PPW) mode.

This occurs since the top (CPW line) and bottom (chuck) metal are not directly contacted, thus a different potential can exist and propagate. The PPW mode can be visualized by plotting the E field intensity below the metal surface, as shown in Figure 4.2(c–d). In the figure the intensity of the E field is acquired on the xy plane placed below the metal plane (i.e., 5 μm). Note, both plots use the same range for the field intensity (blue = minimum, red = maximum) to allow for a direct visual comparison. Conventionally to reduce the PPW mode propagation, an interposer substrate of ferromagnetic material (i.e., providing high losses for the EM waves) is used between the calibration substrate and the metal chuck. Figure 4.2(d) shows a partial reduction of the PPW mode when simulating with the absorber structure. Alternatively, dielectric chucks with a permittivity similar to the one of the calibration substrate can be used to remove the occurrence of the PPW mode.

4.2.2 Surface Wave Modes: TM_0 and TE_1

The overall loss behavior of the CPW structure, including the surface waves when fed by a wafer probe, can be analyzed by using the 3D simulation environment shown in the inset of Figure 4.3(a). A point voltage source with

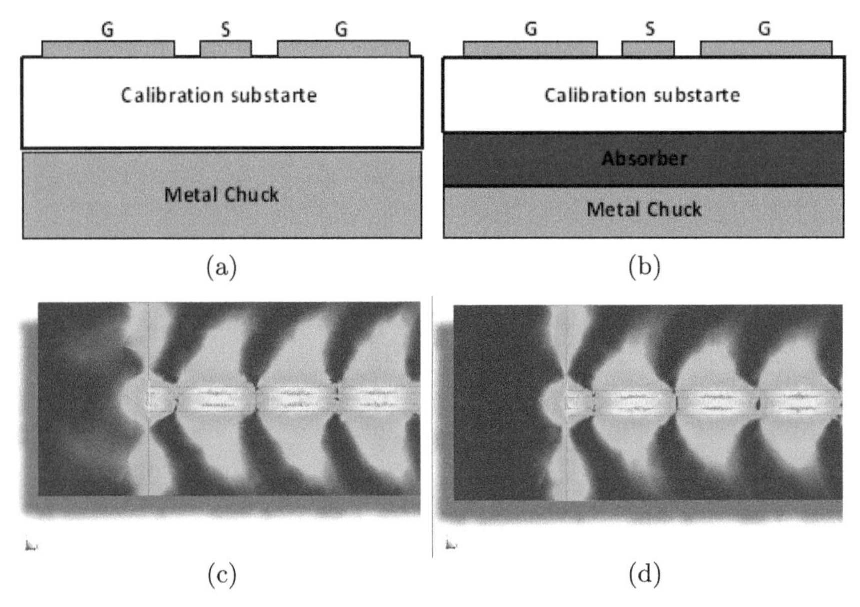

Figure 4.2 Cross section of CPW placed (a) on metal chuck, (b) on absorber. Electrical field intensity below the CPW metal plates (5 μm) for case (c) no absorber and (d) with absorbing boundary conditions in the 3D FEM simulation, both fields were computed at 180 GHz.

a source impedance of 50 Ohm is applied to the bridge to provide a transition similar to a wafer probe. In the 3D simulation environment, the boundary conditions were set to absorbing, thus providing perfect match condition to all the unwanted modes within the structure.

Figure 4.3(a) compares the insertion loss of the CPW structure realized in alumina when the substrate (sub) is enlarged and an air gap (GAP) is applied between the substrate boundary and the radiation boundary of the box, see Figure 4.3(b). Note that the multiple reflections of the unwanted modes within the structure generate an interference pattern (dependent on the distance to the discontinuity) along the trace, as can be seen by the shift of minima and maxima points when the sub-parameter is changed. The simulation does not include conductive or dielectric losses thus the decrease in the transmission parameter S_{21} in Figure 4.3(a) can only be attributed to energy dissipated in the other modes supported by the structure. When considering real structures on alumina substrate (i.e., exhibiting also dielectric losses), it is expected that the lines closer to the edge of the calibration substrate will exhibit a stronger ripple caused by interference with the surface wave mode. In Figure 4.3(b) structures with different distance to the substrate edge where

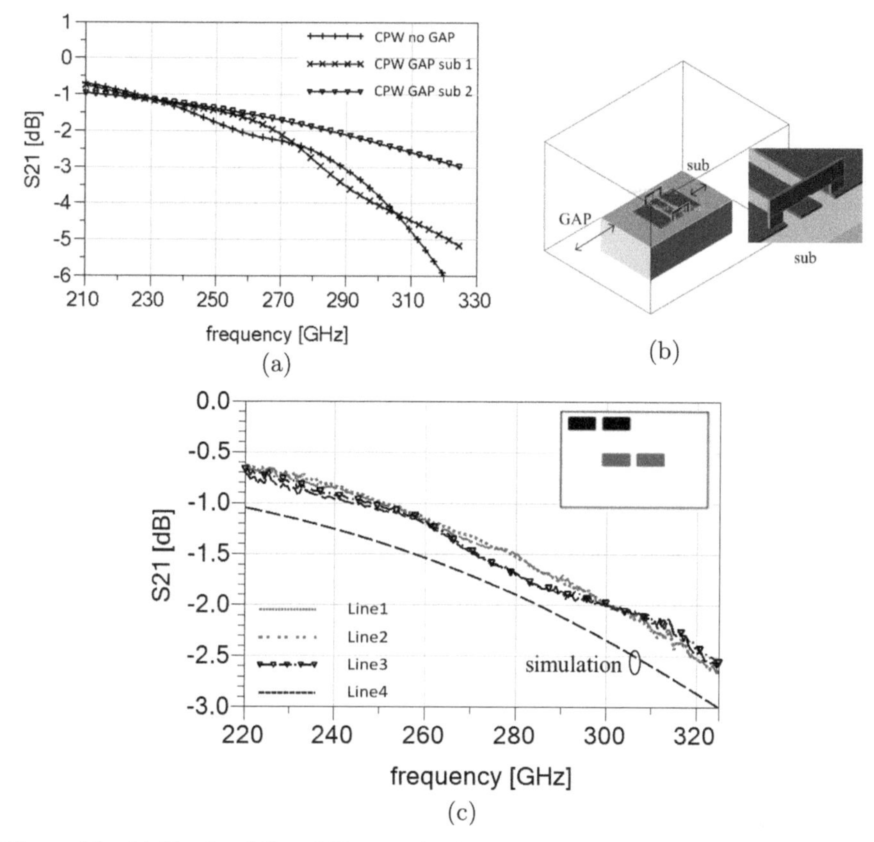

Figure 4.3 (a) Simulated S_{21} of CPW on alumina substrate for various cases: *CPW no GAP* sub = 0 μm GAP = 0 μm, *CPW GAP sub1* sub = 320 μm GAP = 500 μm, *CPW GAP sub2* sub = 420 μm GAP = 500 μm; (b) CPW structure used in the EM simulator with highlight on the lumped bridge configuration; (c) measurement of different (4) thru lines on alumina substrate in different locations of the calibration substrate. Locations (i.e., two middle and two center) identified in the inset on top right.

measured (i.e., center and edge) for the alumina substrate. As can be clearly seen by the figure, the structures at the edge of the substrate exhibit a clear interference pattern, as predicted by the simulation analysis.

4.2.3 Electrically Thin Substrates

Employing lower ϵ_r substrates shifts the occurrence of the TM_1 and TE_1 modes to higher frequencies, and reduces the amount of energy radiated by the CPW common mode, for a given frequency, due to the smaller gap dimension for a given signal width. For these reasons, fused silica ϵ_r can be considered as a good candidate to integrate CPWs to perform TRL calibration in the (sub)mm-wave bands. The same simulation analysis performed for the alumina case in Figure 4.3(a) was carried out for the fused silica substrate, see Figure 4.4(a). As can be seen by the plot a considerably lower amount of energy is transferred to other modes. Moreover, the lower dielectric constant of the substrate provides lower discontinuities when terminated with air, showing close to no-variation when performing a simulation varying the dimension of the parameters sub and GAP, see Figure 4.4(a).

The measured results are then compared with the simulation showing very good agreement in WR3 band, as shown in Figure 4.4(b), confirming also the low loss achieved by the CPW realized on fused silica. Note, that the deviation that can be observed between measured and simulated data above 290 GHz can be explained with reduced sensitivity of the measurement equipment, closer to the edge of the specified band (i.e., WR3 220–325 GHz) and the onset of unwanted modes in the fused silica substrate.

4.2.4 Calibration Transfer

In the previous paragraph the usage of electrically thin substrates was introduced to overcome the limitations exhibited by commercially available calibration devices operating in the mm-wave bands. While using such substrates (i.e., fused silica) improves the calibration quality, an important point is that the measurement quality will also depend on the error introduced by transferring the calibration to the environment where the DUT is embedded.

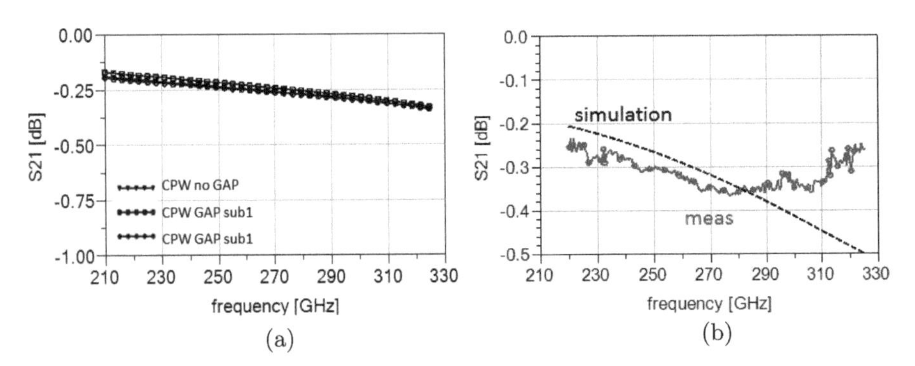

Figure 4.4 (a) Simulated S_{21} of CPW on fused silica substrate for various cases: *CPW no GAP* sub = 0 μm GAP = 0 μm, *CPW GAP sub1* sub = 320 μm GAP = 500 μm, *CPW GAP sub2* sub = 420 μm GAP = 500 μm; (b) measurement versus simulation of a thru line on fused silica substrate.

It is often the case that the DUT is embedded in a different host medium compared to the calibration, i.e., *Si, SiO_2, GaAs*, or other substrate materials. When the measurement is performed on the new host medium, a different probe to substrate interaction will occur, which would not be corrected for by the calibration. This will introduce a residual error that would be a function of the difference in permittivity between the two substrate materials (i.e., calibration and measurement). In a first-order approximation, the calibration transfer effect, associated with the change in the error box, can be seen as a capacitive coupling between the probe tip and the substrate, as schematized in Figure 4.5. This capacitance can be found using a numerical optimizer when the probe geometry is partially known, allowing to minimize the calibration transfer error in the measurement frequency band.

Figure 4.6 shows the results, in terms of worst case of the error bound [Wil92], when transferring the calibration from the primary calibration environment (i.e., alumina and fused silica) to a verification line embedded in the back-end-of-line of a SiGe high-speed process. The calibration quality is evaluated before any optimization is applied (full symbols and solid lines) and after the application of the correction (empty symbols, dotted lines). The maximum value for the error associated with the calibration of alumina decreased from 0.12 to 0.06, with an improvement noticeable over the entire bandwidth. However, no significant improvement is obtained for the fused

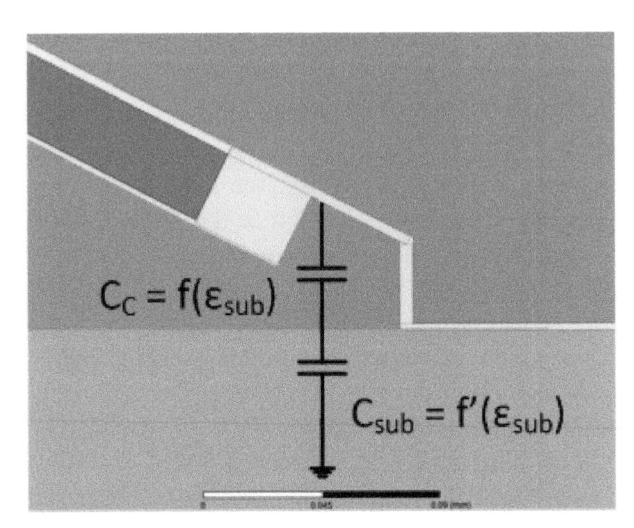

Figure 4.5 Schematic representation of the capacitive coupling between the probe tip and the substrate where the device under test (DUT) is embedded.

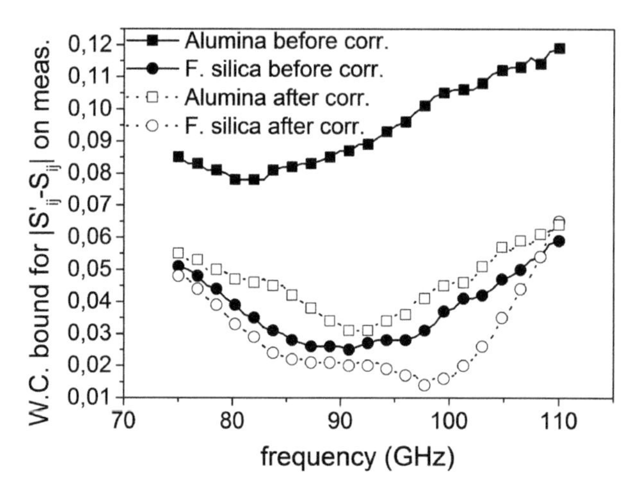

Figure 4.6 Worst case error bound for calibration transfer from fused silica and alumina to SiGe BEOL, before correction (full symbols, solid lines) and after correction (empty symbols, dotted lines), obtained with on-wafer measurements on a 600 μm CPW line manufactured on IHP SiGe 130 nm BiCMOS technology, in the frequency range from 75 to 110 GHz.

silica, where the error associated with the difference in substrate coupling due to calibration transfer is small, due to the similarity of permittivity between the fused silica and the silicon dioxide present in the back-end-of-line of the process.

4.3 Direct On-wafer Calibration

In order to avoid the error arising from the process of transferring the first-tier calibration to another environment, the calibration kit should be implemented in the same environment as that of the DUT. Classical probe-level and on-wafer calibration techniques are based on (partially) known devices and lumped models of the DUT fixture (i.e., SOLT/LRM and lumped de-embedding) or employ distributed concepts (TRL and multi-line TRL). Due to the objective difficulty, especially at higher frequencies, in manufacturing an accurate and predictable resistor in a commercial silicon technology, (multiline)-TRL calibration represents the standard employed technique for *in situ* calibration, as was shown in [Yau10; Yau12; Wil13a; Wil13b; Wil14]. The TRL technique does not require resistors to define the measurement normalization impedance, which is instead set by the characteristic impedance of the lines used during the calibration. Thus, the accurate (frequency-dependent) determination of the calibration lines characteristic impedance becomes a key requirement to allow the correct re-normalization of TRL-calibrated S-parameter measurements.

4.3.1 Characteristic Impedance Extraction of Transmission Lines

To accurately employ TRL techniques in a complex environment such as the BEOL of Silicon-based technologies, a robust approach is required to extract the characteristic impedance of the transmission line. Traditional extraction procedures are based on measurements [Eis92; Mar91a; Wil91a; Wil91b; Mar91b; Wil98], but are only accurate when specific assumptions are verified, such as, low loss substrate, constant capacitance per unit length [Mar91b], and uniform [Eis92] non-inductive pad-to-line transitions [Wil98; Wil01]. For a more extensive analysis of the shortcomings of these methods for (sub)-mm-wave calibration in the BEOL of silicon technologies, the reader is invited to read [Gal17a], where a characterization flow employing 3D EM simulations was developed and validated to accurately extract the Z_0 of transmission lines, excited using waveguide (modal) excitation. The scattering parameters computed during simulation are re-normalized to a given system value (i.e., $Z_{\text{sys}} = 50\ \Omega$) and used in Equation (4.1) to compute the line characteristic impedance [Eis92]:

$$Z_0 = Z_{sys} \cdot \sqrt{\frac{(1 + S_{11}^2) - S_{21}^2}{(1 + S_{11}^2) - S_{21}^2}} \tag{4.1}$$

The approach was validated by benchmarking it with the calibration comparison method using a calibration kit integrated in the BEOL of the IHP SG13G2 130 nm SiGe BiCMOS technology. For the purposes of a fair comparison, the lines were designed to be uniform, with no line-to-pad discontinuities to provide an accurate test case for the calibration comparison method [Wil01]. The calibration kit was designed to allocate different waveguide bands from 75 GHz to 325 GHz. The micro-photographs of the WR-5 (140–220 GHz) structures are displayed in Figure 4.7(a–c), while Figure 4.7(d) shows the schematic line cross section.

The structures were simulated using three different 3D electro-magnetic simulators, Keysight EMPro, Ansoft HFSS, and CST Studio Suite, to check for simulation discrepancies. In the model, the meshed ground planes have been simplified considering a continuous metal connection, both vertically and horizontally. This simplification provides good approximation of the

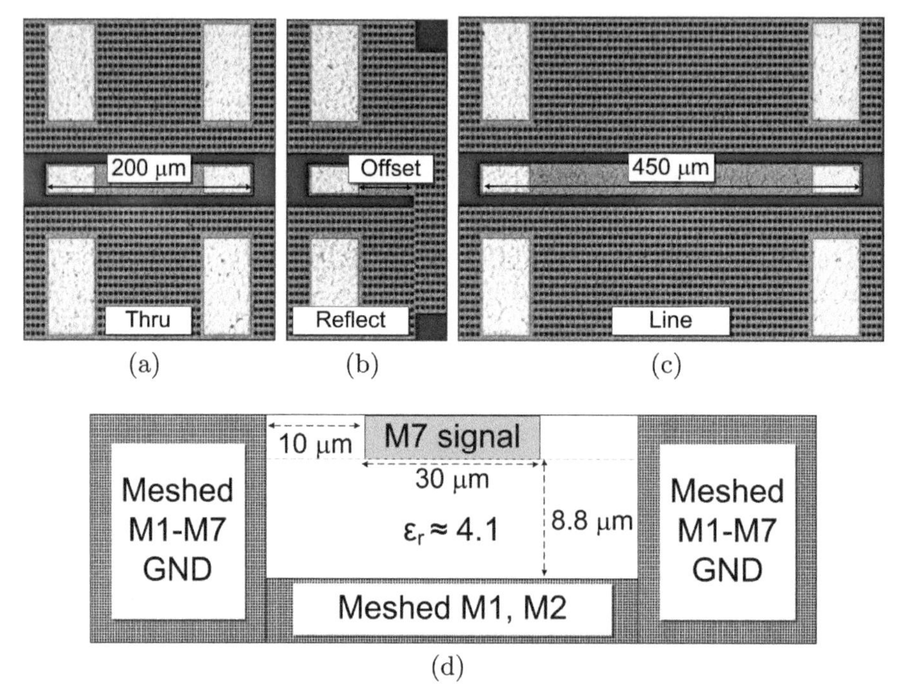

Figure 4.7 Coplanar wave guide CPW calibration structures realized on IHP SiGe 130 nm BiCMOS technology. (a) Microphotograph of the thru line, (b) of the reflect standard and (c) of the transmission line employed for the WR05 calibration kit, (d) schematic cross section of the CPW line.

electrical response of the structure since the openings in the metal mesh are much smaller than the wavelength (maximum aperture is in the order of 2.5 × 2.5 μm^2). The excitation of the CPW lines is provided by means of waveguide (modal) ports. The simulator first solves a two-dimensional eigenvalue problem to find the waveguide modes of this port and then matches the fields on the port to the propagation mode pattern, and computes the generalized (i.e., mode matched) scattering parameters. In all the simulators, the port dimensions are designed using the rules of thumb described in [Wei08], ensuring ideally no fields at port boundaries, as also depicted in Figure 4.8 for two simulator examples.

Absorbing/radiation boundaries are then imposed at the lateral and top faces of the simulation box. The box is defined horizontally by the dimensions (length/width) of the simulated structure, and vertically by the wavelength ($\lambda/4$ at minimum simulation frequency). The bottom face of the simulation

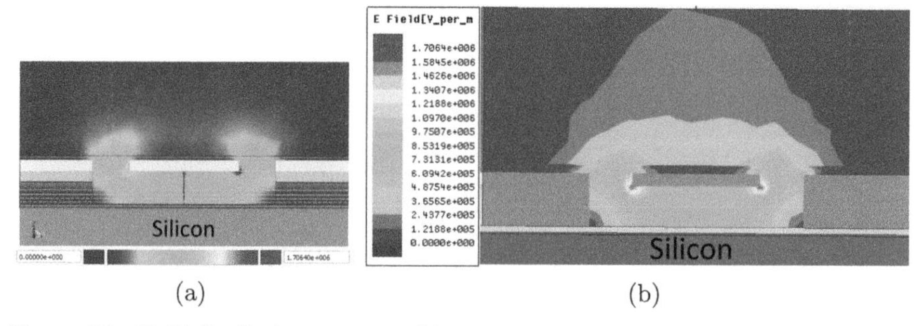

Figure 4.8 Field distribution on waveguide ports at 300 GHz when exciting the structures described in Figure 4.7, for (a) Keysight EMPro and (b) Ansoft HFSS.

box is defined as a perfect electric conductor, simulating the presence of a metallic chuck underneath the structure, as it is the case during measurements. The absorbing boundaries simulate an unperturbed propagation of the EM waves through this boundary. In this respect, the interference with other structures on the wafer is not taken into account in the simulation. Material parameters and lateral dimension are chosen according to the nominal technology values.

Figure 4.9 shows the comparison of the characteristic impedance computed using the proposed method (with different EM simulation tools), and the characteristic impedance extracted with measurements using the calibration comparison method [Wil01] and the Eisenstadt method [Eis92]. Both the measurement based methods are hampered by the (small) discontinuity presented by the probe to line transition, as predicted in [Mar92] and [Wil01]. The EM-based method offers fairly constant (with frequency) characteristic impedance response, as expected. It is interesting to note how simulations performed employing different EM tool, produce slightly different values for the characteristic impedance (max. 1 Ω for the real part and 0.1 Ω for the imaginary part), when applying similar settings in terms of meshing and solving methods. The differences can be attributed to different meshing algorithms, discretization, etc., of the tools which could all be categorized as the intrinsic uncertainty of the proposed method. This comparison shows how the proposed method provides comparable results to the state-of-the-art techniques, when the validity of the latter is still guaranteed by the transmission line design. As the method of [Gal17a] is ideally valid for any kind of transmission line, it can be employed also in situations in which large inductive probe-to-line transitions are present, which is the case when vias are involved.

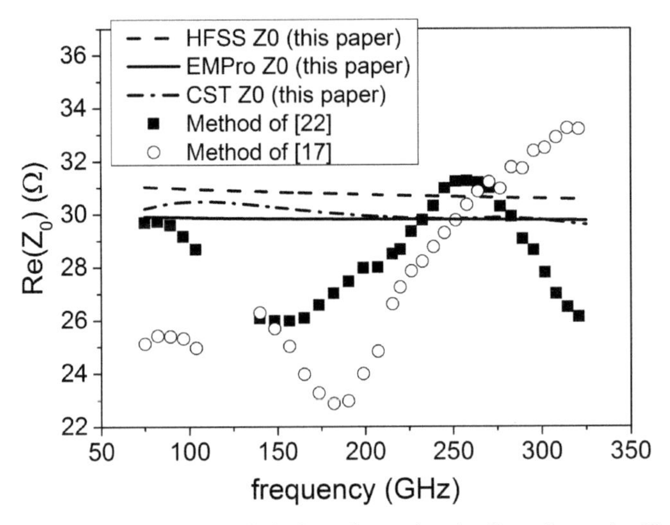

Figure 4.9 Real part of characteristic impedance for the line shown in Figure 4.7(a), computed with the simulation approach described in [Gal17a] (solid lines EMPro, dashed lines HFSS, dashed-dot lines CST), and measured using the method of [Eis92] (empty circles) and the method of [Wil91b] (filled squares).

In order to compare the different calibrations, the method of [Wil92] has been employed, defining an upper bound (UB) error metric as:

$$UB(f) = max\left|S'_{i,j}(f) - S_{i,j}(f)\right| \tag{4.2}$$

Where S' is the reference scattering matrix of the verification line (i.e., 3D simulated S-parameters), $S(f)$ is the frequency-dependent scattering matrix resulting from the investigated calibrations (i.e., LRM on alumina, TRL on fused silica and TRL on BiCMOS) and $i, j \in$ [1,2]. This metric defines the UB of the deviation of the S-parameters measured by one calibration and the reference S-parameters computed using EM simulations. The measurement data used to compute the error bound of Figure 4.10 are based on the same raw data of the verification line, thus removing any measurement variation of the verification structure from the error propagation mechanisms. On these raw data the respective calibration algorithm (with their respectively computed error terms) were applied. In addition, both the methods indicated as TRL on silicon in Figure 4.10 use also the same raw measurement in the calibration procedure (i.e., extraction of error terms), thus confining their difference only to the characteristic impedance values versus frequency, computed with the two different methods.

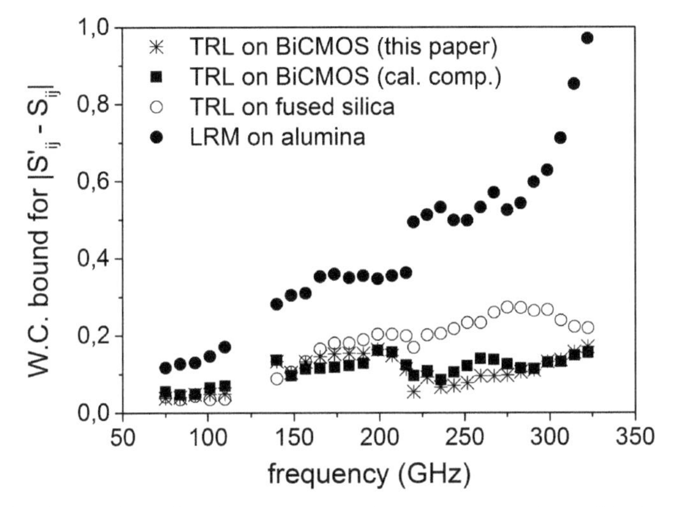

Figure 4.10 Comparison of probe-tips corrected measurements of a verification line manufactured on the SiGe BEOL in the frequency range 75–325 GHz for different calibrations.

As can be seen from Figure 4.10, the calibration performed on SiGe technology is the one that presents smaller deviation from the reference data, with an $UB \leq 0.17$ in the entire frequency band for both characteristic impedance extraction methods considered, i.e., the proposed EM-based method (Figure 4.10, asterisks) and the calibration comparison method (Figure 4.10, filled squares).

4.4 Direct DUT-plane Calibration

The method to derive the characteristic impedance of a transmission line described in the Section 4.3 "Direct On-wafer Calibration" will be applied to extract the Z_0 of transmission lines employed in a TRL calibration/de-embedding kit to perform S-parameters measurements at the lowest metal layer (M1) for direct DUT access. Realizing transmission lines in the lowest metal layers can present several challenges, typically associated with the losses of the underlying substrate (i.e., conductive silicon). One solution was proposed in [Gal17b], where a CPW line realized at M1 was capacitively loaded with a series of floating metal bars (CL-ICPW), realized in a higher metal layer, in order to confine the propagating electromagnetic field in the low loss oxide of the BEOL.

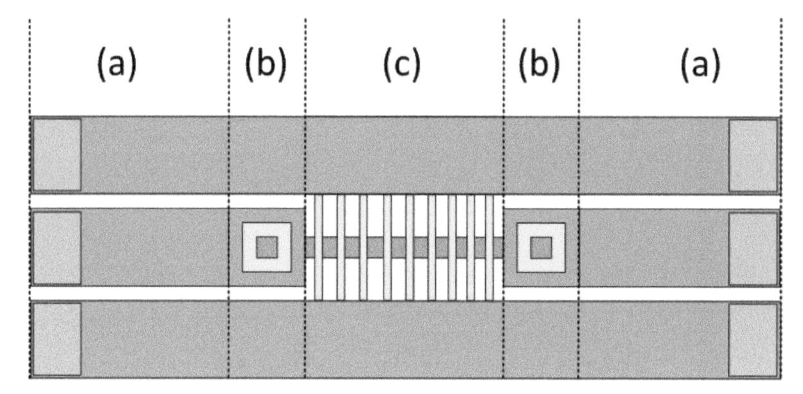

Figure 4.11 Simplified schematic top-view of a generic test-structure realized with CL-ICPW. (a) Input section, (b) M7-M1 vertical transition and (c) DUT stage.

This line topology can be employed in the TRL calibration/de-embedding kit, as depicted in Figure 4.11. The general structure of the fixture features three main sections: an input stage [pad plus launch line, section (a); a transition from top metal to M1, section (b), composed by all metal layers and interconnecting vias; and the final section (c), realized on M1 using CL-ICPWs that can feature a transmission line i.e., thru or line for the TRL de-embedding kit] or an offset short. The calibration/de-embedding kit, was manufactured using the BEOL of *Infineon's* 130 nm SiGe BiCMOS technology B11HFC, featuring seven metal layers. Figure 4.12(a) shows a cross section of Figure 4.11, section (a), where M3 is used as ground shield in order to isolate the CPW from the lossy substrate. The transition from the top metal center conductor of the CPW to the M1 center conductor of the CL-ICPW is realized using a gradual, inverse pyramidal shape. This allows to connect the large top metal conductor (i.e., 30 μm width) with the smaller M1 line, keeping the ground reference at the same metal level (i.e., M3) as shown in Figure 4.12(b).

For the DUT stage, M3 is chosen as the metal layer for the floating shield. This choice allows reducing the losses while guaranteeing a Z_0 of 34 Ω, sufficiently close to the 50 Ω required to minimize the errors arising from reflection losses when measuring in a conventional VNA-based setup [Mub15]. The shield is realized with 2 μm wide metal strips and a fill factor of 50% in order to respect the density rules. The cross section of the final design for the CL-ICPW is shown in Figure 4.12(c). Micro-photographs of the calibration/de-embedding kit for WR-3 (220–325 GHz) waveguide bandwidth are shown in Figure 4.13.

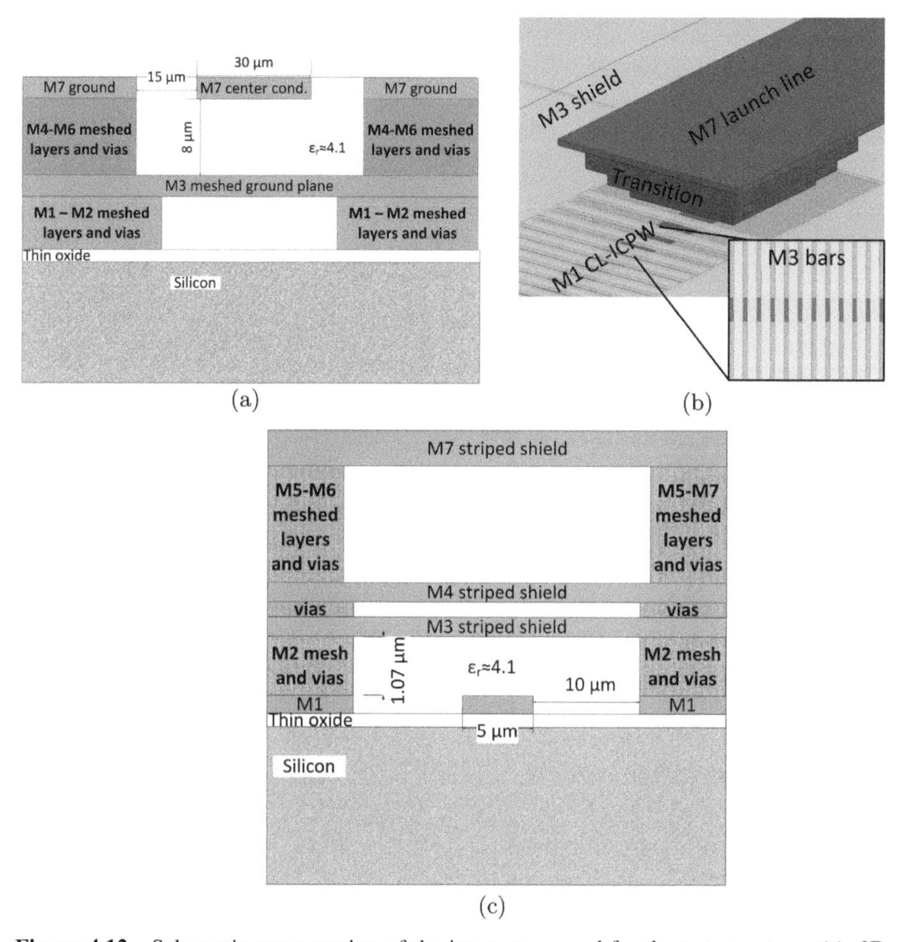

Figure 4.12 Schematic cross section of the input stage used for the test structures (a). 3D model of the vertical transition connecting the central conductor of the input stage in M7 to the CL-ICPW central conductor in M1 (b). Schematic cross section of the CL-ICPW employed in the DUT stage of the calibration kit (c).

The kit employs 130 μm long launch lines. The thru standard is realized by means of a 150 μm CL-ICPW, and it is designed to embed the final DUT (transistor here) in its center reference plane. The de-embedding kit reflects are realized by two symmetric offset shorts, with an offset equal to half the thru length. Further, a longer line with an additional 80 μm length for the CL-ICPW, in respect to the thru, is realized as the line standard. Finally, a test structure consisting of a 310 μm long CL-ICPW has been realized for

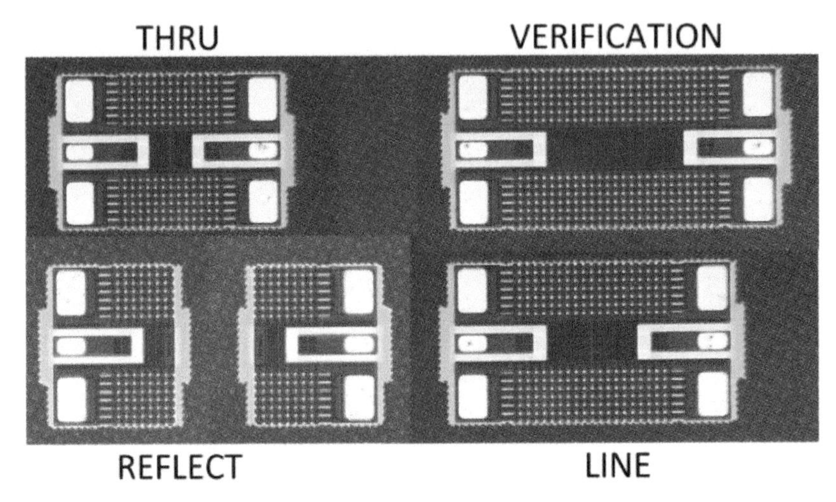

Figure 4.13 Micrograph of the de-embedding kit on Infineon B11HFC technology.

verification. EM simulations are then performed to extract the characteristic impedance of the line. Note that the only structures simulated are the CL-ICPW in Figure 4.12, section (c). For this purpose, the procedure described in Section 4.2 is employed. Once the characteristic impedance is extracted, the proposed kit can be employed for direct calibration at M1. To demonstrate the proposed calibration/de-embedding method in its final application, measurements of a heterojunction bipolar transistor (HBT) featuring two emitter fingers with 5 μm length and 220 nm width were performed. The device was embedded into the test fixture employing CL-ICPW in common-emitter (CE) configuration directly at the calibration reference planes, shown in Figure 4.14. To guarantee proper connection between the CL-ICPW test structure and the transistor (BECEB) modeled in the process design kit (PDK) (i.e., employing a p-type guard ring around the active device, with ground contacts connected to metal level 1) a small bridge at metal 2 (see, Figure 4.14(b) was added. After calibration, EM simulations of these lines are used to de-embed them from the measurements. Note, that the configuration and interconnections (no M1 connections between the emitters and the bases) is only illustrative of the technique. When a different reference plane needs to be defined and different parasitic element included or excluded from the device model this can be achieved by properly setting the reference plane of the calibration through the proper design of the reflect standard and the zero length thru position.

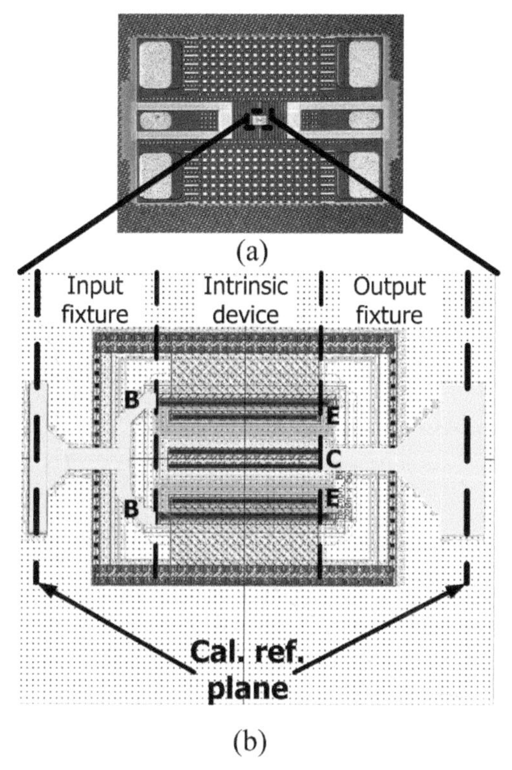

Figure 4.14 Top view of the transistor (BECEB) integrated into the test-structure (a) Detailed view of the layout for the integrated transistor (b), highlighting the input and output fixture (in yellow) required to guarantee connection to the intrinsic device. The base, collector, and emitter contact (B, C, and E, respectively) are marked on the layout.

The device S-parameters have been measured using the direct calibration technique in the frequency range from 220 to 325 GHz, using fixed bias conditions ensuring close to peak f_T, i.e., V_{CE} = 1.5 V and V_{BE} = 0.91 V. The measurement results are then compared with the S-parameters obtained by using the HICUM level 2 model of the device. The device selected in the layout of this work was not supported by a model in the PDK so that an approximate set of parameters had to be generated. Figure 4.15(a) shows the comparison of the magnitude in dB for all the S-parameters of the considered transistor. The measured values for S_{11} and S_{21} agree quite well with the model prediction, with discrepancies in the order of 0.2 dB, while S_{22} shows a bigger error, with a maximum value in the order of 1.1 dB in the entire frequency range. The S_{12} parameter shows the biggest relative

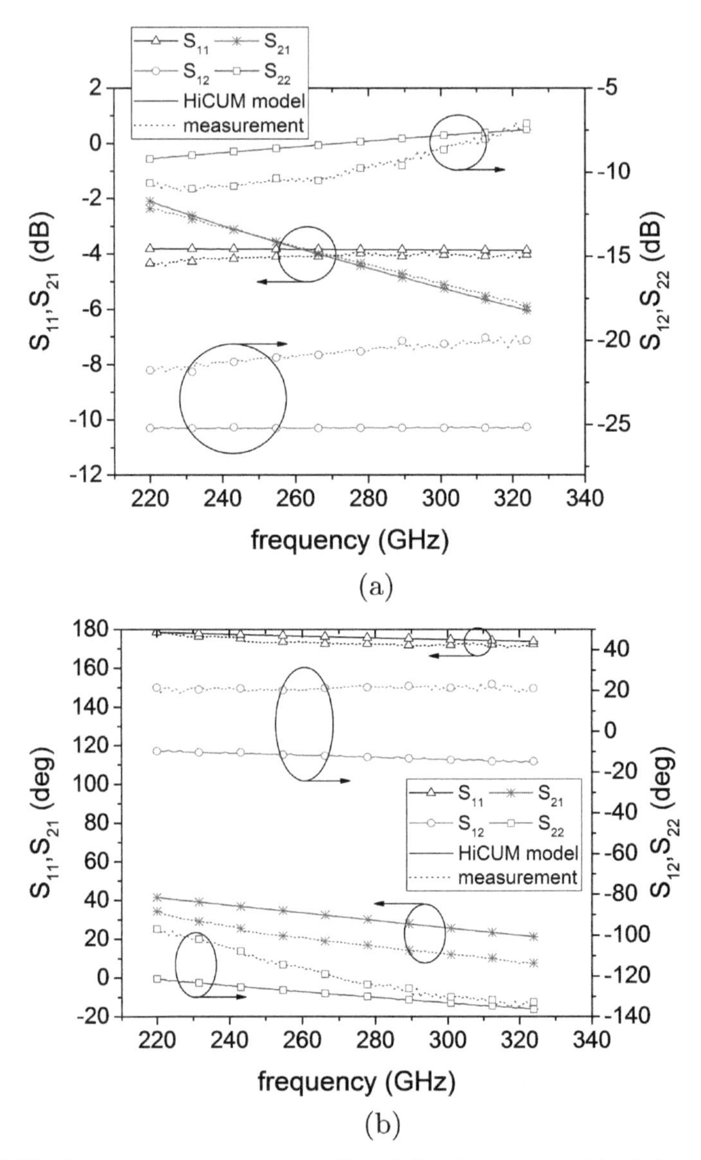

Figure 4.15 S-parameter measurements (dotted lines) versus model of the considered Infineon transistor (solid lines) for both a) Amplitude (in dB) and b) Phase (in degrees).

error in magnitude, due to its small absolute value. Discrepancies between measurements and model are more significant when considering the phase information (see, Figure 4.15(b)) where they can reach 40 degrees for S_{12}.

4.5 Conclusion

In this chapter various concepts and techniques to achieve accurate calibration techniques at (sub)mm-waves for device characterization have been reviewed. The problems of electrically thick substrates have been explained and experimentally validated. Electrically thin substrates with their performance improvement were discussed. The problems and possible error compensations related to substrate transfers are addressed. A complete flow and an EM-based technique to design and characterize TRL-based calibration kits to be embedded in the BEOL of commercial silicon technologies were described. Finally, an approach to realize direct calibration/de-embedding kits capable of measuring the device performance at M1 was presented and experimentally validated.

References

[Dav90] Davidson, A., et al. (1990). "LRM and LRRM calibrations with automatic determination of load inductance," in *Proceedings of the 36th ARFTG Conference Digest*, Monterey, CA, 57–63. doi: 10.1109/ARFTG.1990.323996

[Dio17] Virginia Diodes (2017). *Vector Network Analyzer Extenders.* Available at: https://vadiodes.com/en/products/vector-network-analyzer-extension-modules

[Eis92] Eisenstadt, W. R., et al. (1992). "*S-parameter-based IC interconnect transmission line characterization,*" in *Proceedings of the IEEE Transactions on Components, Hybrids, and Manufacturing Technology*, Monterey, CA, 483–490. doi: 10.1109/33.159877

[Eng79] Engen, G. F., et al. (1979). Thru-reflect-line: an improved technique for calibrating the dual six-port automatic network analyzer. *IEEE Transactions on Microwave Theory and Techniques* 27.12, 987–993. doi: 10.1109/TMTT. 1979.1129778

[Eul88] Eul, H. J., et al. (1988). "Thru-match-reflect: one result of a rigorous theory for de-embedding and network analyzer calibration" in *Proceedings of the 18th European Microwave Conference*, Stockholm, 909–914. doi: 10.1109/EUMA.1988.333924

[Fer92] Ferrero, et al. A. (1992). Two-port network analyzer calibration using an unknown 'thru. *IEEE Microw. Guid. Wave Lett.* 2.12, 505–507. doi: 10.1109/75.173410

[Gal15] Galatro, L., et al. (2015). "Analysis of residual errors due to calibration transfer in on-wafer measurements at mm-wave frequencies," in *Proceedings of the 2015 IEEE Bipolar/BiCMOS Circuits and Technology Meeting – BCTM*, Rome, 141–144. doi: 10.1109/BCTM.2015.7340569

[Gal17a] Galatro, L., et al. (2017). "Millimeter-Wave On-Wafer TRL Calibration Employing 3-D EM Simulation-Based Characteristic Impedance Extraction," in *IEEE Transactions on Microwave Theory and Techniques* 65.4, pp. 1315–1323. doi: 10.1109/TMTT.2016.2609413

[Gal17b] Galatro, L., et al. (2017). "Capacitively Loaded Inverted CPWs for Distributed TRL-Based De-Embedding at (Sub) mm-Waves," in *Proceedings of the IEEE Transactions on Microwave Theory and Techniques*, London, 1–11. doi: 10.1109/TMTT.2017.2727498

[Mar91a] Marks, R. B. (1991). A multiline method of network analyzer calibration. *IEEE Trans. Microw. Theory Techniq.* 39.7, 1205–1215. doi: 10.1109/22.85388

[Mar91b] Marks, et al. R.B. (1991). Characteristic impedance determination using propagation constant measurement. *IEEE Microw. Guided Wave Lett.* 1.6, 141–143. doi: 10.1109/75.91092

[Mar92] Marks, R. B. et al. (1992). "Interconnection transmission line parameter characterization," in *Proceedings of the 40th ARFTG Conference Digest. Institute of Electrical and Electronics Engineers (IEEE)*, San Diego, CA. doi: 10.1109/arftg.1992.327004

[Mub15] Mubarak, F., et al. (2015). "Evaluation and modeling of measurement resolution of a vector network analyzer for extreme impedance measurements," in *Proceedings of the 86th ARFTG Microwave Measurement Conference*, Atlanta, GA, 1–3. doi: 10.1109/ARFTG.2015.7381475.

[Poz04] Pozar, D. M. (2004). *Microwave Engineering*. Hoboken, NJ: Wiley.

[Ryt01] Rytting, D. K. (2001). "Network analyzer accuracy overview" in *Proceedings of the 58th ARFTG Conference Digest*, Atlanta, GA, 1–13. doi: 10. 1109/ARFTG.2001.327486

[Stu09] Stumper, U. (2009). Influence of nonideal calibration items on S-parameter uncertainties applying the SOLR calibration method. *IEEE Trans. Instrument. Meas.* 58.4, 1158–1163. doi: 10.1109/TIM.2008.2006962

[Tep13] Teppati, V., et al. (2013). Modern RF and Microwave Measurement Techniques. The Cambridge RF and Microwave Engineering Series. Cambridge University Press, 2013. isbn: 9781107245181. url: https://books.google.nl/books?id=TEuBKlFGGUUC

[Tie05] Tiemeijer, L. F., et al. (2005). Comparison of the pad-open-short and open-short-load deem bedding techniques for accurate on-wafer RF characterization of high-quality passives. *IEEE Trans. Microw. Theory Tech.* 53.2, 723–729. doi: 10.1109/TMTT.2004.840621

[Wei08] Weiland, T., et al. (2008). A practical guide to 3-D simulation. *IEEE Microw. Magaz.* 9.6, 62–75. doi: 10. 1109/mmm.2008.929772

[Wil01] Williams, D. F., et al. (2001). Characteristic-impedance measurement error on lossy substrates. *IEEE Microw. Wireless Comp. Lett.* 11.7, 299–301. doi: 10.1109/7260.933777

[Wil13a] Williams, D. F., et al. (2013). A prescription for sub-millimeter-wave transistor characterization. *IEEE Trans. Terahertz Sci. Technol.* 3.4, 433–439. doi: 10.1109/TTHZ.2013.2255332

[Wil13b] Williams, D. F., et al. (2013). Calibration-Kit design for millimeter-wave silicon integrated circuits. *IEEE Trans. Microw. Theory Tech.* 61.7, 2685–2694. doi: 10.1109/TMTT.2013.2265685

[Wil14] Williams, D. F., et al. (2014). Calibrations for millimeter-wave silicon transistor characterization. *IEEE Trans. Microw. Theory Tech.* 62.3, 658–668. doi: 10.1109/TMTT.2014.2300839

[Wil91a] Williams, D. F., et al. (1991). Transmission line capacitance measurement. *IEEE Microw. Guid. Wave Lett.* 1.9, 243–245. doi: 10.1109/75.84601

[Wil91b] Williams, D. F., et al. (1991). "Comparison of on-wafer calibrations," in *Proceedings of the 38th ARFTG Conference Digest. Institute of Electrical and Electronics Engineers (IEEE)*, Fort Worth, TX, doi: 10.1109/ arftg.1991.324040

[Wil92] Williams, D. F., et al. (1992). "Calibrating on-wafer probes to the probe tips," in *Proceedings of the 40th ARFTG Conference Digest. Institute of Electrical and Electronics Engineers (IEEE)*, San Diego, CA. doi: 10.1109/arftg.1992.327008

[Wil98] Williams, D. F., et al. (1998). "Accurate Characteristic Impedance Measurement on Silicon," in *Proceedings of the 51st ARFTG Conference Digest. Institute of Electrical and Electronics Engineers (IEEE)*, New York, NY. doi: 10.1109/arftg.1998.327296

[Yau10] Yau, K., et al. (2010). "On-wafer s-parameter de-embedding of silicon active and passive devices up to 170 GHz," in *Proceedings of the 2010 IEEE MTT-S International Microwave Symposium*, Anaheim, CA. doi: 10. 1109/MWSYM.2010.5516659

[Yau12] Yau, K., et al. (2012). Device and IC Characterization Above 100 GHz. *IEEE Microw. Magaz.* 13.1, 30–54. doi: 10.1109/MMM.2011. 2173869

5

Reliability

V. d'Alessandro[1], C. Maneux[2], G. G. Fischer[3], K. Aufinger[4], A. Magnani[1], S. Russo[1] and N. Rinaldi[1]

[1]Department of Electrical Engineering and Information Technology, University Federico II, Italy
[2]Laboratory of Integration of Material to System (IMS), University of Bordeaux, France
[3]IHP, Germany
[4]Infineon Technologies AG, Germany

5.1 Mixed-mode Stress Tests

5.1.1 Introduction to Hot-Carrier Degradation under MM Stress

An important reliability concern in SiGe HBTs is related to long-term degradation (*stress* or *aging*) effects induced by hot carriers (HCs). While in MOSFETs HC mechanisms produce a degradation of drain current and transconductance, as well as a threshold voltage shift [Tya15], in bipolar transistors the HC damage is mainly related to the creation of Si dangling bonds acting as trap states at the semiconductor–insulator interfaces. Interface traps induced during device operation lead to an increased Shockley–Read–Hall (SRH) recombination and hence to an excess non-ideal base current component. Differently from MOSFETs, the collector current remains unaffected, and therefore it can in principle be stated that HC degradation is less critical in bipolar transistors, including the SiGe HBT technology; however, it still entails a number of undesirable consequences, such as current gain reduction (due to the base current growth), noise figure increase, shift of the bias point outside of the functional range, as well as increased power consumption in power amplifiers [Ven00, Cre04, Che09]. Such effects have been traditionally studied under reverse base–emitter stress conditions, where HCs are created by large electric fields across the base–emitter junction (see the early papers [Bur88, Gog00] and the more recent [Sas14a, Sas14b,

Fis15]); this stress test was indeed considered as appropriate for assessing device reliability in BiCMOS operation [Bur88].

Another stress technique has been subsequently proposed and quickly accepted in the literature, which is more representative of device degradation in practical mixed-signal and RF circuit applications; in this technique, referred to as *mixed mode* (MM) [Zha02], the device under test (DUT) is usually operated in common–base (CB) configuration while being simultaneously subjected to large emitter current density ($J_{\mathrm{E,stress}}$) and collector–base voltage ($V_{\mathrm{CB,stress}}$) [the corresponding V_{CE} being higher than the open-base breakdown voltage (BV_{CEO})] [Zhu05, Dio08, Cha15]. Although this biasing condition may seem too severe, the instantaneous operating point of a transistor (e.g., in oscillators and in noise/power amplifiers) can reach either high voltage or high current under large-signal operating mode, thereby gradually increasing HC-triggered damage [Che09, Fis08, Gre09, Fis15]. The high – and continuously applied – stress conditions $J_{\mathrm{E,stress}}$ and $V_{\mathrm{CB,stress}}$ are also denoted as *accelerating factors*, since they give rise to significant MM stress degradation in a relatively short time. Under MM stress tests, the trap creation process involves the following steps [Moe12]:

- the large electric field across the base–collector space-charge region (SCR) first creates primary HCs, and then additional (secondary, tertiary, and so on, depending upon $V_{\mathrm{CB,stress}}$) HCs by impact ionization (II);
- a fraction of the generated HCs can be directed toward the emitter–base oxide spacer (used to separate the Si emitter from the extrinsic base region) or the shallow trench (ST) oxide edge. Along these paths, they lose some energy due to collisions;
- if the HCs reach the oxide interfaces with an energy higher than 1.5 eV, then damage is produced in the form of dissociation of passivated Si–H bonds [Tya15, Tya16]. The damage spectrum depends on the accelerating factors $J_{\mathrm{E,stress}}$ and $V_{\mathrm{CB,stress}}$ for multiple reasons: first, they determine the II rate, but also the device temperature (high temperatures can have a beneficial impact in terms of damage recovery); moreover, they can trigger high-current (Kirk effect) or high-voltage (*pinch-in* effect) phenomena [Che07]. The trap creation at the emitter–base spacer (attributed to hot holes [Van06, Kam17b]) can be monitored by measuring the forward $V_{\mathrm{CB}} = 0$ V Gummel plot, where an SRH-induced growth in the low-V_{BE} base current is observed; the reverse $V_{\mathrm{EB}} = 0$ V Gummel plot is instead used to measure the damage due to HCs hitting the ST interface, since it causes an increase in the reverse-mode base current [Zha02, Zhu05, Che07, Moe12, Cha15].

5.1.2 Long-term MM Stress Characterization on IHP Devices

Mixed mode stress has been investigated in a recent exhaustive work [Fis15], where – differently from previous papers – *long-term* stress tests were performed, plainly showing a decrease in the degradation rate at long times. Moreover, an accurate investigation is presented, which includes the effect of: (i) the accelerating factors $J_{E,stress}$ and $V_{CB,stress}$, (ii) stress temperature, (iii) thermal recovery, and (iv) compact modeling of stress-induced base current components. The results of [Fis15] reported here refer to a packaged single-emitter SiGe:C NPN HBT fabricated by IHP, with effective emitter area[1] $A_E = W_E \times L_E = 0.16\ \mu m \times 0.52\ \mu m$, W_E and L_E being the effective emitter width and length, respectively, featuring peak f_T of 250 GHz at $J_C = 18\ mA/\mu m^2$, peak f_{MAX} equal to 300 GHz, $BV_{CEO} = 1.7$ V, $BV_{CBO} = 5$ V, and mounted in a CB configuration.[2]

The procedure can be described as follows. First, a Gummel plot is measured at $V_{CB} = 0$ V, the HBT being still *fresh* (i.e., stress-unaffected). Then the stress bias (high $J_{E,stress}$ and $V_{CB,stress}$) is applied, and the evolution of the collector and base currents with stress (aging) time are monitored by measuring non-stressing $V_{CB} = 0$ V Gummel plots at chosen time instants and recording their values at $V_{BE} = 0.7$ V. Figure 5.1 shows the CB output

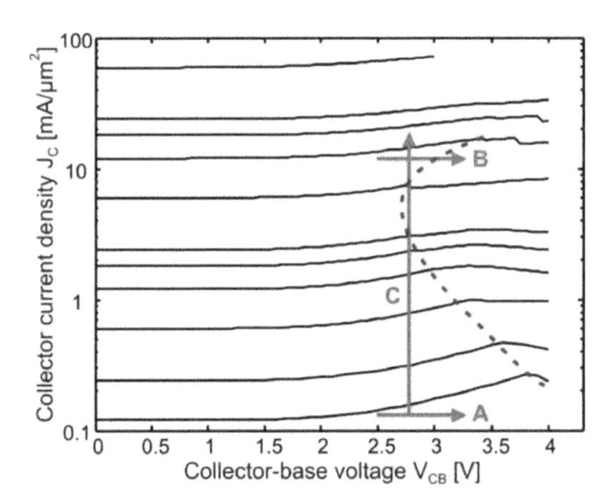

Figure 5.1 CB output characteristics of the DUT manufactured by IHP; also shown are the *pinch-in* locus (red dashed line) and the stress paths **A**, **B**, **C**.

[1]The effective emitter area is the area of the interface between Si emitter and SiGe base, which defines the vertical current flow.

[2]The analysis obviously requires the availability of a number of identical HBTs, one for each stress test to be performed.

characteristics of the DUT; also reported are the locus of *pinch-in* occurrence, which represents the limit of the CB safe operating area (SOA), and the stress conditions: in case **A**, identical transistors are biased with a low $J_{\mathrm{E,stress}}$ (=0.12 mA/μm^2) and different $V_{\mathrm{CB,stress}}$; in experiment **B**, other identical transistors are biased with a high $J_{\mathrm{E,stress}}$ = 12 mA/μm^2 (not far away from the current density at peak f_{T}) and different $V_{\mathrm{CB,stress}}$; in case **C**, $V_{\mathrm{CB,stress}}$ is kept constant at 2.75 V, and different $J_{\mathrm{E,stress}}$ are applied to identical transistors.

The first measurement campaign was conducted by forcing the transistor backside to a temperature T_{B} = 300 K through a thermochuck. In Figure 5.2,

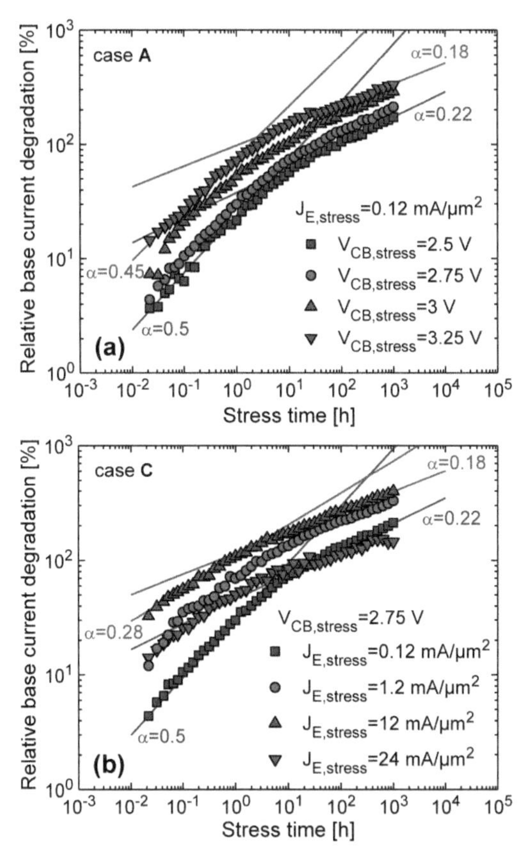

Figure 5.2 Relative base current degradation of the IHP HBT(s) as a function of stress time (a) for $J_{\mathrm{E,stress}}$ = 0.12 mA/μm^2 and various $V_{\mathrm{CB,stress}}$ (series **A**), and (b) for $V_{\mathrm{CB,stress}}$ = 2.75 V and various $J_{\mathrm{E,stress}}$ (series **C**). Also shown is extraction of exponent α at short and long stress times for selected cases.

the relative base current degradation $100 \cdot (I_{\rm Bstress} - I_{\rm Bfresh})/I_{\rm Bfresh}$ due to the enhanced recombination is shown for cases **A** and **C** at different values of $V_{\rm CB,stress}$ (**A**) and $J_{\rm E,stress}$ (**C**); it is found that the damage increases with stress time following a power law dependence ($\sim t^{\alpha}_{\rm stress}$). For short stress times (within a few hours), exponent α is around 0.5 for low/medium currents (consistent with [Che09]); on the other hand, it is found that the degradation rate decreases for longer stress times, where α approaches about 0.2. This is important in terms of device lifetime prediction (e.g., [Pan06]): if data are extrapolated from short-time experiments, the effect of degradation within a, e.g., 10-year timeframe would be largely overestimated. It must be remarked that the measured damage evolution does *not* indicate a trend to saturation within a 1,000-h-long stress time. From Figure 5.2(b) it can also be noted that the damage first increases with stress current, then reaches a maximum and declines at high currents (24 mA/μm^2). This "hump" behavior has also been observed in [Che07] and can be attributed to the higher temperature induced by self-heating (SH): when the temperature exceeds $\sim$350 K, damage indeed reduces as a result of (i) enhanced trap passivation, which starts dominating over trap creation, and (ii) increased carrier scattering, which reduces the number of highly energetic carriers reaching the interface. Moreover, the time exponent α is seen to lower at high currents [Fis13, Fis16]. Although not reported in the figures, this SH-induced reduction in degradation is also observed at high $V_{\rm CB,stress}$ for case **B**.

In order to investigate trap passivation occurring at high temperature, thermal annealing experiments were carried out. A high temperature of the base–emitter junction $T_{\rm j}$ (about 543 K)[3] was reached by increasing $T_{\rm B}$ to 398 K and raising the dissipated power ($P_{\rm D}$) via the application of $V_{\rm CB,anneal}$ = 1.5 V and $J_{\rm E,anneal}$ = 30 mA/μm^2. Figure 5.3 illustrates the relative base current reduction $100 \cdot (I_{\rm Banneal} - I_{\rm Bstress})/I_{\rm Bstress}$ against anneal time for the previously stressed DUTs (series **A** in Figure 5.1). The main findings are: (i) the thermal annealing is more effective in transistors that underwent a heavier stress and (ii) significant current gain recovery is observed, independently of the stress load.

5.1.3 Medium-term MM Stress Characterization on IFX Devices

On-wafer medium-term MM stress tests were performed at University of Naples on single-emitter SiGe:C NPN BEC HBTs manufactured

[3]This value was assessed by a preliminary extraction of the thermal resistance.

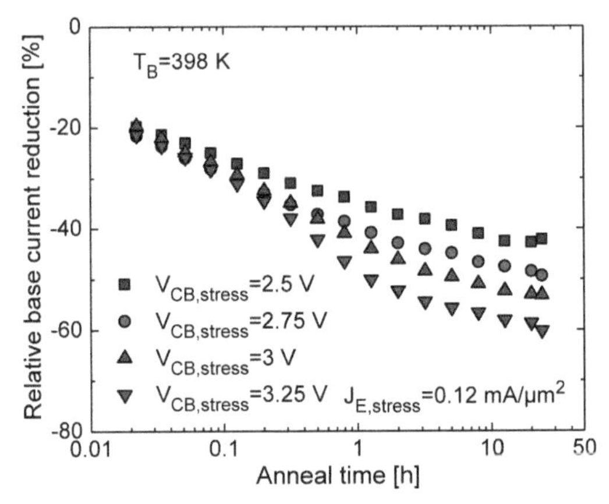

Figure 5.3 Relative base current reduction vs. anneal time obtained by applying $T_B = 398$ K, $V_{CB,anneal} = 1.5$ V, and $J_{E,anneal} = 30$ mA/μm^2 to IHP HBTs previously stressed with the bias conditions of Figure 5.2(a) (also reported in the legend). The monitoring of the base current was performed *in situ*, i.e., at $T_B = 398$ K for $V_{BE} = 0.6$ V and $V_{CB} = 0$ V.

by Infineon Technologies (hereinafter denoted as IFX) [Chev11]. The experiments were conducted on a device with effective emitter area[4] $A_E = W_E \times L_E = 0.13 \times 2.73$ μm^2, exhibiting a peak f_T of 240 GHz, a peak f_{MAX} of 380 GHz at $V_{CB} = 0.5$ V, $BV_{CEO} = 1.5$ V, $BV_{CBO} = 5.5$ V,[5] and mounted in a CB configuration. Similar to the procedure in [Fis15], the stress experiments were conducted by applying high $J_{E,stress}$ and $V_{CB,stress}$ to the DUT, and monitoring the collector and base currents as a function of stress time through measurements of forward $V_{CB} = 0$ V Gummel plots at chosen stress times. A first investigation was carried out by considering different values of $J_{E,stress}$ (namely, 1.4, 7, and 14 mA/μm^2) for the same $V_{CB,stress} = 2$ V. The relative base current degradation evaluated for $V_{BE} = 0.7$ V in the $V_{CB} = 0$ V Gummel plots is illustrated in Figure 5.4; it can be observed that (i) after 10^4 s the damage is still confined below 100% for all cases due to the low $V_{CB,stress}$ applied, and (ii) the highest $J_{E,stress}$ leads to a reduced degradation induced by the high temperature, as will be discussed in the following.

Another analysis was conducted by applying the lowest $J_{E,stress}$ (=1.4 mA/μm^2) and three $V_{CB,stress}$ values, namely, 2, 2.5, and 2.75 V.

---

[4]Further details on the technological definition of effective emitter area for IFX HBTs can be found in [dAl14].

[5]The transistor belongs to set #3 defined in 5.4.1; again, various identical devices were available, one for each stress test.

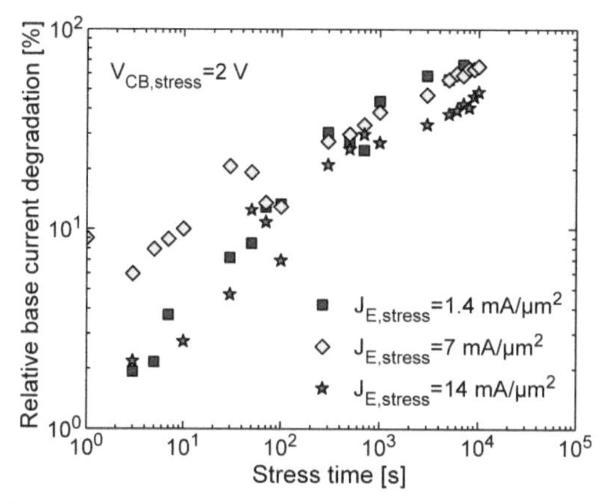

Figure 5.4 Relative base current degradation of the IFX device(s) vs. stress time for $V_{\mathrm{CB,stress}}$ = 2 V and various $J_{\mathrm{E,stress}}$.

Figure 5.5, showing the relative base current degradation at V_{BE} = 0.7 V, evidences that the damage increases with $V_{\mathrm{CB,stress}}$ due to the higher electric field in the base–collector depletion region, which in turn gives rise to a higher number of HCs with an energy higher than 1.5 eV impacting on the interface of the emitter–base oxide spacer and thus creating traps. It is found that the damage exceeds 100% for long times (10^4 s) for $V_{\mathrm{CB,stress}}$ = 2.5 V and even for very short times (300 s) for $V_{\mathrm{CB,stress}}$ = 2.75 V. Consistently with other works, Figure 5.6 witnesses that a power law ($t_{\mathrm{stress}}^{\alpha}$) well describes

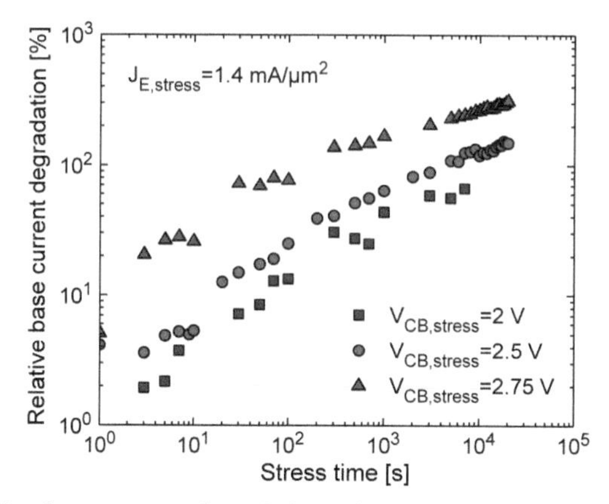

Figure 5.5 Relative base current degradation of the IFX DUT(s) against stress time for $J_{\mathrm{E,stress}}$ = 1.4 mA/µm^2 and various $V_{\mathrm{CB,stress}}$.

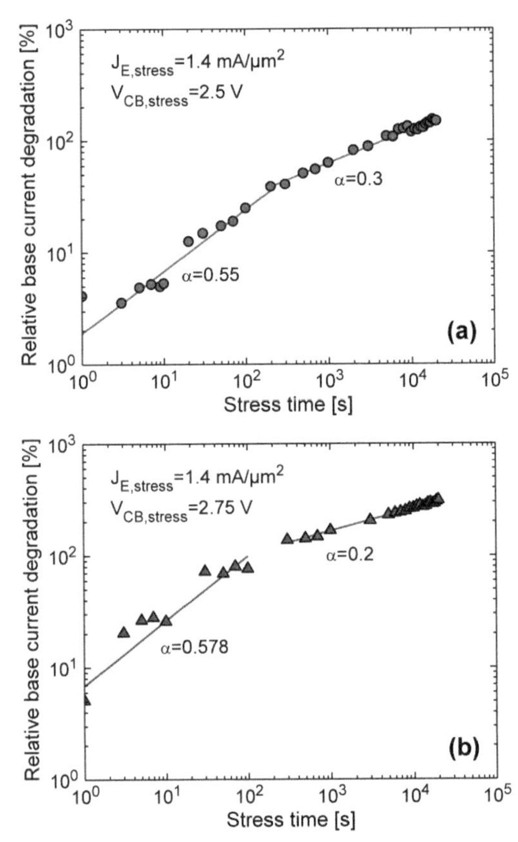

Figure 5.6 Relative base current degradation of the IFX device(s) as a function of stress time for $J_{E,stress}$ = 1.4 mA/μm^2 and (a) $V_{CB,stress}$ = 2.5 V, (b) $V_{CB,stress}$ = 2.75 V. Also shown is the extraction of exponent α at short and medium times.

the evolution of the base current degradation, provided that a different α is considered for short (high α) and medium (low α) stress times.

Following the approach presented in [Van06], an analysis was carried out to gain an in-depth insight into the device behavior under MM stress conditions. In particular, a fresh DUT identical to the stressed ones was measured by sweeping V_{CB} for various assigned J_Es. After a straightforward data processing based on the technique in [Lu89, Zan93] and on the knowledge of the thermal resistance R_{TH} = 7,000 K/W (determined according to the method in the section "Experimental R_{TH} Extraction"), it was possible to obtain the $J_{AV} - J_E$ (J_{AV} being the avalanche current density) and $T_j - J_E$ curves shown in Figure 5.7. It can be inferred that at low J_E the avalanche

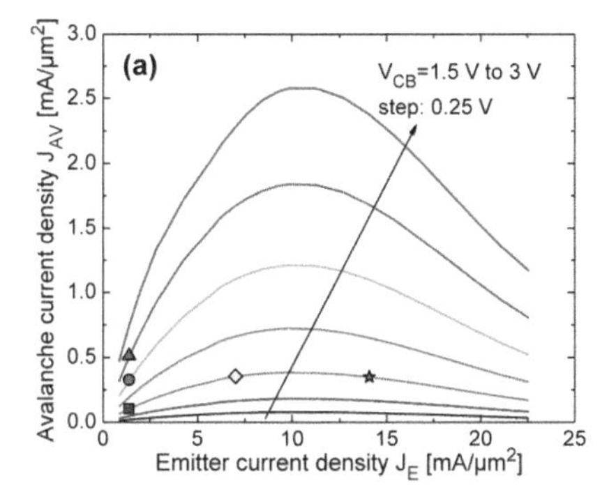

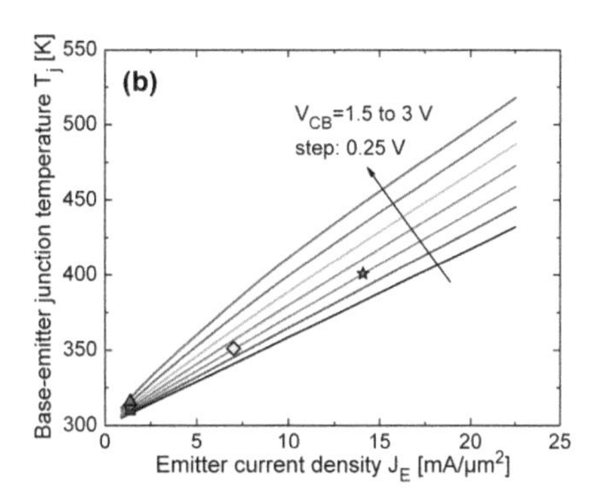

Figure 5.7 (a) Avalanche current density J_{AV} and (b) base–emitter junction temperature T_j as a function of emitter current density J_E for various V_{CB}s. Also shown are the conditions corresponding to the stress tests reported in Figures 5.4 and 5.5 (the same symbols were used for the sake of clarity).

current J_{AV} grows with J_E as a result of the increased II related to the higher number of electrons traveling to the base–collector SCR; conversely, J_{AV} decreases at high J_E due to the concurrent mitigating impact of high-injection (HI) and SH effects. Also identified in Figure 5.7 are the J_{AV}s and T_js corresponding to the MM stress conditions related to Figures 5.4 and 5.5. The main findings are in agreement with the conclusions in [Van06] and can be summarized as follows:

- although the stress test with applied $J_{E,stress}$ = 14 mA/μm^2 (star) shares the same J_{AV} (≈0.3 mA/μm^2) and V_{CB} (=2 V) as the test with $J_{E,stress}$ = 7 mA/μm^2 (rhombus), in the first case the damage is lower due to the higher device temperature (400 K instead of 300 K, as shown in Figure 5.7(b));
- conversely, the tests carried out at the same current density $J_{E,stress}$ = 1.4 mA/μm^2 and different V_{CB}s (square, circle, and triangle) share similar temperatures and different J_{AV}s (the avalanche current increases with V_{CB} due to the higher electric field in the base–collector SCR). As a result, the damage grows with increasing V_{CB}.

5.2 Long-term Stress Tests

The improved frequency performances of state-of-the-art SiGe HBTs have been achieved at the cost of significantly increased operating current densities and lower breakdown voltages [Sch17]. Thus, devices are often operated closer and even beyond the border of the classical SOA; however, this can limit stable device operation due to reliability issues induced by the previously discussed HC degradation. In the following, some dedicated long-term stress tests are carried out to clearly identify the influence of biasing conditions along the SOA limit on the device reliability [Jac15].

5.2.1 Experimental Setup

The stress tests were conducted on devices biased in a common-emitter (CE) configuration under bias conditions close to the SOA border. Since these conditions are not *accelerating* like in conventional MM tests, a long stress time (up to 1,000 h) was required to observe an impact on the electrical characteristics. The transistors are single-emitter SiGe:C NPN CBEBC HBTs fabricated by IFX, with an effective emitter area of $A_E = 0.13 \times 9.93\ \mu m^2$ featuring a peak f_T/f_{MAX} equal to 240/380 GHz and a BV_{CEO}/BV_{CBO} of 1.5/5.5 V [Chev11, Böc15]. In order to observe and record the evolution of the base and collector currents during the tests, non-stressing forward (at $V_{CB} = 0$ V) and reverse (at $V_{EB} = 0$ V) Gummel plots were measured at fixed time instants during the 1,000 h-long experiments [Jac15]. Four stress bias conditions, referred to as P1, P2, P3, and P23, were applied at $T_B = 300$ K along the SOA boundary, as shown in Figure 5.8. The corresponding

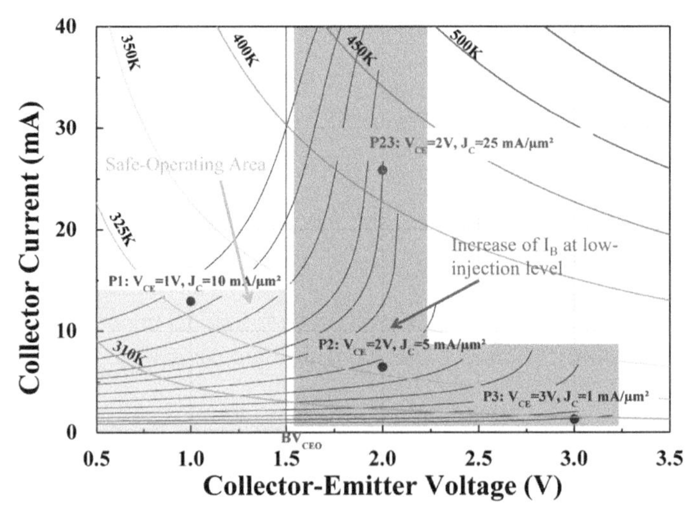

Figure 5.8 Output characteristics of the SiGe HBT under test simulated using HICUM/L2. Also represented are the examined bias conditions (P1, P2, P3, and P23).

voltage (V_{CE}), collector current (I_C), and current density (J_C), as well as the (average) temperature rise ΔT_j over the base–emitter junction (obtained from the thermal resistance R_{TH} = 2,850 K/W determined with the approach in [dAl14]) are summarized in Table 5.1. It must be remarked that P1 is defined below BV_{CEO}, whereas P2, P3, and P23 are beyond BV_{CEO}.

5.2.2 Long-term Degradation Test Results

For P1, P2, and P3, the tests were performed on six HBTs for each bias condition, (that is, 18 identical HBTs were measured). Figure 5.9 shows the forward V_{CB} = 0 V Gummel plots during the stress test at P1, P2, and P3, while Figure 5.10 illustrates the aging-induced I_B growth at V_{BE} = 0.713 V. The following considerations are in order.

- At P3, I_B increases regularly with stress time for low V_{BE}; at V_{BE} = 0.713 V, the variation is 120 nA after 1,000 h.
- At P2, I_B slightly increases for low V_{BE}; at V_{BE} = 0.713 V, the variation amounts to 80 nA after 1,000 h.
- At P1, no sizable degradation is monitored.

In conclusion, the higher V_{CE}, the more the low-injection I_B increases with stress time.

It must be remarked that a slight natural recovery is observed if the device stays *on the shelf* between two stress periods. This recovery is visible at 300 h in Figure 5.10 for the biasing conditions P2 and P3 (the devices were left unstressed for 24 h before the measurement of the Gummel plot).

Concerning P23, the forward V_{CB} = 0 V Gummel plots for each stress time are shown in Figure 5.11, which indicates that I_B increases at low V_{BE}, while remaining almost constant at high V_{BE}. The relative variation of the base current at V_{BE} = 0.65 V is shown in the inset, along with the variation measured at P2 for an identical HBT. The comparison highlights that I_B exhibits similar evolutions at P23 and P2 due to the same collector–emitter voltage (V_{CE} = 2 V), in spite of the four times higher J_C at P3.

Table 5.1 Stress bias conditions and corresponding junction temperatures

	P1	P2	P3	P23
V_{CE} [V]	1 ($<BV_{CEO}$)	2 ($>BV_{CEO}$)	3 ($>BV_{CEO}$)	2 ($>BV_{CEO}$)
I_C [mA]	12.9	6.45	1.29	32.27
J_C [mA/μm^2]	10	5	1	25
ΔT_j [K]	37	37	11	184

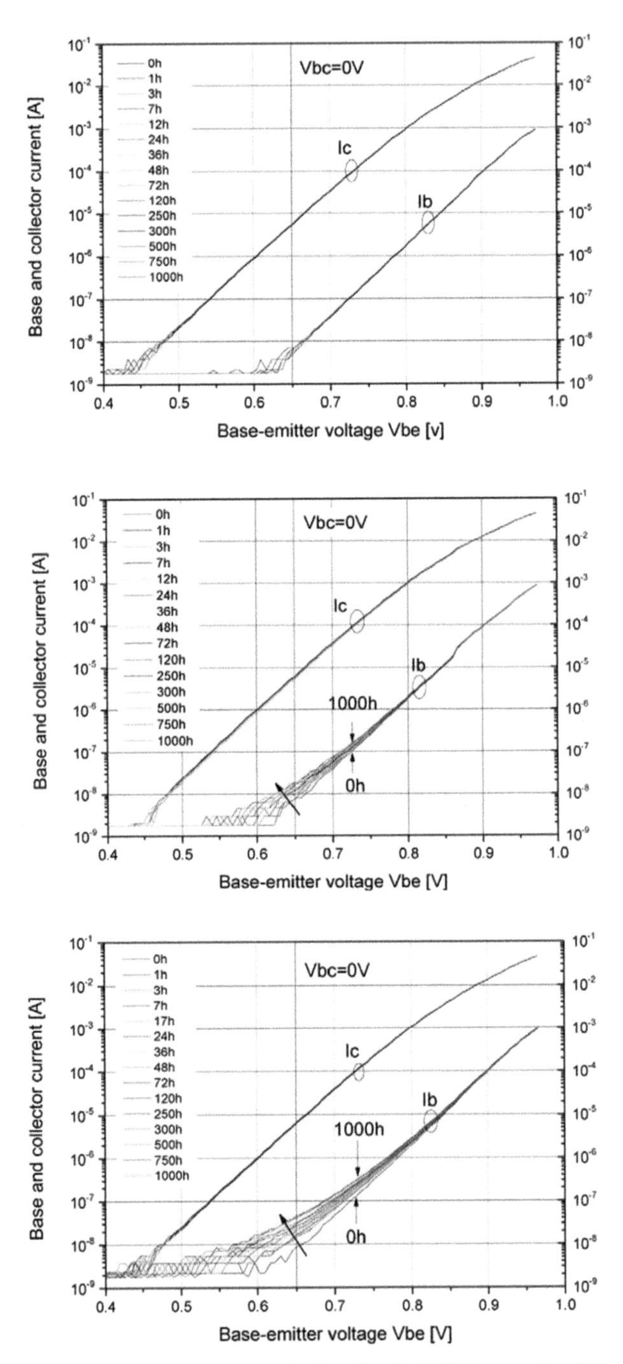

Figure 5.9 Monitoring forward Gummel plots for the DUT stressed at P1, P2, and P3.

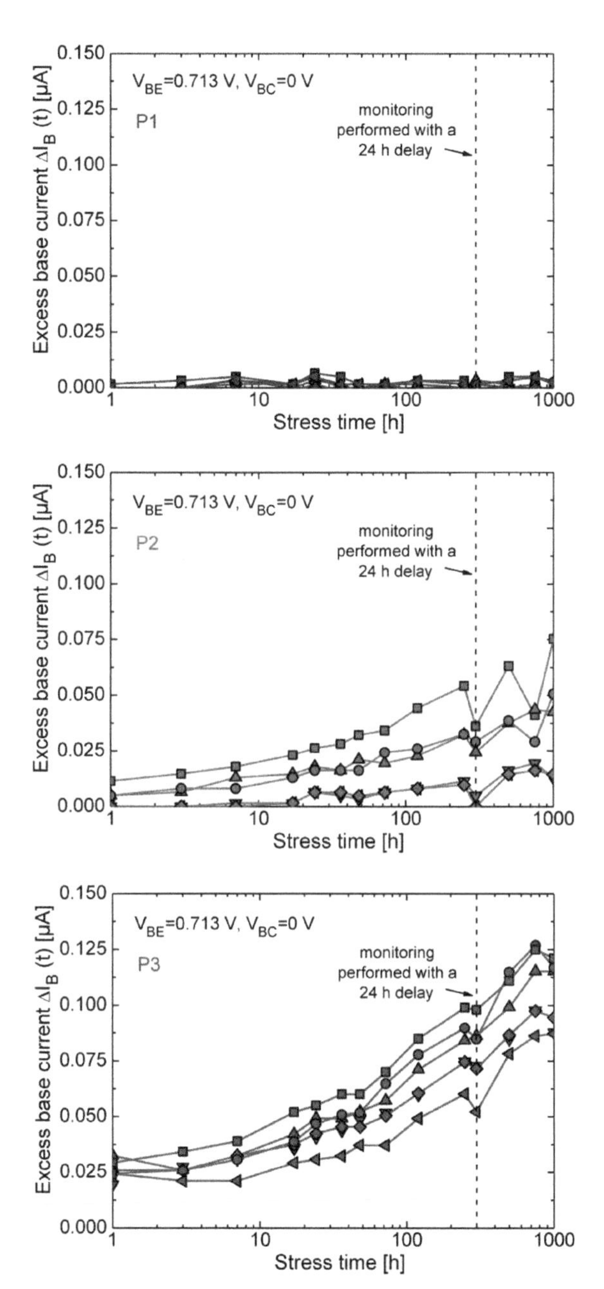

Figure 5.10 Evolution of the excess base current as a function of stress time for six identical HBTs tested at P1, six HBTs tested at P2, and six HBTs tested at P3.

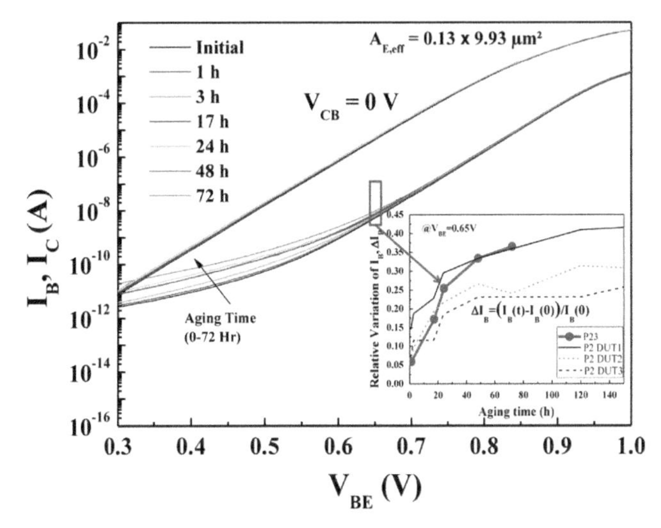

Figure 5.11 Evolution of the forward Gummel plot with stress (aging) time at P23. Shown in the inset is the relative variation of I_B at V_{BE} = 0.65 V.

The reverse V_{EB} = 0 V Gummel plots were also measured for all biasing conditions at chosen time instants during the stress tests. Since the DUTs have the emitter and substrate connected to the ground pad, the base voltage was fixed to 0 V to obtain V_{EB} = 0 V, and a negative collector voltage was swept from 0 V to –1 V to increase V_{BC}. Due to the forward biasing of the substrate–collector junction, a substrate current is added to the emitter current. It was found that the sum of the emitter and substrate currents remains almost constant regardless of the bias point, as can be inferred for the P23 case in Figure 5.12. Different behaviors were instead observed for the $I_B - V_{BC}$ curves: Figure 5.12 also witnesses that at P23 I_B tends to rapidly increase during the first few hours for $V_{BC} < 0.6$ V, eventually saturating after 48 h (as shown in the inset), while remaining constant for $V_{BC} > 0.6$ V. No significant degradation of the reverse Gummel plot was instead observed at P2 (despite the same V_{CE} as P23) and P3 [Jac15]. It is also worth noting that the distortion measured in the reverse V_{EB} = 0 V $I_B - V_{BC}$ plot for low V_{BC} at P23 resembles that of the forward V_{CB} = 0 V $I_B - V_{BE}$ plot at low V_{BE}.

The forward I_B increase observed at low V_{BE} at P2, P3, and P23 has been attributed to hot holes generated by II ($V_{CE} > BV_{CEO}$) at the base–collector SCR and then driven by the electric field to cross the base and hit the edge of the spacer with enough energy to create traps, as determined in

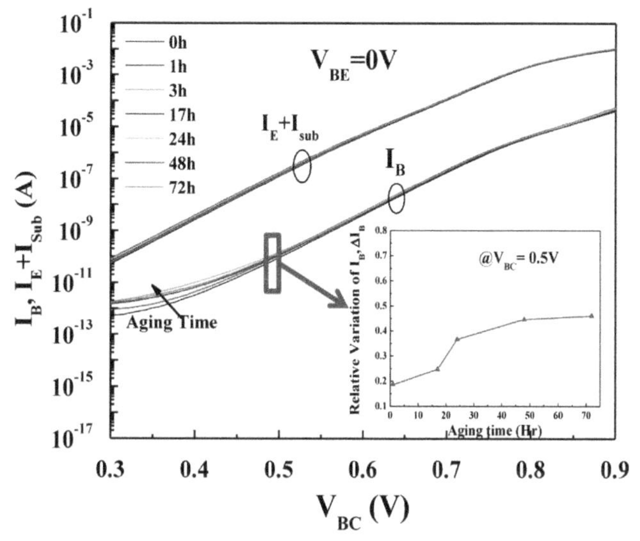

Figure 5.12 Evolution of the reverse Gummel plot with stress (aging) time at P23. Shown in the inset is the relative variation of I_B at V_{BC} = 0.5 V.

[Kam17b] with an advanced TCAD simulation strategy based on the solution of the Boltzmann Transport Equations for electrons and holes through the Spherical Harmonic Expansion approach (see Chapter 2). As clarified in the section "Introduction to hot-carrier degradation under MM Stress," the traps in turn lead to a non-ideal I_B growth via trap-assisted SRH recombination. The higher damage occurring at P3 is due to the higher electric field within the base–collector SCR, which implies a higher concentration of hot holes with enough energy to break the passivated Si–H bonds [Kam17b].

On the other hand, the damage at the ST–Si interface (witnessed by the distorted $I_B - V_{BC}$ plots) at the high-current P23 condition may be associated with both hot holes and hot electrons induced by II, as suggested in [Moe12].

The physical locations of the defects are illustrated in Figure 5.13 with the help of a cross section obtained from a TCAD simulation of the device structure. It is shown that the upper region of the ST–Si interface is more affected by hot holes, whereas the lower region is more affected by hot electrons.

Another numerical analysis was performed using Sentaurus TCAD by Synopsys [Syn] to obtain a deep insight into the degradation mechanism; the hydrodynamic model with optimized parameters reported in [Sas10] was activated. More specifically, simulations were performed to extract the

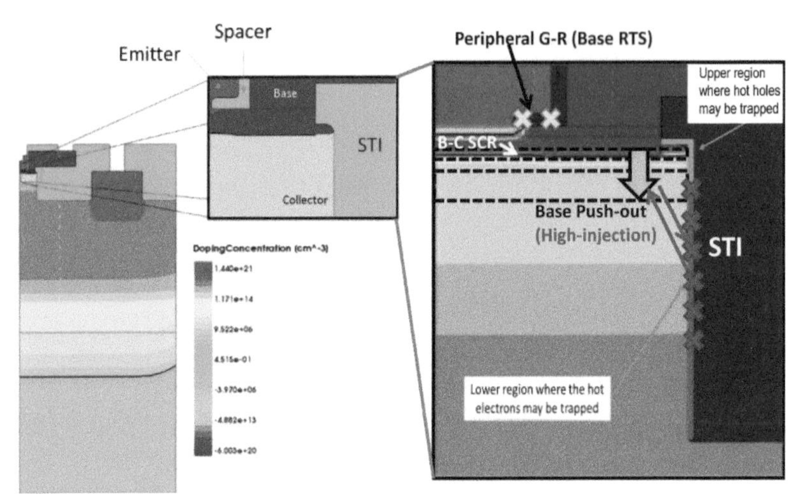

Figure 5.13 Physical origin of the base current degradation represented in a cross section within a TCAD environment.

evolution of the trap density N_t at the spacer interface (P2, P3, and P23) and ST–Si interface (only P23) as follows: for each time instant at which the non-stressing forward and reverse Gummel plots were measured and recorded, N_t (assumed to be at an assigned energy level E_t such as $E_t - E_V = 0.6$ eV, E_V being the valence band limit) was optimized so as to align the simulated plots with the experimental ones. Figure 5.14 reports the matching between measured and computed forward $V_{CB} = 0$ V Gummel plots at P3 after 7 and 750 h. Figure 5.15 illustrates the extracted N_t at the spacer interface as a function of stress time for P2 and P3 [Jac15].

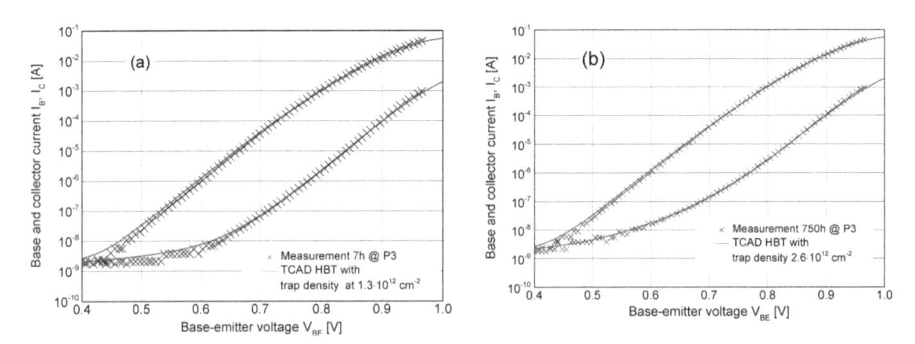

Figure 5.14 Forward Gummel plots at $V_{CB} = 0$ V at P3 after (a) 7 h and (b) 750 h of stress. Measurement results (symbols) are compared with the simulated (solid) counterparts.

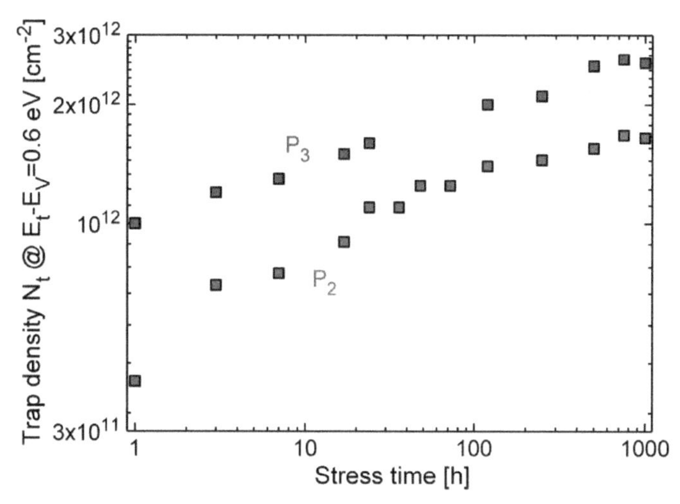

Figure 5.15 Trap density evolution along the interface of the emitter–base oxide spacer vs. stress time at P2 and P3.

5.2.3 Low-frequency Noise Characterization

Noise characterization can be considered as a diagnostic tool for analyzing quality and reliability of bipolar transistors [Vand94, Moh00]. Flicker noise is ubiquitous in almost every electronic device, although its origin is still deep in dispute. Unlike silicon BJTs, current HBTs are often affected by significant generation-recombination (G-R) noise (not so common in large-area devices) at low frequencies, mostly originated in the device external surface and periphery [Cos92, Tut95]. These noise sources may lead to presence of significant random telegraph signal (RTS) noise that can be observed in the time-domain noise signal.

Hereinafter, a comprehensive analysis of the RTS noise in IFX SiGe:C HBTs is presented, in which dominant G-R mechanisms are evidenced at low bias currents in smaller geometries, as confirmed by RTS noise measurements [Muk17]. In larger geometries the RTS noise is not so frequently observed. Eventually, extractions of RTS time constants and their evolution with bias are analyzed to get insight into the active G-R mechanisms. The weak evolution of the RTS noise amplitude with bias indicates that the noise sources are located in the base–emitter peripheral region. Consequently, distinct RTS is observed at the collector side that is activated at high current regimes. This indicates the activation of traps located in trench areas due to fixed imperfections.

The noise characterization setup includes a Keysight E5270B semiconductor parameter analyzer for DC biasing, an HP 35670A dynamic signal analyzer for the measurement of voltage noise spectral density, and a Femto DLPVA-100-F-S low-noise voltage amplifier, which has a variable gain up to 100 dB with a bandwidth of 100 kHz and an 1 TΩ input impedance. The measurements were performed at a gain of 40 dB. The entire on-wafer measurement system is connected through a GPIB interface and is controlled via the ICCAP software. The noise spectral densities of the transistors are measured in V^2/Hz (averaged over 20 spectra). The time-domain RTS noise voltage was measured using the dynamic signal analyzer. The system noise floor was determined to be 2×10^{-17} V^2/Hz. During the biasing, a very high source resistance (R_S) was considered due to the current source for the base biasing, and a 50 Ω resistance was used as load resistance (R_L). The values of transistor parameters β, r_π, R_E, and R_B were extracted from DC measurements. The RTS noise was measured on single-emitter SiGe:C NPN HBTs fabricated by IFX (see the sections "Medium-term MM Stress Characterization on IFX Device" and "Experimental Setup") with various terminal configurations and effective emitter areas, as summarized in Table 5.2. In order to eliminate process variation, the noise was measured on several devices (five to eight) of the same geometry from different dies.

The forward Gummel plot for HBT #3 is shown in Figure 5.16, which depicts the bias range of the noise measurements.

Figure 5.17 shows the base voltage noise spectral density (S_{V_B}) for transistor #1 at V_{BE} = 0.7 V and V_{CE} = 1 V. It can be clearly observed that two distinct G-R mechanisms (referred to as GR1 and GR2) are active. For all the investigated geometries, high G-R noise was observed in the low-frequency noise spectra at lower bias. These G-R noise mechanisms were particularly visible in smaller geometries. The corresponding RTS are shown in the two insets. Interestingly, it is found that one RTS (GR2)

Table 5.2 Details of the DUTs for RTS noise measurements

N°	Configuration	$A_E(=W_E \times L_E)$ [μm^2]
1	BEC	0.13 × 2.71
2	BEBC	0.13 × 4.91
3	CBEBC	0.11 × 9.93
4	BEBC	0.17 × 9.91
5	BEBC	0.25 × 9.91
6	BEBC	0.61 × 9.91
7	BEBC	1.61 × 9.91

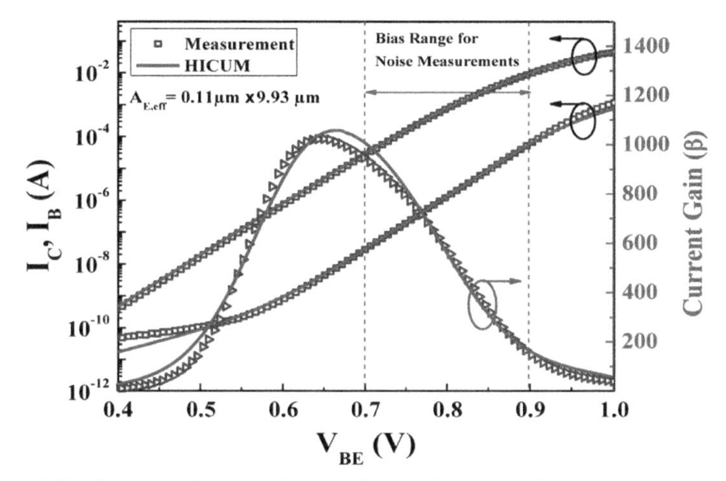

Figure 5.16 Forward Gummel plot showing the bias range for noise measurements.

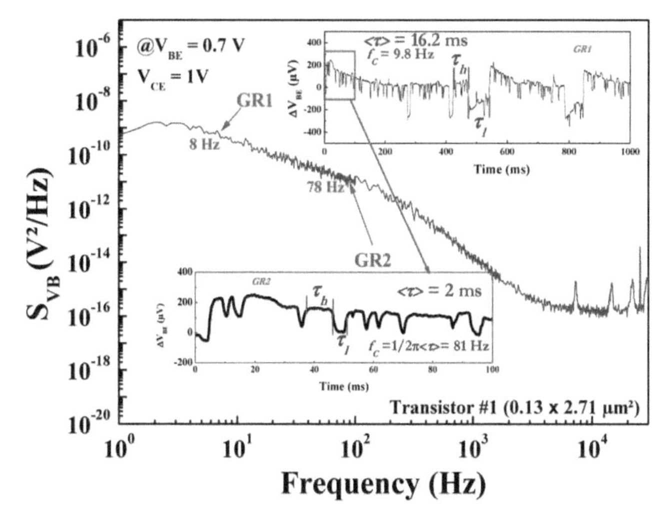

Figure 5.17 S_{V_B} showing different G-R mechanisms and their corresponding RTS in time domain for transistor #1.

is superimposed on the other (GR1). The measured S_{V_B} shows GR1 at a frequency of 8 Hz while the corresponding RTS measurement reveals an average time constant $<\tau>$ of 16.2 ms, equivalent to a G-R cutoff frequency (f_C) of 9.8 Hz (since $\tau = 1/2\pi f_C$). This confirms the existence of GR1. The other RTS (GR2) can be observed in the 0–100 ms range of the time-domain signal superimposed on the principal RTS (GR1) having a smaller time constant of 2 ms (f_C = 81 Hz) that corresponds to the GR2 at 78 Hz. In [Pas04], the observations are quite similar for SiGe HBTs, where a G-R is observed only at low bias and in a frequency range below 100 Hz for smaller devices; it was illustrated from time analysis that this G-R component is related to RTS noise. In our results, the bias dependence shows a weak evolution of the RTS amplitudes. This indicates that such a G-R mechanism is not located in SCRs, and possibly originates at base–emitter periphery [vHa02]. Similar G-Rs were observed in base current noise spectra in earlier stages of this work [Muk16a, Muk16b].

Figures 5.18 and 5.19 illustrate the S_{VB} and the corresponding RTS at different bias conditions for the smallest (transistor #1) and largest (#7) geometries, respectively. Significant G-R contributions are clearly observed with large RTS time constants in #1, whereas #7 does not show significant G-R at low frequencies. For example, at V_{BE} = 0.725 V, transistor #1 exhibits significant GR1 (at 10 Hz) and GR2 components (around 40 Hz), which correspond to time constants of 15 ms and 3 ms (superimposed RTS) in the RTS spectra, respectively. As the bias increases, the RTS time constants become smaller, indicating a faster response from the traps, and at higher bias, such as 0.9 V, the G-R mechanisms completely disappear leading to absence of any RTS in the time-domain noise response. Transistor #7 does not show dominant G-R contribution: a minor G-R contribution can be seen around 105 Hz (time constant of 1.5 ms) that is observed in the RTS at V_{BE} = 0.7 V.

Evidently, the existence of significant RTS noise in smaller geometries is well accepted [Pas04]. In our case, the RTS corresponds to the existence of GR1 in the noise spectral density at low frequency, which we have identified as a contribution due to emitter periphery.

Figure 5.20(a) shows the base RTS noise response of transistor #4 at different bias conditions. Figure 5.20(b) witnesses that the RTS time constants at both the low (τ_l) and high (τ_h) states scale with $1/\exp(qV_{BE}/kT)$, except for the highest V_{BE} where the devices are entering the medium/high injection regime. The capture rate of carriers inversely depends on the available carrier density in the trap position, which can increase with bias. However, a small electric field decrease in the SCR due to a higher V_{BE} is not sufficient to turn into such a rapid roll-off in the characteristic time [vHa02]. This indicates that there must be an additional bias dependence on the trapping and de-trapping mechanisms. In [vHa02], the authors explained their results by tunneling of electrons from the neutral regions across the SCR into traps located in the spacer oxide near the periphery. As V_{BE} increases, the tunneling distance decreases due to reduced SCR width, resulting into faster trap response and therefore reduced RTS time constants (Figure 5.20(b)).

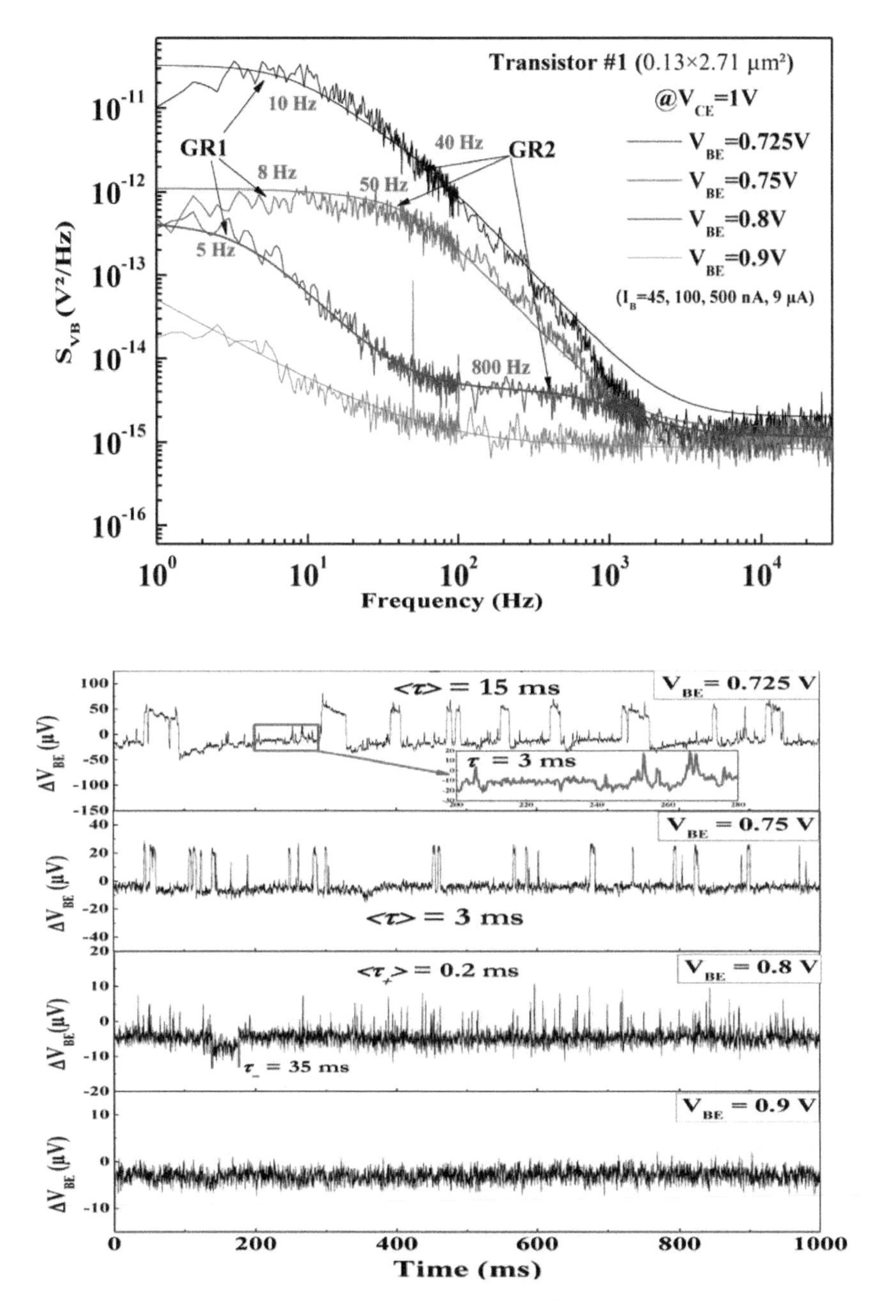

Figure 5.18 S_{V_B} showing different G-R mechanisms at different bias (V_{BE}) conditions and their corresponding RTS in time domain for transistor #1.

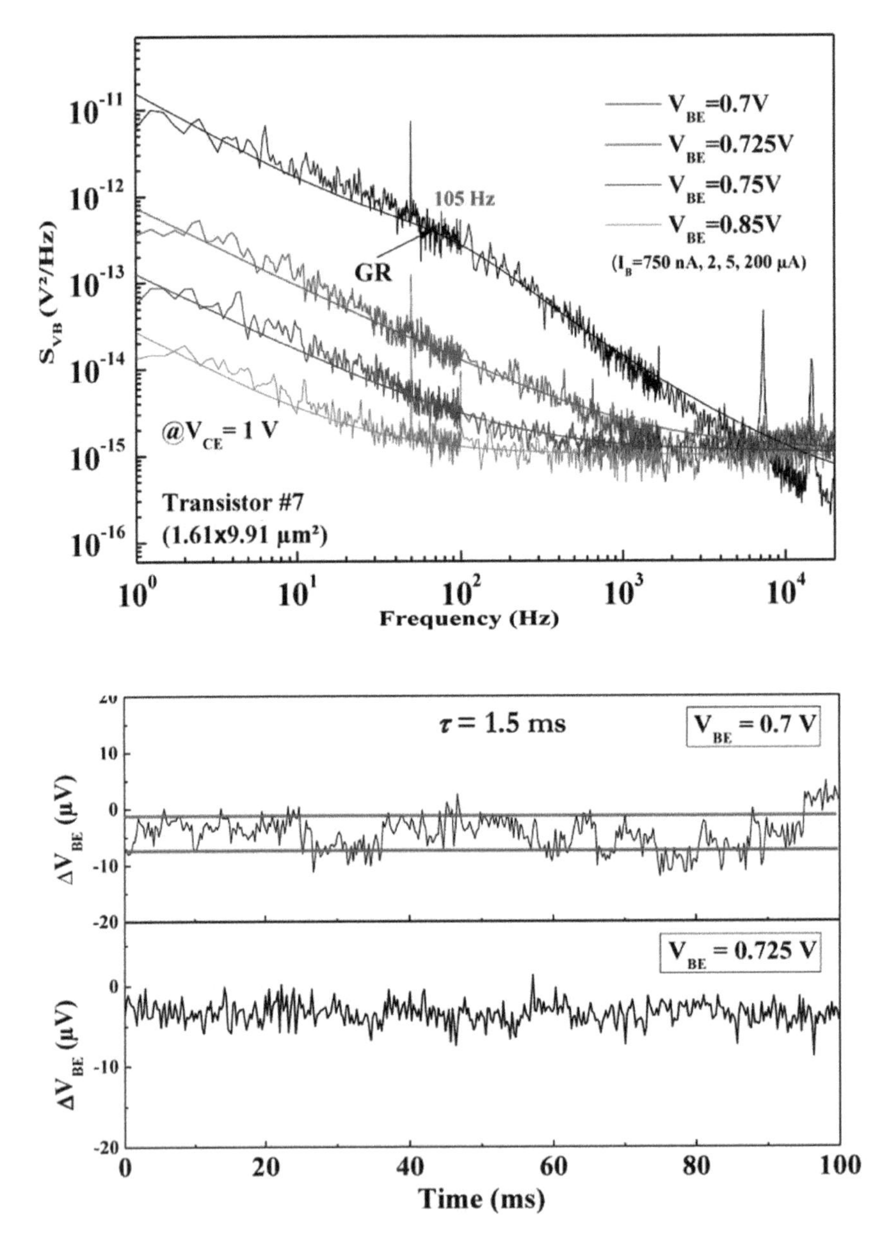

Figure 5.19 S_{V_B} showing different G-R mechanisms at different bias (V_{BE}) conditions and their corresponding RTS in time domain for transistor #7.

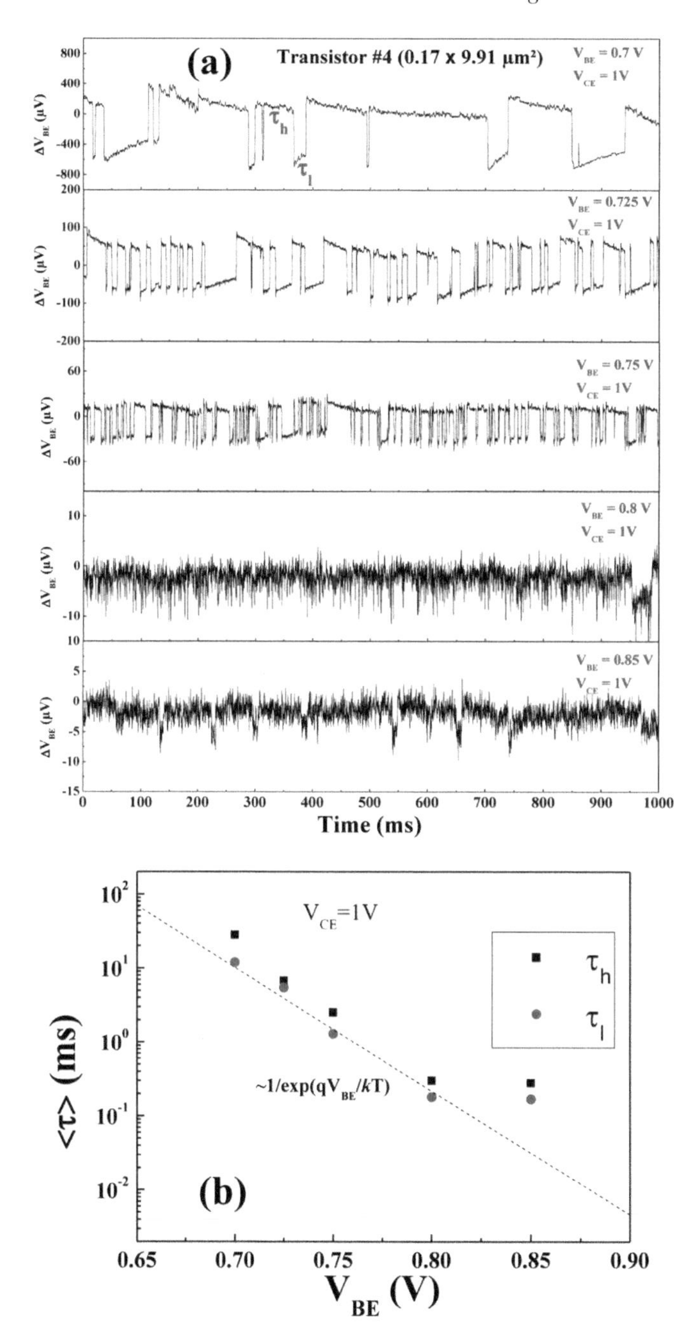

Figure 5.20 (a) Base RTS at different bias conditions, (b) corresponding time constants for the low and the high states as a function of bias (V_{BE}) for transistor #4.

Figure 5.21 shows the RTS noise current amplitude (ΔI_B) of the base noise for different geometries. A very weak bias current dependence ($\sim I_B{}^{0.1}$) is found for smaller transistors, and an almost insignificant dependence is observed for larger geometries. This further corroborates that the RTS noise sources at the base side are located in the emitter–base periphery regions. Also in [vHa02] it was stated that when ΔI_B scales with non-ideal base current component, these fluctuations often originate from noise sources in the spacer oxide at the emitter periphery. In our HBTs a large dispersion in $\Delta I_B/I_B$ ratios was observed (between 0.3% and 3%) from device to device. In larger devices the $\Delta I_B/I_B$ ratio has a relatively higher magnitude, yet this ratio inversely scales with I_B in all geometries. Different $\Delta I_B/I_B$ ratios indicate slightly different physical origins of traps in different geometries.

The bias dependence of the collector RTS is presented in Figure 5.22(a), which shows the RTS at different V_{CB}s for transistor #5. Figure 5.22(b) illustrates the extracted trap time constants as a function of the collector–base bias for two geometries (#4 and #5) at V_{BE} = 0.9 and 0.8 V, respectively. It is expected that the time constants of high and low states are higher in larger geometries at lower bias since the tunneling distance is higher. However, the characteristic times remain almost constant for higher V_{CB}s in both cases, and the steady-state value is reached faster in the case V_{BE} = 0.9 V (onset of high-current effects), whereas a sharp transition at lower V_{CB} is

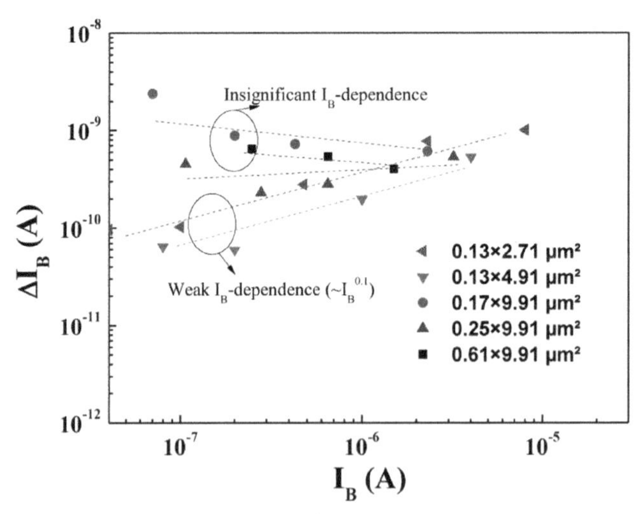

Figure 5.21 Base noise RTS amplitude (ΔI_B) as a function of bias (I_B) for different transistor geometries.

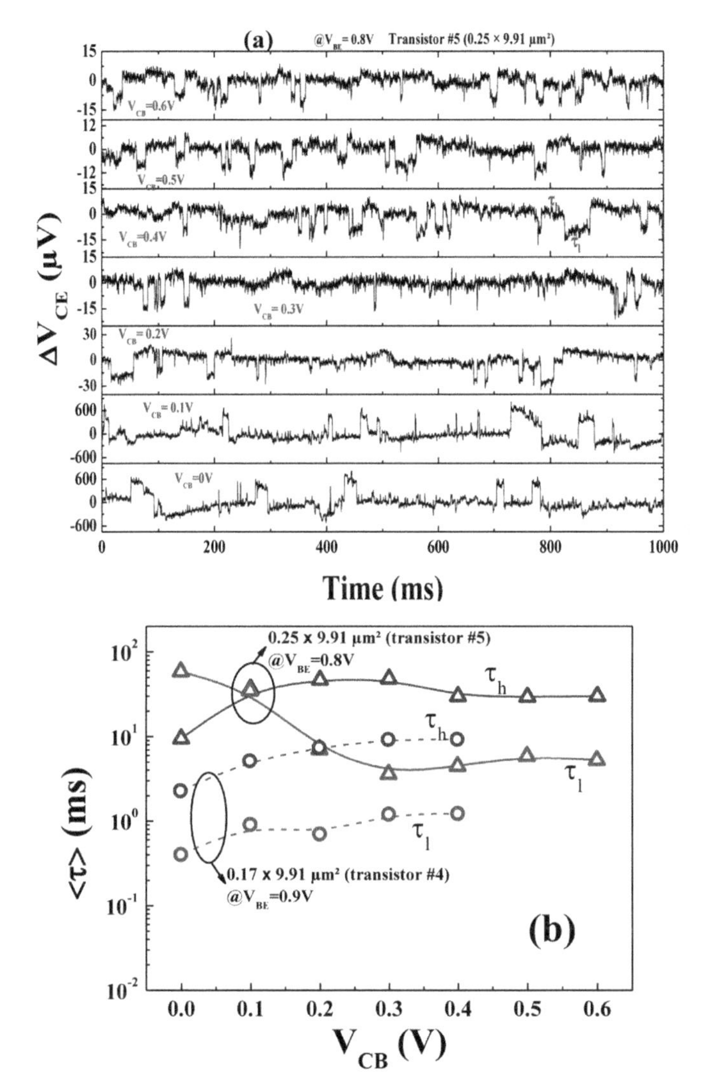

Figure 5.22 (a) Collector RTS at different bias conditions and (b) corresponding RTS time constants as a function of bias (V_{CB}) for transistors #4 and #5.

observed at V_{BE} = 0.8 V. The saturation of trap response at higher V_{CB} indicates that these RTS noise sources are possibly located at the top of the ST walls, and even if the base–collector SCR enlarges, the trap density at the trench sidewalls remains fixed. At V_{BE} = 0.8 V, a sharp transition is observed when the SCR is narrow (V_{CB} ~0.1 V), and with the SCR spreading the tunneling time constants change due to activation of new traps until it reaches saturation. Conversely, due to high-injection (V_{BE} ~0.9 V) and base pushout effects, tunneling times remain constant since all the sidewall traps are already aligned within the SCR area.

In conclusion, this comprehensive analysis of the RTS noise in advanced SiGe:C HBTs demonstrates the evidence of dominant G-R mechanisms at low bias currents in smaller geometries, whereas in larger geometries the RTS noise is not observed. The bias dependence of the RTS reveals a weak evolution in the noise amplitude indicating that the noise sources are located in the base–emitter peripheral region. Distinct RTS is observed at the collector side also near high current regimes, which was attributed to activation of traps located in ST walls. This highlights the applicability of RTS noise characterization to probe these imperfections via non-destructive means.

5.3 Compact Modeling of Hot-Carrier Degradation

The technology selection for the fabrication of integrated circuits is mainly driven by both performance and reliability criteria; in particular, the lifetime is one of the most crucial factors. To evaluate lifetime, the aging behavior of a specific degradation effect can be studied with TCAD simulations focusing mostly on bias-temperature instability and HC injection in MOSFETs, and on MM degradation in bipolar transistors. TCAD simulations can provide some in-depth information on the physical mechanisms. However, these simulations are done at the expense of very long simulation times, thus making them unviable for circuit design. Hence, it is necessary to develop a more practical circuit simulation platform using electrical compact models at transistor level suited to efficiently capture the physics of the degradation through aging laws and subsequently reflect it at circuit level.

5.3.1 Empirical Equations by IHP

Based on the long-term stress results discussed in the section "Long-term Degradation Test Results," IHP developed the following empirical equation for the base current degradation:

$$\Delta I_{\mathrm{B}} = C_{\mathrm{MM}} \cdot f\left(V_{\mathrm{CB,stress}}, J_{\mathrm{E,stress}}\right) \cdot \left(c_{\mathrm{JE}} \cdot t_{\mathrm{stress}}\right)^{\alpha(t_{\mathrm{stress}})} \tag{5.1}$$

where t_{stress} is the stress (aging) time and α is a time-dependent power factor [Fis15, Fis16]. By referring to the early stress stage ($t < 0.1$ h), this general dependence on the stress conditions was extracted [Fis15]:

$$f\left(V_{\text{CB,stress}}, J_{\text{E,stress}}\right) = \exp\left(\mu_0 V_{\text{CB,stress}}\right) \cdot \left(\frac{1\ \text{mA}/\mu\text{m}^2}{J_{\text{E,stress}}} + \frac{J_{\text{E,stress}}}{J_{\text{Ehc}}}\right)^{-0.5} \tag{5.2}$$

where $\mu_0 = 1.5\ \text{V}^{-1}$ and $J_{\text{Ehc}} = 25\ \text{mA}/\mu\text{m}^2$.

The long-term development of the logarithmic aging rate

$$\alpha_{\log} = d\left(\log \Delta I_{\text{B}}\right) / d\left(\log t_{\text{stress}}\right) \tag{5.3}$$

can be approximately fitted to the observed base current degradation by means of the power coefficient

$$\alpha\left(t_{\text{stress}}, J_{\text{En}}\right) = 0.2 + \frac{0.3}{\left(t_{\text{stress}} + J_{\text{En}}^{0.45}\right)^{0.18 \cdot J_{\text{En}}^{0.4}}} \tag{5.4}$$

with pre-factor $c_{\text{JE}} = J_{\text{En}}^{-1}$ and $J_{\text{En}} = J_{\text{E,stress}}/J_{\text{Emin}}$, i.e., the stress current density normalized to the minimum applied density $J_{\text{Emin}} = 0.12\ \text{mA}/\mu\text{m}^2$. This equation is rather complex because $J_{\text{E,stress}}$ has enormous influence on the development of aging rate, as can be seen in Figure 5.23, where aging over a range of stress currents and up to 1,000 h has been successfully simulated by modifying the recombination current of an HBT compact model with the above equations.

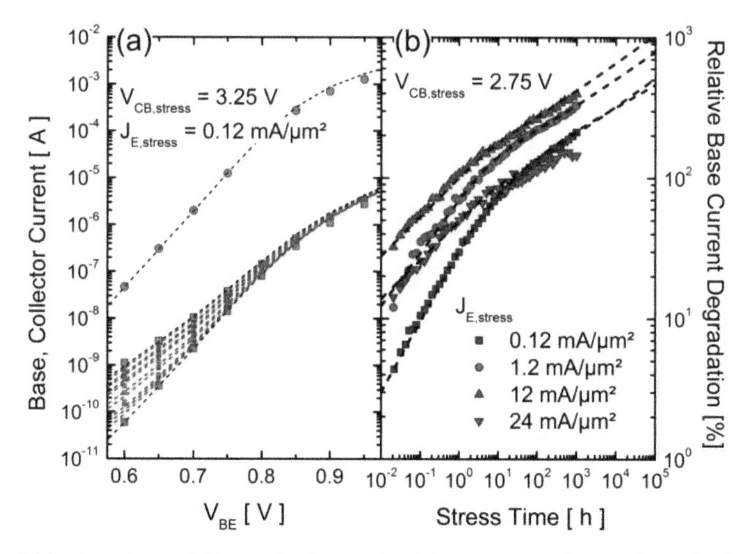

Figure 5.23 (a) Forward Gummel plots and (b) base current degradation of IHP HBTs simulated with an empirical aging function (dashed lines).

5.3.2 HICUM-based Model

Various compact models have been developed for mm-wave circuit applications. One of the most commonly used model for SiGe:C HBTs is referred to as HIgh CUrrent Model (HICUM) [Sch05, Sch10, Sch13], and is based on the General Integral Charge-Control Relation (GICCR) [Sch93]. The GICCR allows taking into account the relevant transport mechanisms through a physical-based approach; this is the reason why HICUM was chosen to implement the aging laws based on the experimental results presented in the section "Long-term Degradation Test Results."

As described before, the degradation (i.e., the base current growth at low V_{BE}) is due to an HC-induced increase in trap density over the interface of the emitter–base oxide spacer. In HICUM the base current in forward mode is subdivided into various components [Sch10], namely, the current I_{jBEi} injected into the intrinsic part of the emitter as a main component, and the peripheral current, in turn composed of the back-injection current across the emitter perimeter junction I_{jBEp} and the recombination current in the perimeter base–emitter SCR I_{REp}; I_{jBEp} and I_{REp} are given by:

$$I_{\mathrm{jBEp}} = I_{BEpS} \cdot \left[\exp\left(\frac{V_{\mathrm{BEjp}}}{m_{\mathrm{BEp}} V_{\mathrm{T}}}\right) - 1\right] \tag{5.5}$$

$$I_{\mathrm{REp}} = I_{\mathrm{REpS}} \cdot \left[\exp\left(\frac{V_{\mathrm{BEjp}}}{m_{\mathrm{REp}} V_{\mathrm{T}}}\right) - 1\right] \tag{5.6}$$

where V_{BEjp} is the peripheral internal (junction) base–emitter voltage, while the saturation currents I_{BEpS} and I_{REpS}, as well as the non-ideality factors m_{BEp} and m_{REp}, are model parameters. In [Gho10, Gho11], it was shown that a possible approach to simulate the I_{B} degradation in InP HBTs is to use I_{REp} given by Equation (5.6). More specifically, the I_{REpS} evolution is expected to follow the N_{t} evolution vs. t_{stress} (extracted as, e.g., in the section "Long-term Degradation Test Results"). In a simplified approach in which I_{REpS} is assumed to saturate for long stress times, the dependence of I_{REp} on time can be accounted for in HICUM through the following differential equation for I_{REpS}:

$$\frac{dI_{\mathrm{REpS}}}{dt} = ATSF \cdot (G - R \cdot I_{\mathrm{REpS}}) \tag{5.7}$$

where G is the generation rate and R (fitting parameter) the annihilation rate of traps, while ATSF is an Aging Time Scale Factor added to shorten the

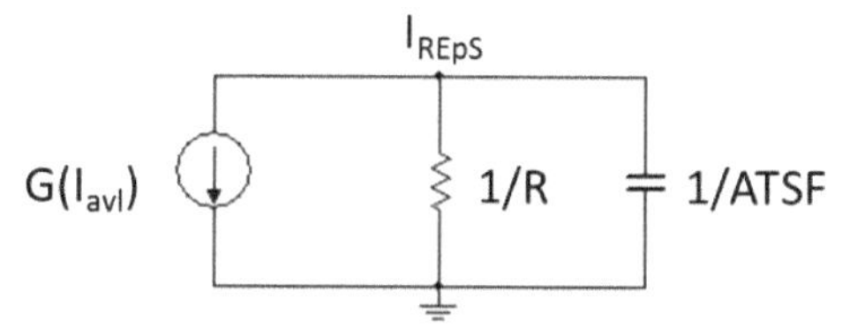

Figure 5.24 New transistor circuit used for aging law implementation in HICUM.

simulation time needed to have a perceptible stress effect to minutes (instead of tens of hours). The generation rate G depends on the bias conditions; in particular, it is an increasing function of the collector–base voltage (V_{CB}) due to the enhanced II current I_{AV} (accounted for in HICUM [Sch10]). The following linear relation was proposed in [Jac15] to include this dependence:

$$G = A \cdot I_{AV} + G_0 \tag{5.8}$$

A and G_0 being fitting parameters. Equation (5.7) with (5.8) was implemented in the Verilog-A code of HICUM/L2 by including the additional circuit shown in Figure 5.24.

5.4 Thermal Effects

Thermal issues have become a serious concern in SiGe HBTs due to the concurrent impact of the following factors: (i) the shrinking of the intrinsic device has induced a growth in power density within the base–collector SCR for a given bias condition; (ii) the trench isolation – exploited to reduce parasitics, crosstalk, and increase f_{MAX} – limits the heat spreading since trenches are filled with materials suffering from low thermal conductivity [Rie05, dAl10, You11, Pet15]. This mechanism is even exacerbated by lateral scaling, which results in a horizontal reduction of the Si volume embraced by trenches; (iii) HBTs are operated at high current densities to boost the frequency performance, which entails a further increase in dissipated power density [Cre13]. Owing to these considerations, thermal effects can be viewed as an undesired, yet unavoidable, by-product of the technology evolution. Unfortunately, the enhanced heat generation (for a given dissipated power) and the reduction in heat removal have pushed the thermal resistances (R_{TH}) of SiGe HBTs into the thousands of K/W [ElR12, Has12, Sah12] and even beyond 10^4 K/W for small emitter windows, as evidenced by recent experimental campaigns conducted on transistors fabricated by STMicroelectronics (hereinafter referred to as STM) [dAl10] and IFX [dAl14]. Thermal effects can lead to a severe distortion of the DC device characteristics (e.g., [LaS09]), and also degrade the low-frequency and high-frequency (since the DC bias is altered) behavior; besides the performance penalty, they may also affect the long-term reliability, and even trigger destructive instability phenomena. Consequently, care must be taken in assessing the impact of the thermal behavior in advanced technology nodes.

5.4.1 Experimental R_{TH} Extraction

Since the steady-state thermal behavior of a device is fully described by the thermal resistance (R_{TH}), a plethora of methods to experimentally extract this critical parameter in bipolar transistors have been developed. Among them, particular interest has been paid to approaches based on DC measurements, for which low effort and relatively cheap instrumentation are required. The most widespread method – presented in slightly different variants in the literature [Daw92, Pfo03, Rie05] – relies on the measurement (i) of the temperature-sensitive base–emitter voltage to employ its temperature coefficient as a *thermometer*, and (ii) of the base–emitter voltage as a function of collector–base voltage (or dissipated power) at an assigned emitter (collector) current. A sticking point of this technique is the thermometer calibration, which can be impacted by SH and thus entail a thermal resistance overestimation. A strategy to purify this procedure from SH has been proposed by Vanhoucke et al. [Van04]. Here an alternative to [Van04] is presented, which suggests a logarithmic law for the current dependence of the temperature coefficient of the internal (junction) base–emitter voltage V_{BEj} and can be explained as follows.

In the absence of HI and II effects, the collector current I_C of a SiGe HBT (which exhibits marginal Early effect) can be described by the simple model

$$I_C = A_E J_{S0} \exp\left[\frac{V_{BEj} + \phi\,(I_E)\,\Delta T_j}{\eta V_{T0}}\right] \tag{5.9}$$

where V_{BEj} is given by:

$$V_{BEj} = V_{BE} - R_E I_E - R_B I_B \tag{5.10}$$

R_B, R_E being the parasitic base and emitter resistances. In Equation (5.9), $A_E = W_E \times L_E$ is the effective emitter area; J_{S0} is the reverse saturation current density, η ($\geq$1) is the ideality factor, and V_{T0} is the thermal voltage, all at temperature T_0 = 300 K; ϕ [V/K] (>0) is the temperature coefficient of

V_{BEj} (in absolute value) and ΔT_{j} is defined as T_{j} - T_0, T_{j} being the (average) temperature over the base–emitter junction. This implies that Equation (5.9) accounts for the temperature dependence of I_{C} ($\approx I_{\mathrm{E}}$) making use of a V_{BEj} shift, by keeping J_{S0}, η, and V_{T0} at their T_0 values (e.g., [Zha96]). The ϕ dependence on I_{C} ($\approx I_{\mathrm{E}}$) can be described with the following logarithmic law [Nen04, dAl10, dAl14, dAl16, dAl17]:

$$\phi\left(I_{\mathrm{C}}\right) = \phi_0 - \eta\frac{k}{q}\ln\frac{I_{\mathrm{C}}}{A_{\mathrm{E}}J_{\mathrm{S0}}} \approx \phi_0 - \eta\frac{k}{q}\ln\frac{I_{\mathrm{E}}}{A_{\mathrm{E}}J_{\mathrm{S0}}} \tag{5.11}$$

Parameters J_{S0} and η in Equation (5.11) can be optimized by invoking the following procedure. First, the $I_{\mathrm{C}} - V_{\mathrm{BE}}$ characteristic of the DUT is measured at various thermochuck temperatures T_{B} under CE conditions by keeping V_{CE} small and sweeping V_{BE} up to values sufficiently low to reasonably neglect SH, HI, II, and resistive effects; as a consequence, the $I_{\mathrm{C}} - V_{\mathrm{BE}}$ curves can be modeled by:

$$I_{\mathrm{C}} = A_{\mathrm{E}}J_{S0}\exp\left[\frac{V_{\mathrm{BE}} + \phi\left(I_{\mathrm{E}}\right)\cdot\left(T_{\mathrm{B}} - T_0\right)}{\eta V_{\mathrm{T0}}}\right] \tag{5.12}$$

which stems from Equation (5.9) by considering $T_{\mathrm{j}} = T_{\mathrm{B}}$ and $V_{\mathrm{BEj}} = V_{\mathrm{BE}}$. Parameters J_{S0} and η are then tailored to match the experimental curve at $T_{\mathrm{B}} = T_0$ with

$$I_{\mathrm{C}} = A_{\mathrm{E}}J_{S0}\exp\left(\frac{V_{\mathrm{BE}}}{\eta V_{\mathrm{T0}}}\right) \tag{5.13}$$

and ϕ_0 is safely (SH is negligible) calibrated so as to ensure good agreement between all the $I_{\mathrm{C}} - V_{\mathrm{BE}}$ characteristics (at different T_{B}s) and the model given by Equation (5.12) with (5.11). Once ϕ_0 is known, Equation (5.11) can be used also at medium current levels, where the extraction of ϕ_0 would be inaccurate due to SH. Further details concerning the derivation of Equation (5.11), as well as the physical meaning of parameter ϕ_0, can be found in [dAl14]. Combining Equations (5.9) and (5.10), it can be obtained that:

$$V_{\mathrm{BE}} = R_{\mathrm{B}}I_{\mathrm{B}} + R_{\mathrm{E}}I_{\mathrm{E}} - \phi\left(I_{\mathrm{C}}\right)\Delta T_{\mathrm{j}} + \eta V_{\mathrm{T0}}\ln\frac{I_{\mathrm{C}}}{A_{\mathrm{E}}J_{\mathrm{S0}}} \tag{5.14}$$

By exploiting the thermal equivalent of Ohm's law, ΔT_{j} is expressed as:

$$\Delta T_{\mathrm{j}} = T_{\mathrm{j}} - T_0 = R_{\mathrm{TH}}P_{\mathrm{D}} + T_{\mathrm{B}} - T_0 \tag{5.15}$$

where P_{D} is the dissipated power. If $T_{\mathrm{B}} = T_0$, Equation (5.15) can be recast as:

$$\Delta T_{\mathrm{j}} = R_{\mathrm{TH}}P_{\mathrm{D}} = R_{\mathrm{TH}}\cdot\left(V_{\mathrm{BE}}I_{\mathrm{E}} + V_{\mathrm{CB}}I_{\mathrm{C}}\right) \approx R_{TH}\cdot\left(V_{\mathrm{BE}} + V_{\mathrm{CB}}\right)I_{\mathrm{E}} \tag{5.16}$$

wherein use has been made of the P_D expression in terms of applied or measurable voltages and currents under CB conditions. By substituting Equation (5.16) into (5.14),

$$V_{\mathrm{BE}} \approx \frac{R_{\mathrm{E}} I_{\mathrm{E}} - \phi\left(I_{\mathrm{E}}\right) R_{TH} V_{\mathrm{CB}} I_{\mathrm{E}} + \eta V_{T0} \ln \frac{I_{\mathrm{E}}}{A_{\mathrm{E}} J_{S0}}}{1 + \phi\left(I_{\mathrm{E}}\right) R_{TH} I_{\mathrm{E}}} \tag{5.17}$$

If a CB measurement is performed at $T_B = T_0$ under a V_{CB} range limited to low values so as to avoid II effects, at a constant I_E sufficiently low to prevent HI and non-linear thermal effects, yet high enough to lead to perceptible SH, the (negative) slope γ of the $V_{BE} - V_{CB}$ characteristic is given by:

$$\gamma = \frac{dV_{\mathrm{BE}}}{dV_{\mathrm{CB}}} = -\frac{\phi\left(I_{\mathrm{E}}\right) R_{\mathrm{TH}} I_{\mathrm{E}}}{1 + \phi\left(I_{\mathrm{E}}\right) R_{\mathrm{TH}} I_{\mathrm{E}}} \tag{5.18}$$

whence the thermal resistance (R_{TH}) can be evaluated as [dAl10, dAl14, dAl16, dAl17, Kam17a]:

$$R_{\mathrm{TH}} = \frac{|\gamma|}{\left(1 - |\gamma|\right) \phi\left(I_{\mathrm{E}}\right) I_{\mathrm{E}}} \approx \frac{|\gamma|}{\phi\left(I_{\mathrm{E}}\right) I_{\mathrm{E}}} \tag{5.19}$$

In [dAl14], this improved approach was applied to about 100 single-emitter SiGe:C NPN BEC HBTs manufactured by IFX. The transistors are divided into three sets corresponding to different technology stages (and scaling strategies), which are hereinafter denoted as #1, #2, and #3. In particular, (i) set #2 is slightly scaled (both laterally and vertically) compared with #1; the collector current of HBTs belonging to #2 at peak f_T is about 30% higher than that of the #1 counterparts with approximately the same emitter area; (ii) set #3 devices underwent an aggressive lateral scaling with respect to #2 ones, while being vertically similar to them. The key figures of the sets are reported in Table 5.3. The thicknesses of the shallow and deep trenches are equal to 0.3 μm and 4.5 μm for all HBTs, respectively. For each set, transistors with several combinations of emitter width/length were available.

Table 5.3 Key figures of the analyzed IFX technology states

	#1	#2	#3
BV_{CBO} [V]	6.5–6.8	5.2–5.9	5.1–5.5
Peak f_T @ V_{CB} = 0 V [GHz]	190	225	235
J_C @ peak f_T, V_{CB} = 0 V [mA/μm^2]	6.5–7.0	9.0–9.5	9.5–10
Peak f_T @ V_{CB} = 0.5 V [GHz]	215	230	240
Peak f_{MAX} @ V_{CB} = 0 V [GHz]	250	310	330
Peak f_{MAX} @ V_{CB} = 0.5 V [GHz]	280	350	380

Figure 5.25 illustrates the experimentally extracted R_{TH}s as a function of L_{E} at assigned widths W_{E} for the three sets. It is shown that R_{TH} (i) significantly increases by reducing L_{E} and (ii) is well above 10^3 K/W and can grow beyond 10^4 K/W for small emitter areas. In particular, the smallest DUTs of sets #1 ($A_{\mathrm{E}} = 0.2 \times 0.57$ μm^2), #2 ($A_{\mathrm{E}} = 0.14 \times 0.39$ μm^2), and #3 ($A_{\mathrm{E}} = 0.11 \times 0.63$ μm^2) suffer from R_{TH} = 14,300, 21,000, and 22,000 K/W, respectively.

Numerical evidence of the accuracy of this technique was provided in [Kam17a], where it was applied to the simulation of an IFX SiGe:C DUT belonging to set #3 through an advanced tool solving the Boltzmann transport equations of electrons, holes, and longitudinal optical phonons, as well as the Energy Balance Equations for the other phonon modes (see Chapter 2).

5.4.2 Thermal Simulation

A viable strategy to assess the impact of technology on the thermal behavior of SiGe HBTs involves the adoption of 3-D finite-element method (FEM) thermal simulations, which are suited to handle structures with arbitrarily complex geometries [Rei01, Wal02].

An interesting contribution has been given in [Sah13], where non-linear steady-state, large signal, and sinusoidal thermal analyses of an STM SiGe:C HBT (with drawn emitter area equal to 0.27 μm × 10 μm) were carried out with Sentaurus; the thermal resistance was found to be in fairly good agreement with the one measured according to the procedure in [Pfo03], although no thermal conductivity degradation mechanisms (e.g., due to high doping) were accounted for. The Back-End-Of-Line (BEOL) structure was found to play a marginal role due to the absence of the metal-via stack above the emitter. This analysis has been recently extended to cover the influence of BEOL on the thermal behavior of multi-finger devices, with emphasis on the coupling among fingers [Dwi16].

In [dAl10], the software package Comsol [Com] was adopted to analyze SH in several STM SiGe:C HBTs. In spite of their geometrical complexity, the devices were reproduced with a very high accuracy up to the emitter, base, and collector contacts, the top surfaces of which were considered adiabatic, that is, the BEOL architecture was not included. Unfortunately, although the upward heat flow was unrealistically suppressed, the numerical R_{TH}s were found to *underestimate* by about 20–25% the experimental values determined through the technique described in the section "Experimental R_{TH} Extraction," independently of technology stage and emitter size. An improved

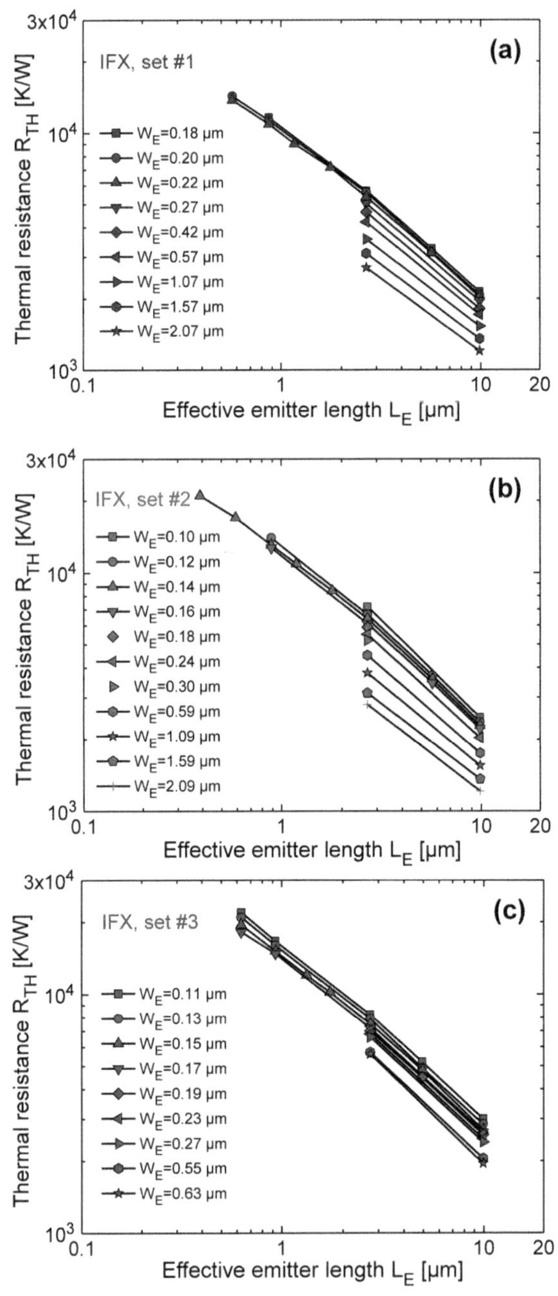

Figure 5.25 Thermal resistance (R_{TH}) as a function of emitter length (L_{E}) for various emitter widths (W_{E}), as experimentally determined for sets (a) #1, (b) #2, and (c) #3.

variant of the approach in [dAl10] was applied to IFX transistors in [dAl16]. The advances with respect to [dAl10] are reported below:

- The whole BEOL structure, comprising five metal (copper) layers and related interconnections (copper vias between metal layers, tungsten contacts between silicon and the lowest metal layer), was taken into account, as well as the external pads, as witnessed by Figure 5.26 reporting the Comsol grid. This allows quantifying the cooling influence due to the upward heat flow (often disregarded in the literature), which is expected to be relevant since – differently from the STM transistor analyzed in [Sah13] – a metal-via stack is located over the emitter in the IFX DUTs.
- In bipolar transistors, the power dissipation occurs at the base–collector SCR. In conventional approaches for thermal simulations, for a rectangular emitter window, such a region is modeled as either a rectangular or a parallelepiped heat source (e.g., [dAl10, Sah13]), both with uniform power density. In [dAl16], the dissipation region is more accurately modeled by resorting to 2-D electrical simulations of the DUTs preliminarily performed with Sentaurus in order to determine a realistic power density distribution; for this aim, the hydrodynamic model with transport parameters optimized for SiGe:C HBTs [Sas10] was used. By referring to the schematic cross section of the DUTs represented in Figure 5.27, the heat sources exploited in the Comsol structures were built with the power density pattern obtained by reproducing the distribution computed by Sentaurus in the (x, z) plane and assuming a

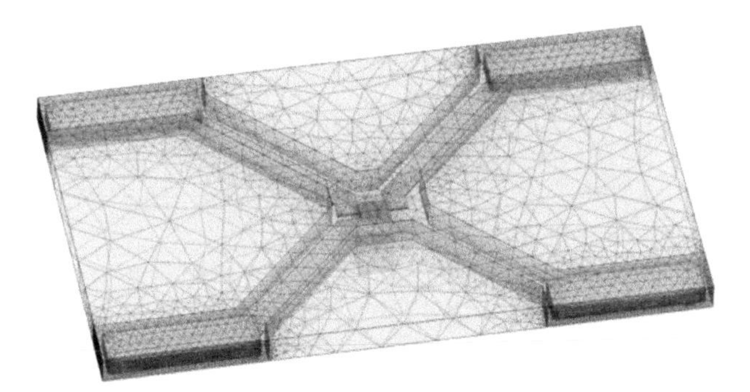

Figure 5.26 Detail of the 3-D Comsol mesh for the IFX transistor with $A_E = 0.13 \times 2.73\ \mu m^2$, composed of 1.35 million tetrahedra of grossly different dimensions, corresponding to 1.8 million degrees of freedom.

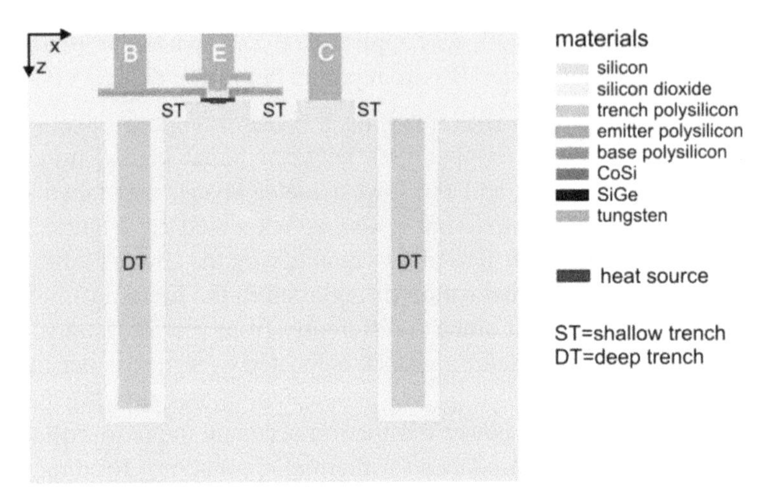

Figure 5.27 Schematic representation (limited to the innermost tungsten contacts) of the typical cross section of the IFX DUTs.

uniform density along the device length (i.e., along the *y*-axis orthogonal to the cross section).

- Thermal simulations are usually performed by setting the thermal conductivities *k* [W/mK] of the materials to values measured from "bulk" samples (listed in Table 5.4). However, in practical cases, many effects concur to reduce *k*, which can be even position-dependent within the same material. In the SiGe alloy, *k* is a function of the z-dependent Ge mole fraction x_{Ge} according to the law [Pal04]

$$k_{\mathrm{SiGe}} = \left[\frac{1 - x_{\mathrm{Ge}}}{k_{\mathrm{Si}}} + \frac{x_{\mathrm{Ge}}}{k_{\mathrm{Ge}}} + \frac{(1 - x_{\mathrm{Ge}})\, x_{\mathrm{Ge}}}{c_{\mathrm{k}}}\right]^{-1} \tag{5.20}$$

Table 5.4 Bulk thermal conductivities

Material	Bulk Thermal Conductivity [W/mK]
Silicon	148
Germanium	60
Silicon dioxide	1.4
Tungsten	177
Copper	390
Emitter polysilicon	40
Base polysilicon	30
Trench polysilicon	20
Cobalt silicide	9.6

where k_{Si} and k_{Ge} are the thermal conductivities of pure Si and Ge, respectively, and c_{k} is a bowing factor equal to 2.8 W/mK. Due to the k lowering imposed by Equation (5.20), the SiGe layer behaves as a barrier for the heat flow from the heat source to the emitter [Pet15]. The thermal conductivity is also adversely impacted by doping due to the enhanced phonon-impurity scattering, as experimentally observed in [Sla64, McC05, Lee12]; a compact formulation to account for this effect is [Lee12]:

$$k_{\mathrm{Si,doped}}\left(k_{\mathrm{SiGe,doped}}\right) = \frac{k_{\mathrm{Si}}\left(k_{\mathrm{SiGe}}\right)}{1 + A \cdot \left(\frac{N}{N_{\mathrm{norm}}}\right)^{\alpha}} \tag{5.21}$$

where N [cm^{-3}] is the position-dependent total doping concentration (acceptors and donors), $N_{\mathrm{norm}} = 10^{20}$ cm^{-3}, while the values of the parameters are $A = 0.74186$, $\alpha = 0.7411$ for boron [Lee12], and $A = 1.698$, $\alpha = 0.8251$ for arsenic, as obtained with a calibration procedure relying on experimental results provided in [McC05]. Lastly, the heat propagation through laterally thin layers can be significantly jeopardized by the phonon scattering with the layer boundaries [Liu05]. In SiGe HBTs, where the heat flow is mostly vertical, scattering mechanisms – expected to be exacerbated in narrow (low-W_{E}) transistors – can take place along device portions like (from the top) emitter tungsten contact, Si emitter, SiGe base, and Si volume surrounded by ST. This deleterious effect can be included by using, e.g., the simple analytical method proposed in [Tor00], which leads to a reduced anisotropic thermal conductivity with x-dependent components given by:

$$\begin{aligned}\frac{k_{\mathrm{y,z}}(x')}{k_{\mathrm{Si,doped}}\left(k_{\mathrm{SiGe,doped}}, k_{\mathrm{contact}}\right)} = \; & 1 - \tfrac{1}{2}\exp\left[-\left(\frac{x'}{x_{\mathrm{charyz}}}\right)^{0.75}\right] \\ & - \tfrac{1}{2}\exp\left[-\left(\frac{1-x'}{x_{\mathrm{charyz}}}\right)^{0.75}\right]\end{aligned} \tag{5.22}$$

$$\begin{aligned}\frac{k_{\mathrm{x}}(x')}{k_{\mathrm{Si,doped}}\left(k_{\mathrm{SiGe,doped}}, k_{\mathrm{contact}}\right)} = \; & 1 - \tfrac{1}{2}\exp\left[-\left(\frac{x'}{x_{\mathrm{charx}}}\right)^{0.95}\right] \\ & - \tfrac{1}{2}\exp\left[-\left(\frac{1-x'}{x_{\mathrm{charx}}}\right)^{0.95}\right]\end{aligned} \tag{5.23}$$

where $x' = x/W$, W being the layer width (along y), $x_{\mathrm{charyz}} = 0.32 \cdot \Lambda/W$ and $x_{\mathrm{charx}} = 0.72 \cdot \Lambda/W$, Λ being the mean free path for phonons (equal to 300 nm in Si and SiGe layers, and to 40 nm in the tungsten emitter contact).

It must be remarked that only Equation (5.20) was accounted for in [dAl10].

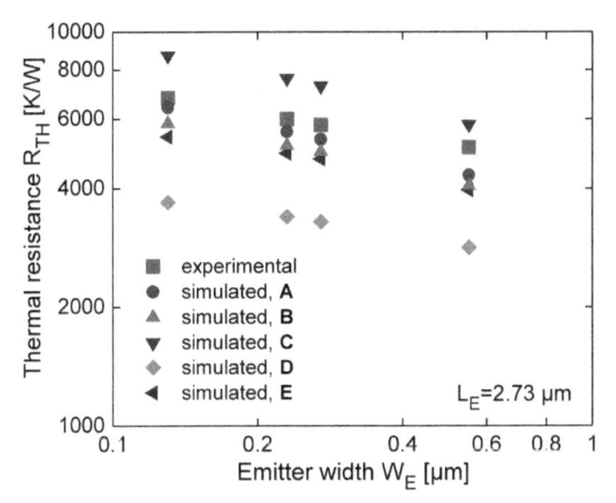

Figure 5.28 Thermal resistances as a function of emitter width for IFX devices sharing L_E = 2.73 µm: experimental (squares) values are compared with those calculated through the simulation approaches **A** (circles), **B** (triangles), **C** (flipped triangles), **D** (rhombi), and **E** (left-oriented triangles).

As discussed in [dAl16], Comsol steady-state simulations were performed by applying an adiabatic boundary condition at the top and lateral faces of the structure, and an isothermal condition on the backside ($T_B = T_0$). The thermal resistance was determined by evaluating the average of the temperature field over the base–emitter junction, which mostly influences the behavior and performance of the device [Zha96], subtracting T_0 and normalizing to the dissipated power (P_D). Results corresponding to DUTs with different W_Es and sharing L_E = 2.73 µm are reported in Figure 5.28, which shows:

- the R_{TH}s determined through the improved experimental technique outlined in the section "Experimental R_{TH} Extraction";
- the R_{TH}s simulated with Comsol by considering the full advanced approach described above (denoted as approach **A**), i.e., by including the BEOL architecture and accounting for the non-uniform power density pattern and the conductivity degradation mechanisms;
- the R_{TH}s calculated with Comsol by modeling the power dissipating region through a standard parallelepiped-shaped source with uniform power density, while considering all other effects and the BEOL structure (approach **B**);
- the R_{TH}s computed with Comsol by accounting for a heat source with non-uniform power density and replacing the metal in the BEOL

architecture with SiO_2 so as to virtually exclude it, while including the first-level tungsten contacts only (approach **C**);

- the R_{TH}s evaluated with Comsol by restoring the BEOL, and considering uncorrected "bulk" values for the thermal conductivities and a standard parallelepiped-based heat source (approach **D**);
- the R_{TH}s computed with Comsol by disregarding the above effects and excluding the BEOL so as to emulate a traditional simulation technique (approach **E**).

By using approach **A**, the R_{TH} of the device with W_E = 0.13 μm was calculated to be 6,437 K/W, which is in fairly good agreement (–5.3%) with the experimental value (6,800 K/W); conversely, a relatively high underestimation (–15%) was obtained for the widest (W_E = 0.55 μm) device, the numerical and measured R_{TH}s being 4,333 K/W and 5,100 K/W, respectively. A post-processing analysis revealed a markedly non-uniform temperature distribution along x over the base–emitter junction compared to low-W_E transistors, which can be ascribed to the concurrent action of the low k_{SiGe} and the narrow tungsten emitter contact (the width of which does not scale with W_E). As a consequence, the evaluation of R_{TH} with a standard geometrical ΔT_j average over the whole junction is likely to be incorrect, and the accuracy should be improved by developing more complex averaging approaches that would lead to a higher FEM R_{TH}. If approach **B** (with the traditional heat source representation) is adopted, the numerical R_{TH} lowers (compared to **A**) from –9% for the HBT with W_E = 0.13 μm to –5.9% for the one with W_E = 0.55 μm, where the base–emitter temperature is non-uniform. Hence, it can be stated that the heat source representation plays a significant role. By making use of the BEOL-free approach **C** (upward heat flow almost annihilated), the FEM R_{TH} of the transistor with W_E = 0.13 μm grows to 8,712 K/W, which corresponds to +28% with respect to the experimental value; this means that, although the low-conductivity SiGe base and Si emitter concur to limit the upward heat flow, the BEOL effectively extracts heat from the emitter. This mechanism is also amplified by the doping-affected conductivity of sub-collector, which counteracts the downward heat propagation. Similar considerations hold for the other narrow HBTs, whereas for the device with W_E = 0.55 μm the lower overestimation (+14%) can be again attributed to the too simple geometrical averaging procedure for the junction temperature field. By exploiting approach **D**, the DUTs enjoy an exacerbated cooling effect dictated by the BEOL architecture and the adoption of the "bulk" thermal conductivity of Si, which favor both the downward and upward heat flow. Consequently, the FEM R_{TH}s are far lower (about –45%) than the experimental counterparts. As expected, employing the traditional approach **E** leads to an underestimation of about –20% regardless of W_{E}, since the deactivation of the k reduction mechanisms (which would imply a heating effect) prevails over the BEOL absence (which would instead cool down the device).

5.4.3 Scaling Considerations

Thermal effects in SiGe HBTs still need to be included in the circuit design process via suitable compact models, which require a geometry-scalable lumped description of the thermal resistance, i.e., an expression of R_{TH} as a function of W_{E} and L_{E} for a given technology stage.

The following simple law was proposed for HICUM/L2 [Sch13]:

$$R_{\mathrm{TH}} = \frac{R_{\mathrm{TH0}}}{1 + a_{\mathrm{W}} W_{\mathrm{E}} + a_{\mathrm{L}} L_{\mathrm{E}}} \tag{5.24}$$

where R_{TH0} [K/W], a_W [$\mu\mathrm{m}^{-1}$], a_{L} [$\mu\mathrm{m}^{-1}$] are fitting parameters. Another formulation, conceived for Mextram504, relies on the preliminary knowledge (from experiments) of the thermal resistance (R_{THref}) of a reference transistor, and three dimensionless fitting parameters (b_A, b_{W}, b_{L}) to be calibrated [Wu06a, Wu06b]:

$$R_{\mathrm{TH}} = \frac{R_{\mathrm{THref}}}{1 + b_{\mathrm{A}}\left(\frac{W_{\mathrm{E}} L_{\mathrm{E}}}{W_{\mathrm{Eref}} L_{\mathrm{Eref}}} - 1\right) + b_{\mathrm{W}}\left(\frac{W_{\mathrm{E}}}{W_{\mathrm{Eref}}} - 1\right) + b_{\mathrm{L}}\left(\frac{L_{\mathrm{E}}}{L_{\mathrm{Eref}}} - 1\right)} \tag{5.25}$$

Lastly, a more sophisticated model was developed by resorting to the following procedure. The exact closed-form solution to the heat transfer equation for a rectangle-shaped indefinitely-thin heat source (THS) with area $W_{\mathrm{E}} \times L_{\mathrm{E}}$ located on the adiabatic top surface of a semi-infinite homogeneous “bulk” domain (with thermal conductivity k) is given by [Rin00]

$$R_{\mathrm{TH}} = \frac{1}{2\pi k}\left[\frac{1}{L_{\mathrm{E}}}\ln\left(\frac{L_{\mathrm{E}} + \sqrt{W_{\mathrm{E}}^2 + L_{\mathrm{E}}^2}}{-L_{\mathrm{E}} + \sqrt{W_{\mathrm{E}}^2 + L_{\mathrm{E}}^2}}\right) + \frac{1}{W_{\mathrm{E}}}\ln\left(\frac{W_{\mathrm{E}} + \sqrt{W_{\mathrm{E}}^2 + L_{\mathrm{E}}^2}}{-W_{\mathrm{E}} + \sqrt{W_{\mathrm{E}}^2 + L_{\mathrm{E}}^2}}\right)\right] \tag{5.26}$$

It is worth noting that the width and length of the THS were assumed to coincide with the emitter ones, which is a reasonable assumption. Unfortunately, Equation (5.26) with k = 148 W/mK (thermal conductivity of Si, as can be seen in Table 5.4) revealed to be unsuited for SiGe HBTs: the R_{TH} values were found to be about 65–75% lower than the experimental counterparts (addressed later) although the cooling effect due to the upward heat flowing to the BEOL structure is not modeled. This means that the heating effect caused by the shallow/deep trenches filled with low thermal conductivity materials – not included in Equation (5.26) as well – plays a role more important than BEOL. Equation (5.26) can be recast in the form:

$$R_{\mathrm{TH}} = \frac{1}{2\pi k}\left[\frac{1}{L_{\mathrm{E}}}\ln\left(\frac{1+\sqrt{\left(\frac{W_{\mathrm{E}}}{L_{\mathrm{E}}}\right)^2+1}}{-1+\sqrt{\left(\frac{W_{\mathrm{E}}}{L_{\mathrm{E}}}\right)^2+1}}\right)+\right.$$
$$\left.\frac{1}{W_{\mathrm{E}}}\ln\left(\frac{\frac{W_{\mathrm{E}}}{L_{\mathrm{E}}}+\sqrt{\left(\frac{W_{\mathrm{E}}}{L_{\mathrm{E}}}\right)^2+1}}{-\frac{W_{\mathrm{E}}}{L_{\mathrm{E}}}+\sqrt{\left(\frac{W_{\mathrm{E}}}{L_{\mathrm{E}}}\right)^2+1}}\right)\right] \tag{5.27}$$

If $W_{\mathrm{E}}/L_{\mathrm{E}} \ll 1$, the square root can be approximated with a first-order Taylor series expansion

$$\sqrt{\left(\frac{W_{\mathrm{E}}}{L_{\mathrm{E}}}\right)^2+1} \approx 1+\frac{1}{2}\left(\frac{W_{\mathrm{E}}}{L_{\mathrm{E}}}\right)^2 \tag{5.28}$$

By substituting Equation (5.28) into (5.27) and neglecting the second-order terms, it is found that:

$$R_{\mathrm{TH}} \approx \frac{1}{2\pi k}\left\{\frac{2}{L_{\mathrm{E}}}\ln\left(2\frac{L_{\mathrm{E}}}{W_{\mathrm{E}}}\right)+\frac{1}{W_{\mathrm{E}}}\left[\ln\left(1+\frac{W_{\mathrm{E}}}{L_{\mathrm{E}}}\right)-\ln\left(1-\frac{W_{\mathrm{E}}}{L_{\mathrm{E}}}\right)\right]\right\} \tag{5.29}$$

Finally, by expressing also the logarithms with a first-order Taylor series expansion, after some algebra,

$$R_{\mathrm{TH}} \approx \frac{1}{\pi k L_{\mathrm{E}}}\left[\ln\left(2\frac{L_{\mathrm{E}}}{W_{\mathrm{E}}}\right)+1\right] \tag{5.30}$$

Equation (5.30) represents a good approximation of the *exact* (5.26) for heat sources with medium/high aspect ratio $W_{\mathrm{E}}/L_{\mathrm{E}}$, while slightly losing accuracy when $W_E \rightarrow L_E$. It was empirically demonstrated that Equation (5.30) can be extended to a wider range of W_E values by introducing a correction term equal to 0.12 in the logarithm argument, which leads to:

$$R_{TH} = \frac{1}{\pi k L_E}\left[\ln\left(0.12 + 2\frac{L_E}{W_E}\right) + 1\right] \quad (5.31)$$

In order to potentially predict the L_E and W_E dependence of the R_{TH} for SiGe HBTs of a specific technology stage, Equation (5.31) was further generalized to [dAl14]:

$$R_{TH} = \frac{1}{\pi k L_E}\left[\ln\left(c + c_R\frac{L_E}{W_E}\right) + 1 + \frac{c_W}{W_E}\right] \quad (5.32)$$

where c, c_R, c_W and the thermal conductivity (k) are fitting parameters. In particular, the term c_W/W_E was introduced to ensure a good fitting over a broad W_E span. It must be remarked that considering k as a fitting parameter has physically sense in SiGe HBTs, since the heat emerging from the dissipation region propagates through various materials with different thermal conductivities (e.g., Si, SiGe, poly, oxide, and tungsten). Models (5.24), (5.25), and (5.32) with optimized parameters (see Table 5.5) were compared to experimental data determined with the approach described in the section "Experimental R_{TH} Extraction" on set #2 devices for various W_Es and three emitter lengths in Figure 5.29.

Results can be summarized as follows: law (5.24) relying on three fitting parameters (the calibrated R_{TH0} is well above the range of the experimental R_{TH}s, and thus cannot be interpreted as a real thermal resistance) and (5.25) based on three fitting parameters plus a "reference" (measured) thermal resistance are suited to offer a fairly good matching with experimental data within a wide range of L_E and W_E values. Excellent agreement is provided by (5.32), which is an extended version of a formulation derived for homogeneous "bulk" domains, and makes use of four fitting parameters, one of which is thermal conductivity. Interestingly, it was found that the optimized k value

Table 5.5 Optimized parameters of the scalable R_{TH} models

Model	Parameters			
(5.24)	R_{TH0} = 34,893 K/W		a_W = 4.46 μm^{-1}	a_L = 1.44 μm^{-1}
(5.25)	R_{THref} = 3,570 K/W ($W_E \times L_E$ = 0.14 × 5.69 μm^2)	b_A = 0.034	b_W = 0.045	b_L = 0.792
(5.32)	c = 0.97	c_R = 1.265	c_W = 0.0166	k = 80 W/mK

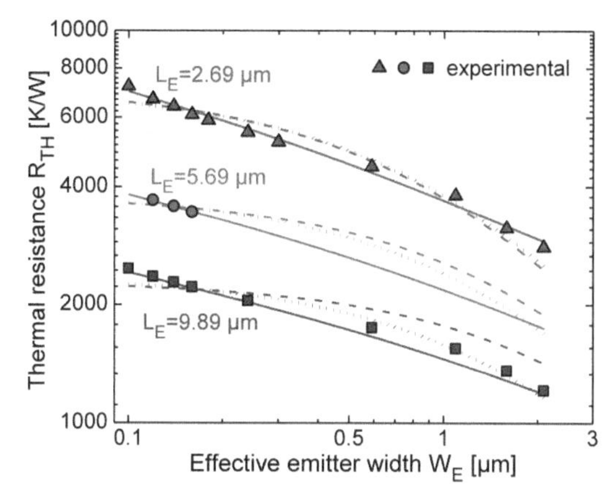

Figure 5.29 Comparison between scalable models (5.24) (dashed lines), (5.25) (dotted), (5.32) (solid) with calibrated parameters, and experimental R_{TH}s (symbols) for set #2 transistors.

(80 W/mK) is lower than the Si counterpart, which is physically reasonable since the lateral heat propagation is mostly influenced by shallow/deep trenches filled with the low-conductivity materials like poly and oxide.

References

[Böc15] Böck, J., Aufinger, K., Boguth, S., Dahl, C., Knapp, H., Liebl, W., et al. (2015). "SiGe HBT and BiCMOS process integration optimization within the DOTSEVEN project," in *Proceedings of the IEEE BCTM*, 121–124.

[Bur88] Burnett, J. D., and Hu, C. (1988). Modeling hot-carrier effects in polysilicon emitter bipolar transistors. *IEEE Trans. Electron Devices* 35, 2238–2244.

[Cha15] Chakraborty, P. S., and Cressler, J. D. (2015). "Hot-carrier degradation in Silicon-Germanium heterojunction bipolar transistors", *Hot Carrier Degradation in Semiconductor Devices*, ed. T. Grasser (Cham: Springer International Publishing), 371–398.

[Che07] Cheng, P., Zhu, C., Appaswamy, A., and Cressler, J. D. (2007). A new current-sweep method for assessing the mixed-mode damage spectrum of SiGe HBTs. *IEEE Trans. Device Mater. Reliab.* 7, 479–487.

[Che09] Cheng, P., Grens, C. M., and Cressler, J. D. (2009). Reliability of SiGe HBTs for power amplifiers–Part II: underlying physics and damage modeling. *IEEE Trans. Device Mater. Reliab.* 9440–448.

[Chev11] Chevalier, P., Meister, T. F., Heinemann, B., Van Huylenbroeck, S., Liebl, W., Fox, A., et al. (2011). Towards THz SiGe HBTs," in *Proceedings of the IEEE BCTM*, 57–65.

[Com] Comsol Multiphysics (2016). *User's Guide, Release 5.2a*. Stockholm: COMSOL.

[Cos92] Costa, D., and Harris, J. (1992). Low-frequency noise properties of n-p-n AlGaAs/GaAs heterojunction bipolar transistors. *IEEE Trans. Electron Devices* 39, 2383–2394.

[Cre04] Cressler, J. D. (2004). Emerging SiGe HBT reliability issues for mixed-signal circuit applications. *IEEE Trans. Device Mater. Reliab.* 4, 222–236.

[Cre13] Cressler, J. D. (2013). "A retrospective on the SiGe HBT: what we do know, what we don't know, and what we would like to know better," in *Proceedings of the IEEE SiRF*, New Orleans, LA, 81–83.

[dAl10] d'Alessandro, V., Marano, I., Russo, S., Céli, D., Chantre, A., Chevalier, P., et al. (2010). "Impact of layout and technology parameters on the thermal resistance of SiGe:C HBTs," in *Proceedings of the IEEE BCTM*, 137–140.

[dAl14] d'Alessandro, V., Sasso, G., Rinaldi, N., and Aufinger, K. (2014). Influence of scaling and emitter layout on the thermal behavior of toward-THz SiGe:C HBTs. *IEEE Trans. Electron Devices* 61, 3386–3394.

[dAl16] d'Alessandro, V., Magnani, A., Codecasa, L., Rinaldi, N., and Aufinger, K. (2016). Advanced thermal simulation of SiGe:C HBTs including back-end of line. *Microelectron. Reliab.* 67, 38–45.

[dAl17] d'Alessandro, V. (2017). Experimental DC extraction of the thermal resistance of bipolar transistors taking into account the Early effect. *Solid State Electron.* 127, 5–12.

[Daw92] Dawson, D. E., Gupta, A. K., and Salib, M. L. (1992). CW measurement of HBT thermal resistance. *IEEE Trans. Electron Devices* 39, 2235–2239.

[Dio08] Diop, M., Revil, N., Marin, M., Monsieur, F., Schwartzmann, T., and Ghibaudo, G. (2008). "Coupled approach for reliability study of fully self aligned SiGe:C 250GHz HBTs," in *Proceedings of the IEEE IIRW*, 77–80.

[Dwi16] Dwivedi, A. D. D., Chakravorty, A., D'Esposito, R., Sahoo, A. K., Frégonèse, S., and Zimmer, T. (2016). Effects of BEOL on self-heating and thermal coupling in SiGe multi-finger HBTs under real operating condition. *Solid State Electron.* 115, 1–6.

[ElR12] El Rafei, A., Saleh, A., Sommet, R., Nébus, J. M., and Quéré, R. (2012). Experimental characterization and modeling of the thermal behavior of SiGe HBTs. *IEEE Trans. Electron Devices* 59, 1921–1927.

[Fis08] Fischer, G. G., and Gliŝić, S. (2008). "Temperature stability and reliability aspects of 77 GHz voltage controlled oscillators in a SiGe:C BiCMOS technology," in *Proceedings of the IEEE SiRF*, 171–174.

[Fis13] Fischer, G. G., Molina, J., and Tillack, B. (2013). "Comparative study of HBT ageing in a complementary SiGe:C BiCMOS technology," in *Proceedings of the IEEE BCTM*, 167–170.

[Fis15] Fischer, G. G., and Sasso, G. (2015). Ageing and thermal recovery of advanced SiGe heterojunction bipolar transistors under long-term mixed-mode and reverse stress conditions. *Microelectron. Reliab.* 55, 498–507.

[Fis16] Fischer, G. G. (2016). "Analysis and modeling of the long-term ageing rate of SiGe HBTs under mixed-mode stress," in *Proceedings of the IEEE BCTM*, 106–109.

[Gog00] Gogineni, U., Cressler, J. D., Niu, G., and Harame, D. L. (2000). Hot electron and hot hole degradation of UHV/CVD SiGe HBT's. *IEEE Trans. Electron Devices* 47, 1440–1448.

[Gho10] Ghosh, S., Marc, F., Maneux, C., Grandchamp, B., Koné, G. A., and Zimmer, T. (2010). Thermal aging model of InP/InGaAs/InP DHBT. *Microelectron. Reliab.* 50, 1554–1558.

[Gho11] Ghosh, S., Grandchamp, B., Koné, G. A., Marc, F., Maneux, C., Zimmer, T., et al. (2011). Investigation of the degradation mechanisms of InP/InGaAs DHBT under bias stress conditions to achieve electrical aging model for circuit design. *Microelectron. Reliab.* 51, 1736–1741.

[Gre09] Grens, C. M., Cheng, P., and Cressler, J. D. (2009). Reliability of SiGe HBTs for power amplifiers–Part I: large-signal RF performance and operating limits. *IEEE Trans. Device Mater. Reliab.* 9, 431–439.

[Has12] Hasnaoui, I., Pottrain, A., Gloria, D., Chevalier, P., Avramovic, V., and Gaquière, C. (2012). Self-heating characterization of SiGe:C HBTs by extracting thermal impedances. *IEEE Electron Device Lett.* 33, 1762–1764.

[Jaq15] Jacquet, T., Sasso, G., Chakravorty, A., Rinaldi, N., Aufinger, K., Zimmer, T., et al. (2015). Reliability of high-speed SiGe:C HBT under electrical stress close to the SOA limit. *Microelectron. Reliab.* 55, 1433–1437.

[Kam17a] Kamrani, H., Jabs, D., d'Alessandro, V., Rinaldi, N., Aufinger, K., and Jungemann, C. (2017). A deterministic and self-consistent solver for the coupled carrier-phonon system in SiGe HBTs. *IEEE Trans. Electron Devices* 64, 361–367.

[Kam17b] Kamrani, H., Jabs, D., d'Alessandro, V., Rinaldi, N., Jacquet, T., Maneux, C., et al. (2017). Microscopic hot-carrier degradation modeling of SiGe HBTs under stress conditions close to the SOA limit. *IEEE Trans. Electron Devices* 64, 923–929.

[LaS09] La Spina, L., d'Alessandro, V., Russo, S., Rinaldi, N., and Nanver, L. K. (2009). Influence of concurrent electrothermal and avalanche effects on the safe operating area of multifinger bipolar transistors. *IEEE Trans. Electron Devices* 56, 483–491.

[Lee12] Lee, Y., and Hwang, G. S. (2012). Mechanism of thermal conductivity suppression in doped silicon studied with nonequilibrium molecular dynamics. *Phys. Rev. B* 86, 075202–1–075202–6.

[Liu05] Liu, W., and Asheghi, M. (2005). Thermal conduction in ultrathin pure and doped single-crystal silicon layers at high temperature. *J. Appl. Phys.* 98, 123523–1–123523–6.

[Lu89] Lu, P.-F., and Chen, T.-C. (1989). Collector-base junction avalanche effects in advanced double-poly self-aligned bipolar transistors. *IEEE Trans. Electron Devices* 36, 1182–1188.

[McC05] McConnell, A. D., and Goodson, K. E. (2005). Thermal conduction in silicon micro- and nanostructures. *Annu. Rev. Heat Trans.* 14, 129–168.

[Moe12] Moen, K. A., Chakraborty, P. S., Raghunathan, U. S., Cressler, J. D., and Yasuda, H. (2012). Predictive physics-based TCAD modeling of the mixed-mode degradation mechanism in SiGe HBTs. *IEEE Trans. Electron Devices* 59, 2895–2901.

[Moh00] Mohammadi, S., Pavlidis, D., and Bayraktaroglu, B. (2000). Relation between low-frequency noise and long-term reliability of single AlGaAs/GaAs power HBT's. *IEEE Trans. Electron Devices* 47, 677–686.

[Muk16a] Mukherjee, C., Jacquet, T., Chakravorty, A., Zimmer, T., Böck, J., Aufinger, K., and Maneux, C. (2016). Low-frequency noise in advanced SiGe:C HBTs–Part I: analysis. *IEEE Trans. Electron Devices* 63, 3649–3656.

[Muk16b] Mukherjee, C., Jacquet, T., Zimmer, T., Maneux, C., Chakravorty, A., Böck, J., et al. (2016). "Comprehensive study of random telegraph noise in base and collector of advanced SiGe HBT: bias, geometry and trap locations," in *Proceedings of the IEEE ESSDERC*, 260–263.

[Muk17] Mukherjee, C., Jacquet, T., Chakravorty, A., Zimmer, T., Böck, J., Aufinger, K., et al. (2017). Random telegraph noise in SiGe HBTs: reliability analysis close to SOA limit. *Microelectron. Reliab.* 73, 146–152.

[Nen04] Nenadović, N., d'Alessandro, V., Nanver, L. K., Tamigi, F., Rinaldi, N., and Slotboom, J. W. (2004). A back-wafer contacted silicon-on-glass integrated bipolar process–Part II: a novel analysis of thermal breakdown. *IEEE Trans. Electron Devices* 51, 51–62.

[Pal04] Palankovski, V., and Quay, R. (2004). *Analysis and Simulation of Heterostructure Devices.* Vienna: Springer-Verlag.

[Pan06] D. Panko, T. Vanhoucke, R. Campos, and G. A. M. Hurkx, "Time-to-fail extraction model for the "mixed-mode" reliability of high-performance SiGe bipolar transistors," in *Proc. IEEE AIRPS*, 2006, pp. 512–515.

[Pas04] Pascal, F., Chay, C., Deen, M. J., G-Jarrix, S., Delseny, C., and Penarier, A. (2004). Comparison of low-frequency noise in III-V and Si/SiGe HBTs. *IEE Proc. Circ. Devices Syst.* 151, 138–147.

[Pet15] Petrosyants, K. O., and Torgovnikov, R. A. (2015). "Electro-thermal modeling of trench-isolated SiGe HBTs using TCAD," in *Proceedings of the IEEE SEMI-THERM*, 172–175.

[Pfo03] Pfost, M., Kubrak, V., and Brenner, P. (2003). "A practical method to extract the thermal resistance for heterojunction bipolar transistors," in *Proceedings of the IEEE ESSDERC*, 335–338.

[Rei01] Reid, A. R., Kleckner, T. C., Jackson, M. K., Marchesan, D., Kovacic, S. J., and Long, J. R. (2001). Thermal resistance in trench-isolated Si/SiGe heterojunction bipolar transistors. *IEEE Trans. Electron Devices* 48, 1477–1479.

[Rie05] Rieh, J.-S., Greenberg, D., Liu, Q., Joseph, A. J., Freeman, G., and Ahlgren, D. C. (2005). Structure optimization of trench-isolated SiGe HBTs for simultaneous improvements in thermal and electrical performances. *IEEE Trans. Electron Devices* 52, 2744–2752.

[Rin00] Rinaldi, N. (2000). Thermal analysis of solid-state devices and circuits: an analytical approach. *Solid State Electron.* 44, 1789–1798.

[Sah12] Sahoo, A. K., Frégonèse, S., Weiβ, M., Malbert, N., and Zimmer, T. (2012). A scalable electrothermal model for transient self-heating effects in trench-isolated SiGe HBTs. *IEEE Trans. Electron Devices* 59, 2619–2625.

[Sah13] Sahoo, A. K., Frégonèse, S., Weiβ, M., Maneux, C., Malbert, N., and Zimmer, T. (2013). "Impact of Back-End-Of-Line on thermal impedance in SiGe HBTs," in *Proceedings of the IEEE SISPAD*, 188–191.

[Sas10] Sasso, G., Rinaldi, N., Matz, G., and Jungemann, C. (2010). "Analytical models of effective DOS, saturation velocity and high-field mobility for SiGe HBTs numerical simulation," in *Proceedings of the IEEE SISPAD*, 279–282.

[Sas14a] Sasso, G., Maneux, C., Böck, J., d'Alessandro, V., Aufinger, K., T. Zimmer, and Rinaldi, N. (2014). "Evaluation and modeling of voltage stress-induced hot carrier effects in high-speed SiGe HBTs," in *Proceedings of the IEEE CSICS*.

[Sas14b] Sasso, G., Rinaldi, N., Fischer, G. G., and Heinemann, B. (2014). "Degradation and recovery of high-speed SiGe HBTs under very high reverse EB stress conditions," in *Proceedings of the IEEE BCTM*, Miami, FL, 41–44.

[Sch93] Schröter, M., Friedrich, M., and Rein, H.-M. (1993). A generalized integral charge-control relation and its application to compact models for silicon-based HBT's. *IEEE Trans. Electron Devices* 40, 2036–2046.

[Sch05] Schröter, M. (2005). "Advanced compact bipolar transistor models – HICUM," in *Silicon Heterostructure Handbook*, ed. J. D. Cressler (New York, NY: CRC Press), 807–823.

[Sch10] Schröter, M., and Chakravorty, A. (2010). *Compact Hierarchical Bipolar Transistor Modeling with HICUM*. Singapore: World Scientific Publishing Co. Pte. Ltd.

[Sch13] Schröter, M., Pawlak, A., Krause, J., and Sakalas, P. (2013). Latest developments of HICUM/L2 for mm-wave applications. *Paper Presented at the IEEE BCTM Open Workshop*, Bordeaux.

[Sch17] Schröter, M., Rosenbaum, T., Chevalier, P., Heinemann, B., Voinigescu, S. P., Preisler, E., et al. (2017). SiGe HBT technology: future trends and TCAD-based roadmap. *Proc. IEEE* 105, 1068–1086.

[Sla64] Slack, G. A. (1964). Thermal conductivity of pure and impure silicon, silicon carbide, and diamond. *J. Appl. Phys.* 35, 3460–3466.

[Syn] Synopsys TCAD (2010). *User's Guide, Release 2010.12*. Synopsys TCAD, CA: Synopsys TCAD.

[Tor00] Tornblad, O., Sverdrup, P., Yergeau, D., Yu, Z., Goodson, K. E., and Dutton, R. W. (2000). "Modeling and simulation of phonon boundary scattering in PDE-based device simulations," in *Proceedings of the IEEE SISPAD*, 58–61.

[Tut00] Tutt, M. N., Pavlidis, D., Khatibzadeh, A., and Bayraktaroglu, B. (1995). Low-frequency noise characteristics of self-aligned AlGaAs/GaAs heterojunction bipolar transistors. *IEEE Trans. Electron Devices* 42, 219–230.

[Tya15] Tyaginov, S. (2015). "Physics-based modeling of hot-carrier degradation", in *Hot Carrier Degradation in Semiconductor Devices*, ed. T. Grasser (Cham: Springer International Publishing), 105–150.

[Tya16] Tyaginov, S., Jech, M., Franco, J., Sharma, P., Kaczer, B., and Grasser, T. (2016). Understanding and modeling the temperature behavior of hot-carrier degradation in SiON nMOSFETs. *IEEE Electron Device Lett.* 37, 84–87.

[Vand94] Vandamme, L. K. J. (1994). Noise as a diagnostic tool for quality and reliability of electronic devices. *IEEE Trans. Electron Devices* 41, 2176–2187.

[Van04] Vanhoucke, T., Boots, H. M. J., and van Noort, W. D. (2004). Revised method for extraction of the thermal resistance applied to bulk and SOI SiGe HBTs. *IEEE Electron Device Lett.* 25, 150–152.

[Van06] Vanhoucke, T., Hurkx, G. A. M., Panko, D., Campos, R., Piontek, A., Palestri, P., et al. (2006). "Physical description of the mixed-mode degradation mechanism for high performance bipolar transistors," in *Proceedings of the IEEE BCTM*.

[Ven00] Vendrame, L., Pavan, P., Corva, G., Nardi, A., Neviani, A., and Zanoni, E. (2000). Degradation mechanisms in polysilicon emitter bipolar junction transistors for digital applications. *Microelectron. Reliab.* 40, 207–230.

[vHa02] von Haartman, M., Sandén, M., Östling, M., and Bosman, G. (2002). Random telegraph signal noise in SiGe heterojunction bipolar transistors. *J. Appl. Phys.* 92, 4414–4421.

[Wal02] Walkey, D. J., Smy, T. J., Reimer, C., Schröter, M., Tran, H., and Marchesan, D. (2002). Modeling thermal resistance in trench-isolated bipolar technologies including trench heat flow. *Solid State Electron.* 46, 7–17.

[Wu06a] Wu, H.-C., Mijalković, S., and Burghartz, J. N. (2006). "A referenced geometry based configuration scalable Mextram model for bipolar transistors," in *Proceedings of the IEEE BMAS Workshop*, 50–55.

[Wu06b] Wu, H.-C. (2006). *A Scalable Mextram Model for Advanced Bipolar Circuit Design*. Ph.D. dissertation, Delft University of Technology, Delft.

[You11] You, S., Decoutere, S., Van Huylenbroeck, S., Sibaja-Hernandez, A., Venegas, R., and De Meyer, K. (2011). "Impact of isolation scheme on thermal resistance and collector-substrate capacitance of SiGe HBTs," in *Proceedings of the IEEE ESSDERC*, 243–246.

[Zan93] Zanoni, E., Crabbé, E. F., Stork, J. M. C., Pavan, P., Verzellesi, G., Vendrame, L., et al. (1993). Extension of impact-ionization multiplication coefficient measurements to high electric fields in advanced Si BJT's. *IEEE Electron Device Lett.* 14, 69–71.

[Zha96] Zhang, Q. M., Hu, H., Sitch, J., Surridge, R. K., and Xu, J. M. (1996). A new large signal HBT model. *IEEE Trans. Microw. Theory Tech.* 44, 2001–2009.

[Zha02] Zhang, G., Cressler, J. D., Niu, G., and Joseph, A. J. (2002). A new "Mixed-Mode" reliability degradation mechanism in advanced Si and SiGe bipolar transistors. *IEEE Trans. Electron Devices* 49, 2151–2156.

[Zhu05] Zhu, C., Liang, Q., Al-Huq, R. A., Cressler, J. D., Lu, Y., Chen, T., et al. (2005). Damage mechanisms in impact-ionization-induced mixed-mode reliability degradation of SiGe HBTs. *IEEE Trans. Device Mater. Reliab.* 5, 142–149.

6

Millimeter-wave Circuits and Applications

A. Mukherjee[1], W. Liang[1], M. Schröter[1,2], U. Pfeiffer[3], R. Jain[3], J. Grzyb[3] and P. Hillger[3]

[1]Chair for Electron Devices and Integrated Circuits, Technische Universität Dresden, Germany
[2]Department of Electrical and Computer Engineering, University of California at San Diego, USA
[3]Institute for High-Frequency and Communication Technology (IHCT), University of Wuppertal, Germany

6.1 Millimeter-wave Benchmark Circuits and Building Blocks

A. Mukherjee, W. Liang and M. Schröter

The continuous progress of SiGe:C HBT BiCMOS process technology paves the way for high-volume low-cost mm-wave and sub-mm-wave applications. The design of the corresponding high-frequency (HF) integrated circuits requires accurate compact models for both active and passive devices. Especially, the compact models for active devices must cover many physical effects occurring in advanced process technologies and address a wide bias, temperature, and geometry range as well as high-frequency (HF) effects such as non-quasi-static delay and substrate coupling. Devices used in HF circuits typically operate at 3 to 10 times the circuit speed due to the harmonics generated within the circuits that ultimately determine the signal shape. The verification of compact models at such a high speed has become a major issue since device measurement capability has not kept pace with process and circuit development. While there has been some effort toward extending small-signal (S-parameter) measurement capability toward several 100 GHz, direct experimental verification of compact models for *large-signal* operation at mm- and sub-mm-wave frequencies still appears illusive. For instance, load-pull measurements beyond 50 GHz are not only difficult and expensive but also do not provide any phase shift information, which is important for describing time-dependent large-signal switching correctly.

The demand for model accuracy to ensure one-pass success for saving R&D cost in mm-wave and sub-mm-wave circuit design forces compact models of transistors to undergo tests in a vast range of operating conditions instead of merely verifying typical device characteristics. Therefore, model verification has been extended to small circuits in which transistor operation can be tested under realistic application-relevant conditions. These circuits comprise benchmark blocks and small building blocks of larger systems.

Benchmark circuits on the one hand have to be sufficiently simple so as to avoid masking compact transistor model deficiencies by other effects, but should on the other hand resemble the typical transistor operation in related larger circuit building blocks. A well-selected set of benchmark circuits should allow the transistors (and their associated models) to be exercised in application-relevant operating modes beyond the typical standard device characteristics measured in a characterization lab. In addition, the same benchmark circuits can also be employed for evaluating process performance and for detecting processing issues in terms of the targeted applications during the process development phase.

The circuit building blocks are concerned with practical needs toward, e.g., lowering power consumption or utilizing the transistor non-linearity for harmonic power generation in mm-wave circuits. Here, transistor operation in extreme regions is of interest, e.g., at low collector–emitter voltages (i.e., at significantly forward-biased base–collector junction) or beyond the open-base breakdown voltage. The related building blocks target competitive figures of merit (FoMs) and serve also for demonstrating the process technology's capability.

Using these relatively small circuits for the above-mentioned purposes has so far been hampered by various factors. On one side, modeling and process engineers lack the necessary circuit design expertise and on the other side circuit designers have little interest in designing, from their perspective, relatively simple circuits. In DOTSEVEN, for the first time, an attempt was started to better bridge these two worlds by fabricating a set of circuit blocks partially designed by the modeling community of the project. The experimental results of the various circuits were then compared with simulations in order to establish a solid understanding of the accuracy of the compact models under circuit-relevant constraints. Several examples are presented below.

6.1.1 Benchmark Circuits

A. Broadband amplifier using a Darlington pair

The broadband amplifier (BBA) is an integral part of both wireless and wireline communication systems. Figure 6.1 depicts two variants of the BBA

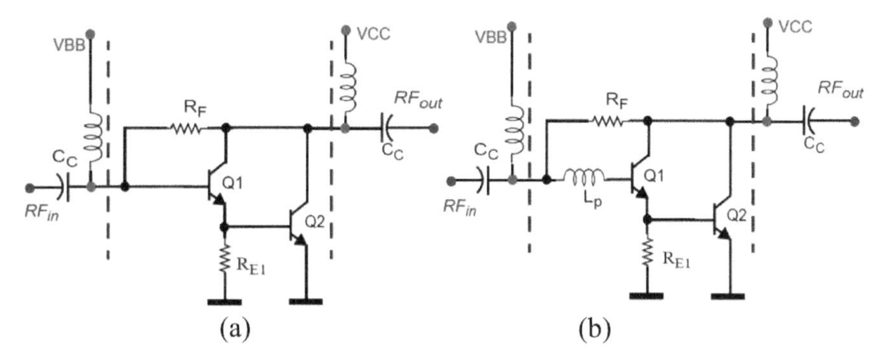

Figure 6.1 Schematic of the broadband Darlington amplifier (a) without and (b) with peaking inductor (L_{p}).

schematic, with their input and output matched to the 50 Ω system impedance. This type of amplifier generally shows a low-pass behavior, i.e., it provides its maximum gain at low frequencies. Here, the BBA topology is derived from the basic Darlington configuration [Mukh16, Gray08, Vera13, Vera14], but uses a modified Darlington pair consisting of an emitter follower and a common emitter transistor.

The degeneration resistance (R_{E1}) actually increases the terminal impedance of the stage and can at the same time be used to bias the transistor Q1. The advantage of the Darlington configuration over a simple single degenerated stage is that an appropriate choice of R_{E1} yields a current gain bandwidth which can approach twice that of a single stage [Armi89].

The operating points of the transistors are adjusted, as shown in the schematic, through external bias-Tees and the resistor (R_{E1}). The feedback resistance (R_{F}) allows adjusting the gain flatness. One of the main aspects of benchmark circuits is the need to be able to quickly design them, preferably by modeling or process engineers. This requires the development of a generic design procedure [Mukh16], which may not achieve world-record performance but guarantees a working circuit that meets the circuit purposes. Such a procedure is given below:

1. The design starts with choosing V_{CC} at or near BV_{CEO} as specified by the corresponding process design kit (PDK) documentation.
2. Both the transistors are biased at $J_{\mathrm{C}}(f_{\mathrm{T,peak}})$ according to the f_{T} – J_{C} plot (cf. Figure 6.2). The corresponding V_{BE} values determine $V_{\mathrm{BB}} = V_{\mathrm{BE1}} + V_{\mathrm{BE2}}$.
3. As no explicit input and out matching network is used in the circuit, the emitter length of Q1 is adjusted so as to make the real part of the input

impedance 50 Ω. For both transistors the minimum emitter width should be used.

4. Since Q1 is operated as emitter follower, the current through R_{E1} is much larger than the base current into Q2 so that $R_{E1} \approx V_{BE2}/I_{C1}$.
5. The initial emitter length of Q2 can then be chosen similar to Q1, but needs to be adjusted according to its larger V_{CE} to maintain operation at $J_C(f_{T,peak})$.
6. The initial value of R_F can be obtained from, $R_F/[1 - S_{21}(f = 0)] = 50\ \Omega$.
7. To enhance the bandwidth of the amplifier, a peaking inductor (L_P) can be added in series to the input of Q1. The value of L_P can be calculated from the resonance condition at the input of Q1, knowing its input impedance, and at the 3 dB frequency of the gain of the BBA without peaking inductance. The value can be calculated by $L_p = \mathrm{Imag}(Z_{in})/(2\pi f_{3dB})$, where f_{3dB} is the original 3 dB frequency of the amplifier.
8. Further optimization of the circuit is required after EM simulation of the entire circuit.

Two variants, with and without the peaking inductor, of the BBA were fabricated in IHP's first DOTSEVEN technology run. The backend of that process offered seven metal layers with five thin metals and two thick top metal layers [IHP03]. Important transistor characteristics along with the comparison to the compact model data are shown in Figure 6.2.

The "topmetal2" of the process has been used to realize the inductor and the other required transmission line interconnects. Both amplifier versions were configured for on-wafer measurements using GSG probes along with DC biasing through bias-Tees. The die micrographs of the two amplifiers

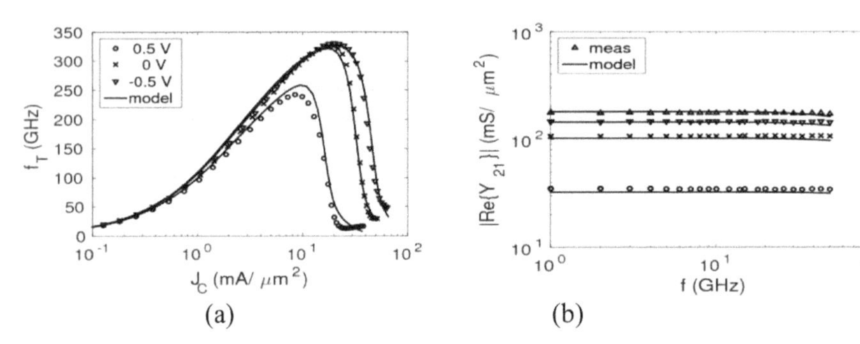

Figure 6.2 Comparison between measured data (symbols) and compact model HICUM/L2 results (lines). (a) Transit frequency f_T vs. J_C for different V_{BC} values. (b) Transconductance g_m vs. frequency for V_{BC} = 0V and at different J_C = (1, 5, 10, 20) mA/μm^2.

are shown in Figure 6.3. The total die area of the individual amplifiers is just 0.05 mm^2 and fits into regular HF GSG pads also used for transistor characterization. The two amplifiers were biased with a single supply voltage of 1.8 V.

The S-parameters were measured using a Keysight PNA-L5235A with 110 GHz extenders. The measurement includes effects of pad parasitics, on-chip transmission lines connecting input and output and other components of the amplifier layout, i.e., no de-embedding of those elements was performed. The small signal gain S_{21} in the frequency range of 0.5–110 GHz along with a comparison with post-layout simulation is shown in the Figure 6.4.

The measurement shows a bandwidth of 78 GHz for the BBA without peaking inductor and of 109 GHz for the amplifier with peaking inductor. The measured stability factors for both amplifier versions are shown in

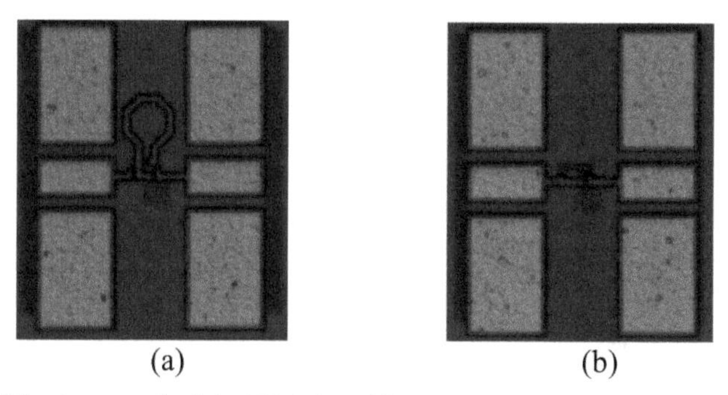

Figure 6.3 Die photograph of the BBA (a) with and (b) without peaking inductor. The chip size in both cases is (0.245×0.18) mm^2.

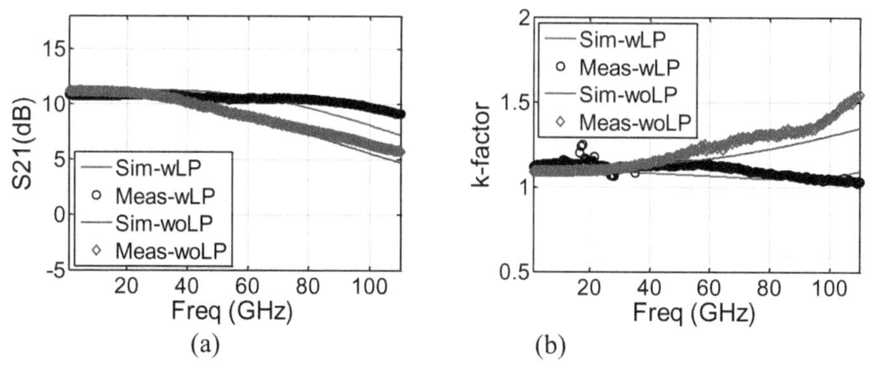

Figure 6.4 (a) Small-signal gain and (b) stability factor of the BBA with and without peaking inductor: comparison between simulation (lines) and measurement (symbols).

Figure 6.4(b) and ensure unconditional stability. The simulations agree quite well with the measurements.

Transistor models for advanced SiGe HBTs have to cover many physical effects in order to achieve the desired accuracy for enabling first-pass design. The resulting model complexity makes it difficult to understand the impact of each physical effect or model parameter on circuit performance under realistic operating conditions. Knowing this dependence is important for model developers, circuit designers, and process engineers. Therefore, a sensitivity analysis was performed by analyzing the changes of the relevant circuit FoMs with respect to variations in model parameters. For the BBA here, the gain (S_{21}), the input and output reflection coefficients S_{11} and S_{22}, the stability factor k, and the bandwidth were selected as the important FoMs. The model parameters of the two transistors (Q1 and Q2) were varied separately to identify the model parameters that are most influential on the above-mentioned FoMs. The Figure 6.5(a) displays the maximum relative sensitivity of the above mentioned FoMs for those model parameters of Q2 that cause changes of more than 5% in at least one of the FoMs as a response to a ±20% change in the respective model parameter. Figure 6.5(b) shows for Q1 the three model parameters that have the highest impact in at least one of the FoMs.

It can be observed in Figure 6.5 that the emitter resistance (*re*), the low-current transit time (*t0*), and the thermal resistance (rth) of Q2 have the biggest impact. *re* causes a large change in input return loss as a small change in its value directly affects the corresponding BE-voltage of the transistor. The emitter resistance (*re*) of Q1 also shows considerable impact on input return loss but its contribution is masked by the large value of R_{E1} in series.

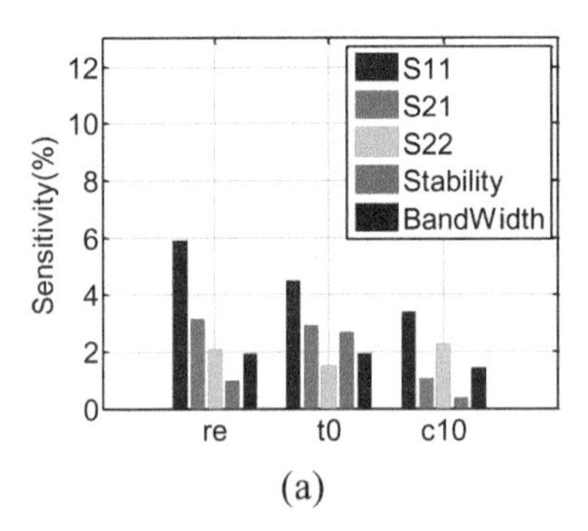

(a)

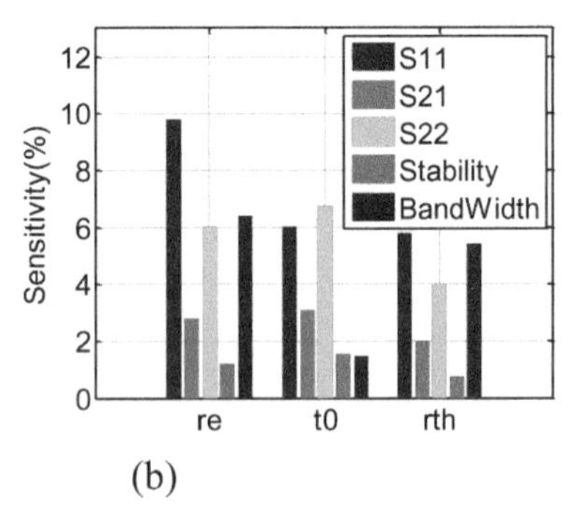

(b)

Figure 6.5 Sensitivity of the series peaked BBA performance parameters with respect to HICUM model parameters for the transistors (a) Q1 and (b) Q2.

The influence of *t0* is caused by the associated diffusion capacitance (S_{11}), which dominates the input capacitance, and the base–collector voltage-dependent mobile charge in the transfer current, impacting the output conductance and thus S_{22}. Interestingly, the bandwidth is mostly impacted by *rth* of the second transistor Q2. The sensitivity analysis of the BBA without series peaking inductance shows a very similar trend and thus is not shown here.

B. W-band low-noise amplifier (LNA)

In this section, the design and implementation of a wide-band LNA for the frequency range of 90–110 GHz is described. The architecture employed here includes a shunt–shunt feedback resistor along with an input matching LC π-network and a post-cascode series peaking inductor [Lin07]. The basic idea of the π-network is to add a low-pass filter at the input side of the LNA [Lin07]. It enables the input impedance of the LNA to be matched with the source impedance (50 Ω) when the input frequency is less than the cutoff frequency of the low-pass filter,

$$f \leq \frac{1}{2\pi\sqrt{L_B C_1}} \tag{6.1}$$

Considering the equivalent circuit (EC) as shown in Figure 6.6(b), the input impedance (Z_{in}) can be written as,

$$Z_{in} = \frac{s^2 R_f C_\pi L_B + sL_B + R_f}{s^3 R_f C_{in} C_\pi L_B + s^2 L_B C_{in} + sR_f\left(C_{in} + C_\pi\right) + 1}. \tag{6.2}$$

Setting Z_{in} = 50 Ω and with $s = j\omega$, the above equation can be solved for two frequencies with perfect input impedance matching,

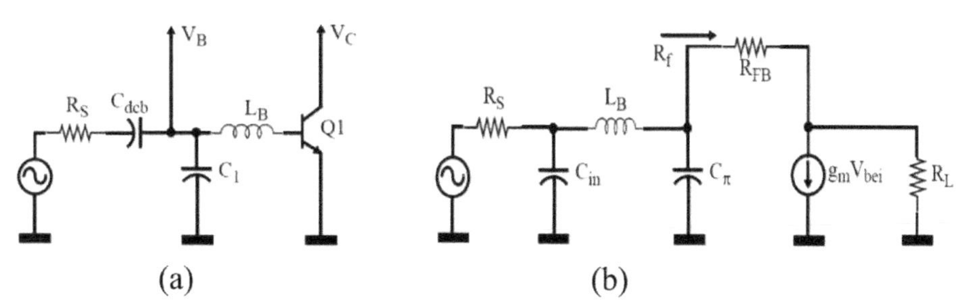

Figure 6.6 (a) LNA with π-network for input matching, (b) small signal equivalent circuit of the LNA with feedback resistance R_{FB} and input matching elements.

$$\omega_1 = 0 \tag{6.3}$$

$$\omega_2 = \sqrt{\frac{2}{L_B C_\pi} - \frac{1}{Z_0^2 C_\pi^2}} \tag{6.4}$$

For the design of the LNA the cascode configuration is adopted due to its better reverse isolation and higher small-signal gain. The optimum DC bias voltage for the common-emitter transistor was found from the $NF_{\min}$ vs. J_C measurement (cf. Figure 6.7) of a separately measured single-emitter transistor.

Figure 6.8 shows the schematic of the implemented wide-band LNA.

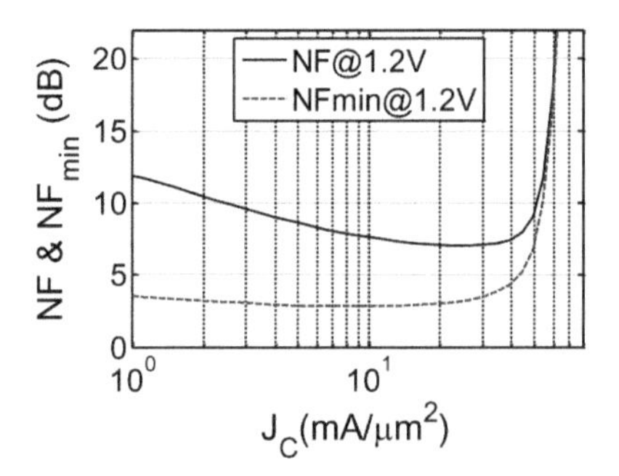

Figure 6.7 NF and $NF_{\min}$ versus J_C for a SiGe HBT with A_{E0} = 0.7 × 0.9 μm² for V_{CE} = 1.2 V at f = 90 GHz.

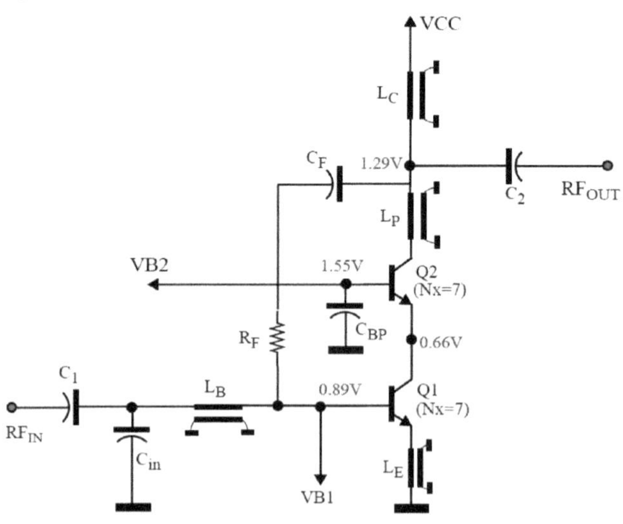

Figure 6.8 Schematic of the single-stage wide-band (90–110 GHz) LNA.

At mm-wave frequencies, the layout of the circuit plays a pivotal role for the circuit performance. Initially both transistors were contacted up to top metal. The insertion of the series inductor L_{B} and parallel capacitor C_{in} compensates for the effect of the input capacitance C_{π}. This strategy produces a third-order ladder type low-pass filter network which can reduce the imaginary part of Y_{11} and hence increase the input-matching bandwidth [Lin07]. The initial values of L_{B} and C_{in} were calculated with the help of Equation (6.4) and can be further adjusted for optimized performance. C_{in} was implemented using the MIM capacitor between metal5 and topmetal1. At the desired high frequencies, L_{B} was realized using the available topmetal2 of the process. A small degeneration inductor L_{E} ($\sim$20 pH), implemented as a metal line, was added to ensure good linearity and better stability [Ko96, Afsh06]. After several simulation iterations, the value of R_{F} was fixed to 415 Ω.

The output matching network consists of L_{C} and C_2, the values of which were carefully chosen to provide S_{22} matching over the frequency range of interest. A small peaking inductor L_{P} was added to achieve a better small-signal gain (S_{21}) flatness. All interconnects between the passive elements were realized with topmetal2 and were EM-simulated to include the effects of the layout parasitics. The base biases of the two transistors were fed through 3 kΩ resistors that use unsalicided, p-doped gate polysilicon as resistor material [IHP03].

Depending on the biasing of this circuit different results are obtained. One goal here was to verify the compact model for low-power applications operating in saturation. Therefore, the DC bias values at the base terminals were chosen as V_{B1} = 0.94 V and V_{B2} = 1.6 V, respectively. Together with a supply voltage (V_{CC}) of 1.4 V this ensured that both transistors work in the "saturation region", i.e., with positive external V_{BC} of about 0.23 V.

The on-wafer S-parameter measurements were performed using a Keysight PNA-L5235A with 110 GHz extenders. The noise of the amplifier was measured at the IMS lab in Bordeaux, France. Figure 6.9 shows the S-parameters of the fabricated amplifier along with circuit simulation. The moderate performance of S_{22} up to 80 GHz affects the output reflection coefficient of the measurement equipment. Generally, the agreement between measurement and simulation is quite satisfactory though.

The noise measurement was performed from 75 GHz to 90 GHz, which was the highest frequency range for which a noise source and respective measurement equipment were available. The measured *NF* at 90 GHz is 5 dB,

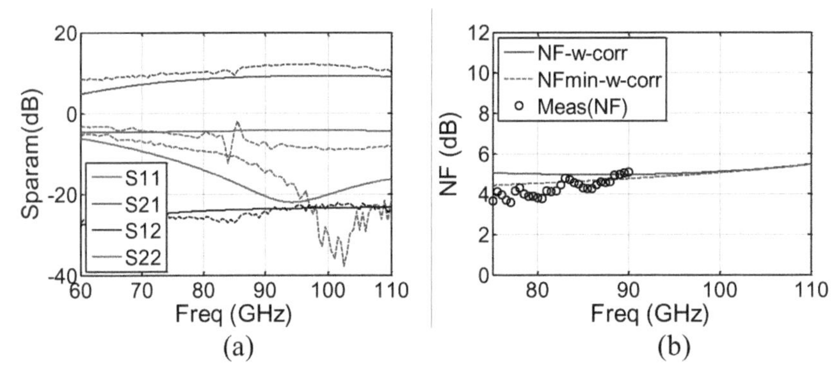

Figure 6.9 (a) Small-signal results of the wide-band LNA: comparison between measurement (dashed lines) and simulation with HICUM/L2 (solid lines). (b) Corresponding frequency-dependent noise figure NF: comparison between measurements (symbols) and simulation of (blue line) and $NF_{\rm min}$ (red dashed line).

which is slightly less than simulated. The simulation was performed with the noise-correlation model turned on.

Figure 6.10 shows the results of a sensitivity analysis for the designed LNA where the model parameters of the two transistors Q1 and Q2 were varied separately. In case of transistor Q1, the impacts of only those model parameters are shown that cause the maximum relative sensitivity of the relevant FoMs to vary at least 4% in response to a parameter variation in the range of ±20%. In case of Q2 the three most influential model parameters are shown.

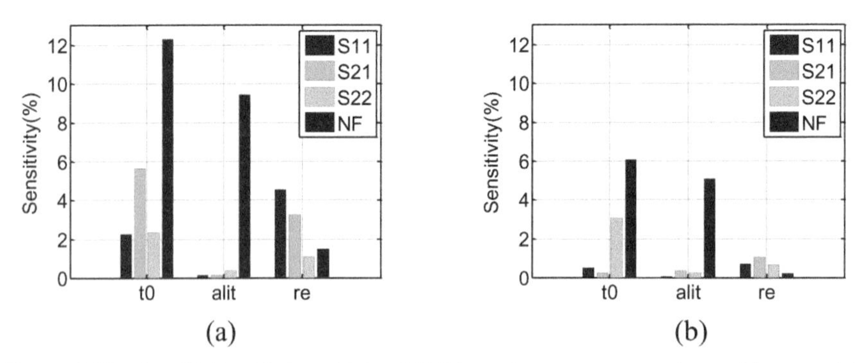

Figure 6.10 Sensitivity of the wideband LNA performance parameters with respect to HICUM model parameters for the transistors (a) Q1 and (b) Q2.

It can be observed from the above figure that the low-current transit time (*t0*) and the associated delay time of the transfer current (*alit*t0*) of Q1 and Q2 have the biggest impact on mainly the noise figure (NF). The delay time enters the sensitivity through the noise correlation. The gain of the cascode amplifier is basically controlled by the CE transistor (Q1) rather than the transistor in CB mode (Q2). The emitter resistance (*re*) of Q1 mostly impacts the input reflection and gain. Generally, the FoMs are less sensitive though to the parameters of Q2 compared to those of Q1.

It must be mentioned here that in this sensitivity study the model parameters are varied individually, i.e., their correlation through process and structural parameters of the transistor were ignored [Schr05]. A study including the correlations can be done using a special transistor scaling tool [Schr99].

The performance of LNAs can be compared through the following FoM,

$$FoM_{LNA} = \frac{Bandwidth\ (GHz) \times Gain\ (dB)}{(F_{avg} - 1) \times P_{DC}(mW)},$$

where, F_{avg} is the average noise factor within the band and P_{DC} is the DC power dissipation of the circuit. In this *FoM*, the gain in decibels used as the power consumption is proportional to gain in decibels [Sato10]. Table 6.1 compares the performance of this LNA with other state-of-the-art broadband mm-wave LNAs reported recently. Despite operation in saturation, the performance compares reasonably to the other designs and its FoM is only exceeded by an 80 nm HEMT amplifier.

6.1.2 Circuit Building Blocks

So far, some results going along this direction have been reported. In [Seth11, Inan14], LNAs for the 8–12 GHz and 10–22 GHz bands were designed with reduced supply voltages. At a higher frequency, a 65 GHz LNA was implemented in a 130 nm SiGe HBT process in [Agar14]. A 53.5 GHz SiGe HBT oscillator with only 0.5 V supply voltage was reported in [Sah14]. In the following context, the design of a W-band low-power LNA is presented, which uses an ultra-low supply voltage (V_{CC} = 0.5 V). This work aims to give an example showing how far the DC power consumption can be reduced while maintaining meaningful circuit performance for a mm-wave LNA with HBT transistors biased in the saturation region. The impact of varying transistor series resistances on voltage gain, minimum noise figure

Table 6.1 Comparison of LNA related FoMs for different technologies and topologies (*simulation; #78–110 GHz estimated)

Reference	Tech f_T/f_{MAX} (GHz)	Topology	3 dB BW (GHz)	Gain (dB)	*NF* (dB)	OP_{1dB} (dBm)	DC (mW)	FOM
[Kiss10]	SiGe:C HBT 220/285	2-stage CE	55–77 (33%)	20	5.8 (est)	3	40	3.92
[Gilr11]	0.18μ BiCMOS 200/200	5-stage CE	69–95 (31%)	20	<12	–	63	0.917
[May10]	0.12μ BiCMOS 200/265	5-stage CE	82–100 (20%)	27	8*	–	27.6	3.31
[Chen12]	0.18 μ BiCMOS 200/180	4-stage cascode	86–106 (21%)	25	<9	–	–	–
[Sato10]	80 nm InP HEMT 380/283	3-stage CG	68–110 (47%)	18	3.5	–4	12	50.85
[Koch10]	100 n InAlAs mHEMT 200/300	4-stage CS	115–150 (26%)	15	5–6 (est)	–	35–40	5.42
[Zhan12]	0.13 μ BiCMOS	4-stage	132–160 (19%)	21	<9.5*	–	14.5	5.84
[Liu13]	0.25 μ BiCMOS 180/220	2-stage cascode	47–77 (48%)	22.5	<7.2	4.5*	52	4.35
[Liu13]	0.13 μ BiCMOS 250/300	2-stage cascode	70–140* (66%)	25	<7# <9*	1*	54	6.10
Thiswork	0.13 μ BiCMOS 505/720	1-stage cascode	67–117 (54%)	12	<9.6* 5 @90 GHz	0.49*	12	12.5

(NF_{min}), and third-order input intercept point (IIP3) is also investigated for this low-power LNA.

Furthermore, the design of a W-band frequency tripler with 0.5 V supply voltage is presented, aiming at an output signal at 96 GHz from a 32 GHz input signal with as low as possible DC power consumption. This frequency tripler could be used as a candidate for generating W-band signals, together with a fundamental-tone oscillator located at a much lower frequency, to alleviate the problem of directly designing a W-band oscillator with satisfactory performance. Comparison between simulated and experimental results is given to verify the accuracy of HICUM model parameters at W-band when transistors are used to design non-linear mm-wave circuits with a reduced supply voltage.

6.1.2.1 W-band low-noise amplifier (LNA) with 0.5 V supply voltage

IHP SG13G2 SiGe HBT technology was used for designing the LNA in this section. Figure 6.11(a) shows that a collector current density of more than 10 mA/μm^2 is needed to bias the HBT from this technology at its peak transit frequency (f_{T}). The corresponding base–emitter DC bias voltage (V_{BE}) can be obtained from Figure 6.11(b), which shows that at least 0.85 V is needed to achieve such a collector current density. Therefore, if the design of an LNA with ultra-low supply voltage is targeted (like V_{CC} = 0.5 V), then the transistor has to be biased in the saturation region ($V_{\text{BC}} > 0.35$ V). Figure 6.11(a) also implies that the HBT in this technology can still provide

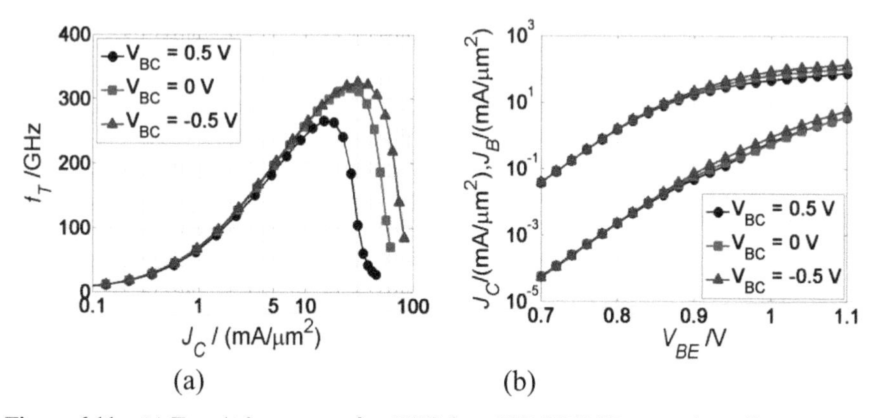

Figure 6.11 (a) Transit frequency of an HBT from IHP SG13G2 versus its collector current density with V_{BC} = –0.5, 0, and 0.5 V. (b) Corresponding transfer characteristics.

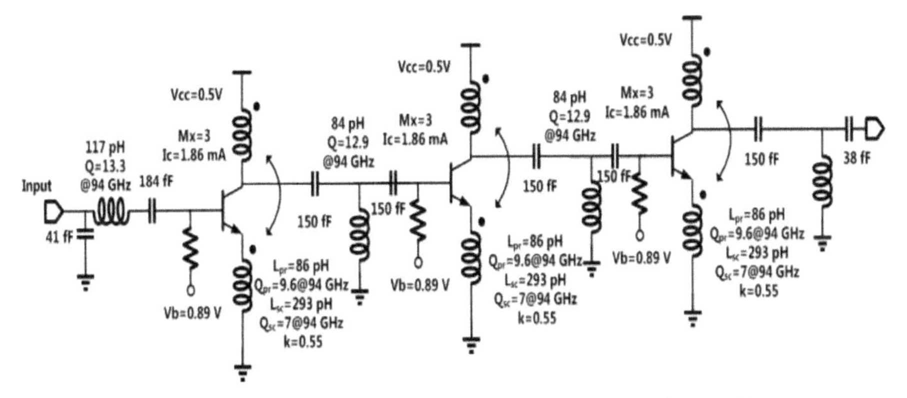

Figure 6.12 Schematic of the three-stage common-emitter LNA.

an acceptable value for peak f_T when V_{BC} equals 0.5 V (around 260 GHz, compared with the value of around 320 GHz with –0.5 V V_{BC} in the normal forward-active case). In other words, with a supply voltage as low as 0.5 V, this transistor still retains a decent speed for designing mm-wave circuits.

The topology of the LNA, shown in Figure 6.12, consists of three stages of common-emitter configuration with emitter–collector transformer feedback to improve reverse isolation and stability at high frequencies. Besides stability, the emitter series inductor also serves as part of the impedance matching network. The amplifier is biased with V_b = 0.89 V and V_{cc} = 0.5 V, while the total power consumption of the three stages is only 2.79 mW (1.86 mA for each stage). The topmost metal layer provided by the technology (TopMetal 2) is used to fabricate the transmission lines, whereas the lower metal layers (TopMetal 1, and Metal 5/4/3/2) are used for transitions going through different low-level layers. The bottom metal layer (Metal 1) is used as the ground plane all over the layout of the circuit. The LNA is designed with the aid of constant available power gain circles and constant NF circles of each stage.

Figure 6.13 illustrates the constant available power gain circles and constant NF circles of the transistor used in the first stage of the amplifier (without transformer feedback) at 94 GHz. The source impedance posed to the base of the transistor (transformed from a 50-Ω signal source by the input matching network) is selected as close to the center of the constant available power gain circles as possible for higher power gain, while the source impedance is also chosen as close to the center of the constant NF circles as possible for lower noise mismatch (leading to lower noise contribution by the

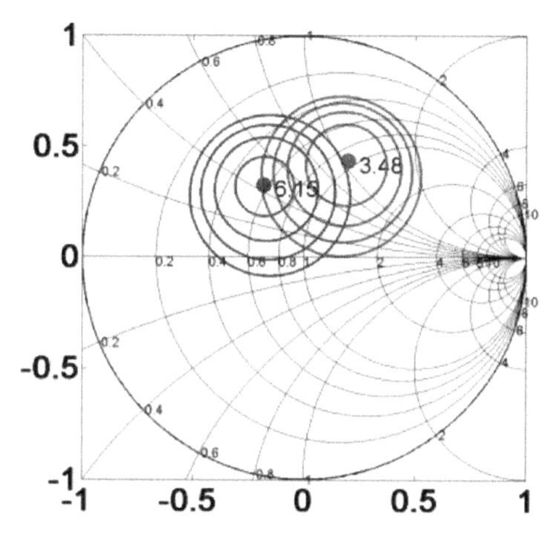

Figure 6.13 Constant available power gain circles (blue curves, from 6.15 dB to 5.35 dB with a 0.2 dB step) and constant NF circles (red curves, from 3.48 dB to 4.28 dB with a 0.2 dB step) for the first stage of the amplifier at 94 GHz.

corresponding amplifier stage). Therefore, in practice a compromise has to be made between power matching and noise matching in an LNA design.

The measured and simulated S-parameter results of the three-stage amplifier are shown in Figure 6.14(a). With only 0.5 V collector supply voltage, this LNA can still provide 14.38 dB peak power gain at 91 GHz and more than 10 dB power gain over a frequency range from 86 GHz

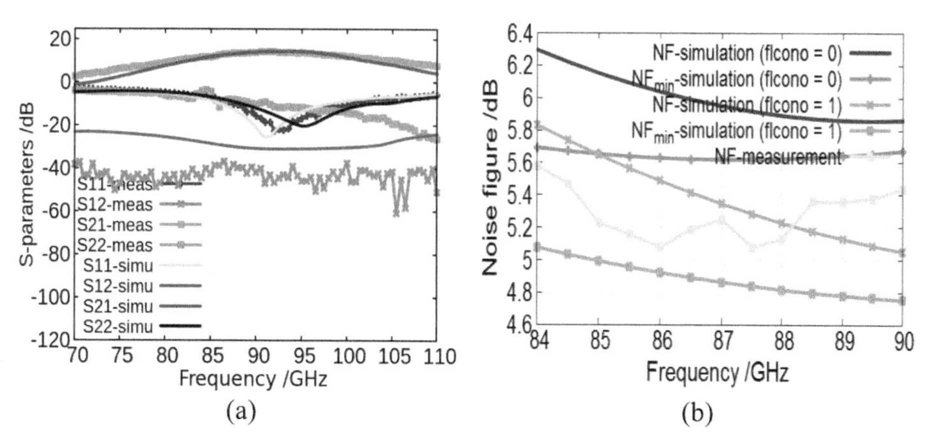

Figure 6.14 Measured (symbols) and simulated (lines) results of the W-band ultra-low power three-stage LNA: (a) S-parameters and (b) noise figure.

to 100 GHz. The measured and simulated NFs of the LNA are shown in Figure 6.14(b), where the correlated noise has been turned on (flcono = 1) and off (flcono = 0), respectively. The measured NF at 90 GHz is 5.44 dB. The measured input-referred 1 dB compression point is –21 dBm at 91 GHz. Fairly good agreement between measurement and simulation (especially for the S-parameter results) has been achieved, which verifies the accuracy of the compact model (HICUM/L2) with a forward-biased BC junction at high frequencies (W-band). Figure 6.14(b) also shows that including noise correlation is not negligible, when there is a demand to accurately capture the noise performance of amplifiers at W-band. Note that noise correlation increases with frequency.

To investigate the impact of transistor series resistances on the circuit performance of this LNA (S_{21}, NF_{min}, and IIP3), a sensitivity analysis was performed at 92 GHz for the emitter, base, and collector series resistances (±30% variation) as shown in Figure 6.15. The corresponding model

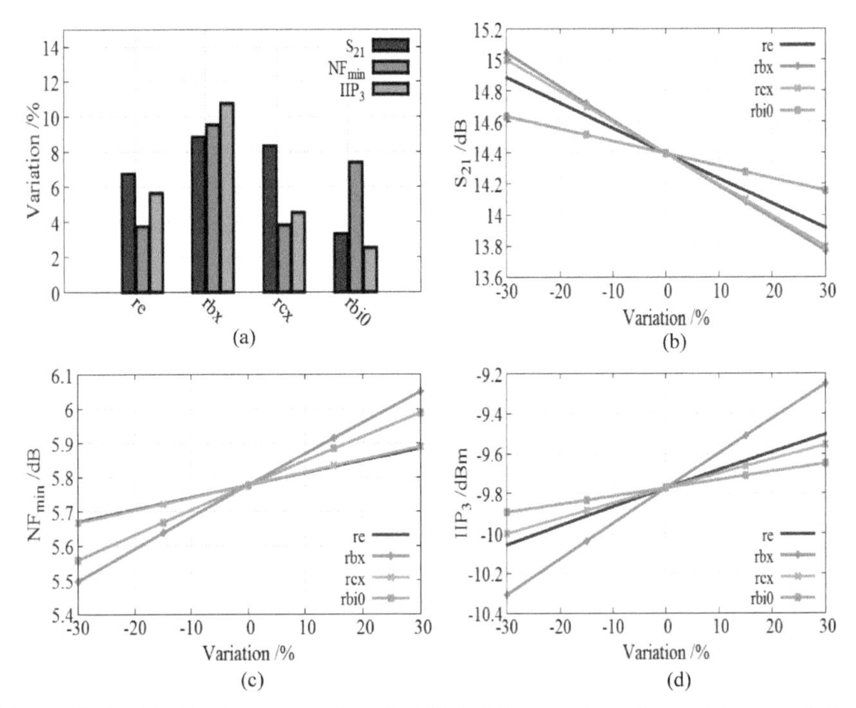

Figure 6.15 (a) Absolute sensitivity of LNA FoMs w.r.t. to series resistance variation. Detailed variation of (b) S_{21}, (c) minimum noise figure, (d) input referred third-order intercept point, all w.r.t. to series resistance variation.

Table 6.2 Performance summary of the W-band LNAs

Reference	Technology	Freq (GHz)	$20*\log_{10}\|S_{21}\|$ (dB)	NF (dB)	IIP_3 (dBm)*	P_{DC} (mW)
[Ceti12]	45 nm CMOS	95	10.7	6	14.6	52
[Vigi16]	28 nm CMOS	90	28	7	–2.7	31.3
[Sev10]	130 nm SiGe	95	9	8.6	–5.3	13
[May10]	120 nm SiGe	95	23	8	N/A	28
[Yang13]	90 nm SiGe	90	19	5.1	–10.4	43
[Ina14]	90 nm SiGe	94	10	4.2	–1.9	8.8
This work	130 nm SiGe	90	14.3	5.44	–9.1	2.79

*The listed IIP_3 results are estimated from the reported input-referred 1 dB compression points.

parameters are the zero-bias internal base resistance r_{Bi0}, the external base resistance r_{Bx}, the emitter resistance r_E, and the external collector resistance r_{Cx}. Regarding the sensitivity of the input-referred IIP3, one would expect r_E to have the largest impact on the linearity of an amplifier due to the series negative feedback introduced by this resistance. A reason for the less-than-expected change of IIP3 may be that the impact of r_E (8.16 Ω) is masked by that of the inductor in series at the emitter (cf. schematic in Figure 6.12), which has an impedance ($\omega*L$) of 8 Ω at 92 GHz. The detailed variations of S_{21}, NF_{min}, and IIP3 with regard to the variations of series resistances are shown in Figures 6.15(b–d). The fact that r_{Bx} has the biggest impact on the FoMs confirms that, at least for the considered process technology, the maximum oscillation frequency is the more relevant standard device FoM.

The performance of this LNA, along with the comparison with other reported LNAs operating around 90 GHz, is summarized in Table 6.2. The results of this work clearly imply the option of operating transistors with very low collector supply voltage (0.5 V) while maintaining a reasonable power gain and NF performance at W-band.

6.1.2.2 W-band low-power frequency tripler

In this section, the design of a W-band low power frequency tripler is introduced. This frequency tripler is also designed in IHP SG13G2 SiGe HBT technology. The schematic of the core part of this frequency tripler is shown in Figure 6.16. The transistors used in the core harmonic generation cells have an emitter size of 0.07 μm × 0.9 μm × 3. The total DC power consumption (including the buffer amplifier) is 4.66 mW with a 0.5 V supply voltage [Lia17].

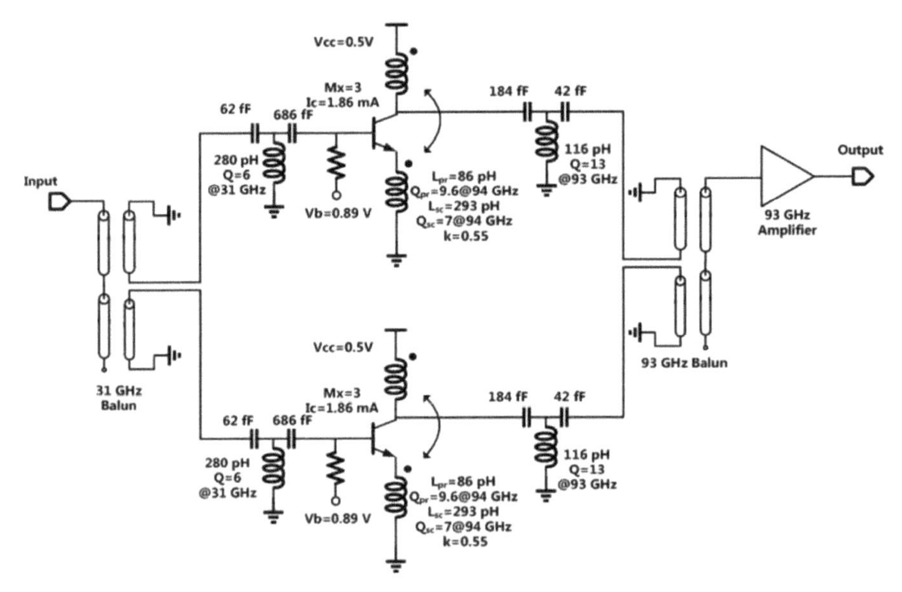

Figure 6.16 Schematic of the W-band low-power frequency tripler.

The topology of the tripler consists of two parts: the harmonic generation part and the output buffer amplifier part. Differential configuration is used in the harmonic generation part to suppress the even-order harmonic signal, with the use of on-chip baluns for single-ended-to-differential conversion. Extensive electromagnetic (EM) simulation was performed during this design as shown in Figure 6.17. The small-signal input/output return loss results are measured from one break-up harmonic generation cell as shown in Figure 6.18(a), which implies that the strongest output signal occurs at around

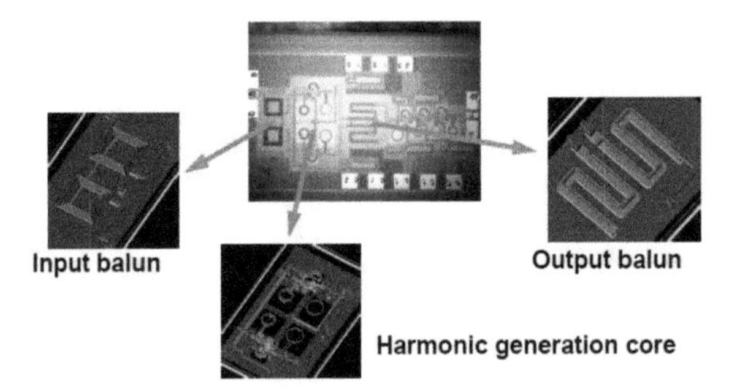

Figure 6.17 Layout and corresponding EM simulation views of the frequency tripler.

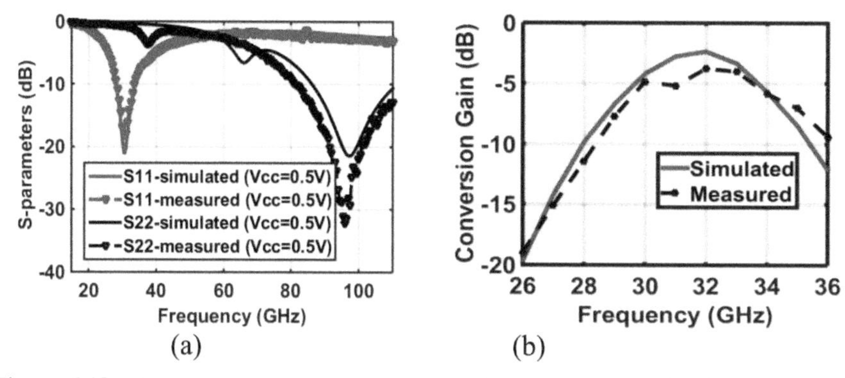

Figure 6.18 (a) Input and output return losses of the core part of the frequency tripler; (b) Conversion gain (actually loss) of the frequency tripler.

96 GHz with an input at around 32 GHz, which is as expected for the correct function of a W-band frequency tripler. The simulated and measured conversion loss results of the frequency tripler are shown in Figure 6.18(b), which shows a minimum conversion loss of 3.79 dB when generating a 96 GHz output signal. These conversion loss results are measured with only –10 dBm input signal over the frequency range of 26–36 GHz.

The performance of this frequency tripler is summarized in Table 6.3 along with the performance of some other reported W-band frequency triplers. The work in [Yeh13] has also demonstrated an ultra-low power frequency tripler using the injection-locking mechanism, but the required input signal power can be as high as 6 dBm, which will impose significant additional power consumption and design effort on the preceding circuit blocks. Looking at Table 6.3, it seems that the design of the frequency tripler in this work has proved the potential for utilizing high-speed SiGe HBT technology biased in the saturation region to implement a competitive

Table 6.3 Performance summary of the W-band frequency triplers

Reference	Technology	Freq (GHz)	Pin (dBm)	Peak Conv. Gain (dB)	Harmonic Rejection (dB)	P_{DC} (mW)
[Chen10]	65 nm CMOS	85–95.2	0	–13.5	>30	19.8
[Wang12]	180 nm SiGe	96	0	–7	>20	75
[Hung10]	150 nm mHEMT	72–114	14.5	–20	–	120
[Vish12]	90 nm CMOS	90–115	8	–2	–	17
[Yeh13]	90 nm CMOS	94	–1	–	>20	3
This work	130 nm SiGe	88.5–103.5	–10	–3.79	>30	4.66

W-band frequency tripler while greatly reducing the DC power consumption. Also note the relaxed and thus cost-efficient process node (130 nm) of the SiGe technology used here.

6.2 Millimeter-wave and Terahertz Systems

U. Pfeiffer, R. Jain, J. Grzyb and P. Hillger

With DOTSEVEN technology it becomes conceivable to realize high-speed circuits operating up to fundamental frequencies of 300 GHz and with utilization of higher harmonics (sub-harmonic operation) even beyond the intrinsic cutoff frequency of the active device. This is the portion of the electromagnetic spectrum, where millimeter-wave and terahertz-systems meet, and where advanced SiGe HBT technologies have a wide-range potential. For instance, the RF bandwidth in communication systems is typically in the order of 10% of the carrier frequency, at 300 GHz, this provides a wide absolute bandwidth of 30 GHz, enabling data-rates in the order of tens of gigabits per second. Similarly, future high-precision radars will profit from the abundant bandwidth at frequencies above 200 GHz and terahertz 3D computed tomography (CT) imagers can be entirely implemented in a silicon process technology. The design, simulation, and performance of this emerging application space are described in the following.

The section "240 GHz SiGe Chipset" describes a 240 GHz SiGe chipset for ultra-high data-rate communication at frequencies above 200 GHz. The high f_{MAX} achieved in IHPs DOTSEVEN technology enabled the design of high-performance fundamentally operated 240 GHz transmitter (Tx) and receiver (Rx) chip-set fully packaged including an on-chip primary antenna coupled to a secondary low-loss hyper-hemispherical silicon lens antenna. A record data-rate of 40 Gbps for QPSK modulation was demonstrated.

The section "210–270 GHz Circularly Polarized Radar" describes a 240 GHz circularly polarized FMCW radar demonstrator in IHPs DOTSEVEN technology. It shows the highest operational bandwidth and range resolution reported for any silicon-based radar system. The proposed circular polarization concept additionally increases the SNR by 6 dB when compared to conventional radar implementations.

The process improvements in Infineon's DOTSEVEN technology made it possible to implement an all-silicon terahertz 3D imager demonstrator presented in the section "0.5 THz Computed Tomography." The main driving motor for this development was to showcase the potential of free-running triple-push oscillator source at around 500 GHz for high-quality absorption measurements of hidden objects. The sources have been used together with custom asymmetric terahertz detectors to build a 3D terahertz CT system. This demonstrator is able to reconstruct 3D volume renders of hidden objects with an optically limited voxel resolution of around 2 mm $\times$ 2 mm. Contrary to previously demonstrated terahertz CT systems that typically use bulky and expensive III–V sources, the demonstrator is comprised solely of hardware fabricated in SiGe HBT technology from the DOTSEVEN project.

6.2.1 240 GHz SiGe Chipset

From an application perspective, the frequency upscaling above 200 GHz comes with a lot of benefits. A higher fractional bandwidth and a finer diffraction-limited spatial resolution both benefit the numerous applications ranging from high-data rate communication, RADAR imaging, and even spectroscopic characterization of the materials. The implementation of such systems requires wideband RF front-end components and wideband on-chip antennas. In this section, we present a generic 240 GHz Tx and Rx chipset which was developed under the DOTSEVEN project. This differential chipset operates in the quadrature mode, and the frequency of 240 GHz refers to the center frequency of the local oscillator (LO) signal, which was designed to be very wideband and tunable to make the chipset useful for a plethora of applications.

The block diagrams for both the Tx and Rx are shown in Figure 6.19 [Sarm16, Sarm16b]. The LO generation network consists of an active balun, a $\times$16 frequency multiplier followed by a three-stage power amplifier (PA), and a differential 90° hybrid. The active balun is used for single-ended to differential conversion of the single-ended low-frequency signal (13.75–17.25 GHz) applied from an external frequency synthesizer which drives the succeeding $\times$16 stage. The $\times$16 frequency multiplier circuit forms the core of the LO generation network and it consists of four cascaded frequency-doubler stages, which are staggered tuned in frequency to increase the operational bandwidth [Sarm13, Sarm14]. The LO signal thus generated is amplified with a three-stage PA and then passed through a passive wideband 90° hybrid coupler to generate the quadrature signal. In the Tx, the quadrature LO signal is mixed with external quadrature IF signal to generate a wideband RF which is boosted in power with a four-stage PA and is then radiated through an on-chip ring antenna into a hyper-hemispherical silicon lens and subsequently to the free space. Similarly, at the Rx, the RF signal travels from the lens-antenna to a three-stage PA and subsequently to IF down-conversion mixers. The quadrature LO generation network at the Rx is similar to that

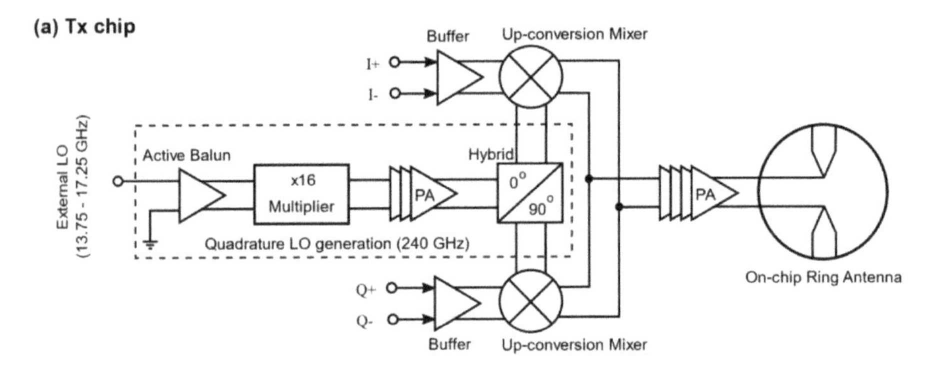

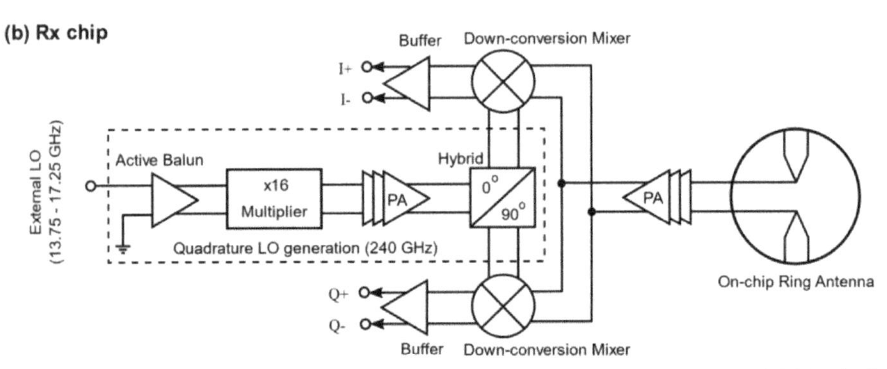

Figure 6.19 Block diagram of 240 GHz quadrature Tx (a) and Rx (b) chipset with identical on-chip ring antenna, after [Sarm16, Sarm16b].

of the Tx, and both Tx and Rx use on-chip 50-Ω differential buffers for the external IF interface [Sarm16b]. The bandwidth is further optimized with a specially designed high-speed printed circuit board (PCB), which will also be discussed later. Each of these circuit blocks is discussed individually in the subsequent sections. We also demonstrate an end-to-end communication system with a measured data rate of 30 Gbps with an EVM of 26% and 50 Gbps with an EVM of 29% for BPSK and QPSK modulation respectively, without applying any channel equalization or error correction techniques [Pedro17, Grz17b].

6.2.1.1 Wideband LO signal generation

Let us start discussing the design details for this chipset, starting with the LO generation circuitry which forms the core for both the Tx and the Rx. In this design, the frequency multiplication technique is used instead of a HF voltage-controlled oscillator (VCO) for LO-signal generation above 200 GHz. This preference was made based on the following reasons:

1. Frequency multipliers offer higher tuning range, higher usable bandwidth, and a flexible phase noise performance compared to the VCOs. At high-frequencies, the overall VCO tuning range is limited by the vector parasitics [Chi13].
2. A wideband tunable LO is also needed to realize a generic chipset which can be used across a spectrum of applications such as high-speed communication, material characterization, imaging, and frequency-modulated continuous wave (FMCW) RADAR [Sarm16, Grz16].
3. The often stated and major drawback of multiplier chains is that they have an overall higher power consumption. However, VCO-based LO sources also need additional frequency dividers, which increase their overall power consumption as well. A free running VCO is otherwise limited to the on–off keying (OOK) modulation with a poor spectral efficiency [Sarm14].

An expanded block diagram of the LO generation network is shown in Figure 6.20. The ×16 multiplier chain is composed of four cascaded doubler stages, each of which is based on the common Gilbert-cell topology where the RF and LO ports are supplied with the same signal for in-phase multiplication to extract the second harmonic. The three-stage PA is composed of pseudo-differential cascode topology which shall be discussed later.

The LO generation circuitry for a generic transceiver chipset must fulfill the following performance requirements:

1. To ensure the generic nature of the chipset, the LO signal must have a large bandwidth. While a system based on fixed or narrowband LO along with wideband mixers can amplifiers is suitable enough for communication applications, other applications such as FMCW radar require a wideband tunable LO source [Grz16].

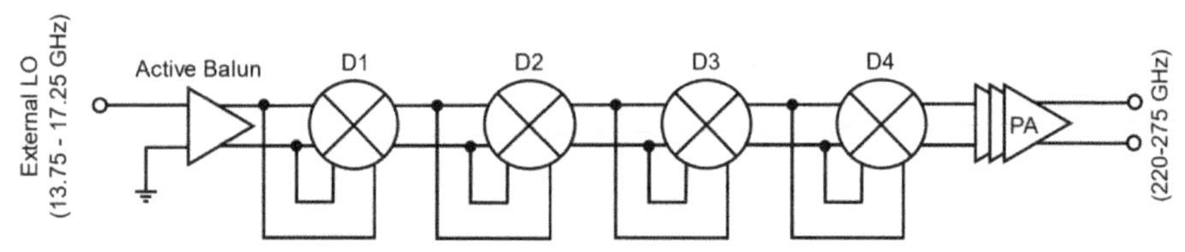

Figure 6.20 Block diagram of LO signal source consisting ×16 frequency multiplication over four cascaded frequency doublers (D1–D4) and a wideband three-stage PA, after [Sarm16].

2. As mentioned above, the power-hungry nature of the multiplier chains is often a major concern. The design must limit the power dissipation of the multiplier chain to as low as possible.
3. As the LO generation is based on harmonic extraction; the spectral purity of the multiplier chain is very important. Any spurious tones from the doubler stages reaching the mixer may start corrupting the IF thereby limiting the IF bandwidth.
4. The mixers need a minimum LO drive power of around 1 mW (0 dBm). Therefore, the generated LO signal power must be high enough to manage this power level along with the additional losses in the passive hybrid coupler.

For these reasons, the LO generation sub-system was designed to generate a power of at least 5 dBm over a 3 dB bandwidth of 40 GHz [Sarm14]. To understand the design further, we need to have a look at the circuit description of each component individually.

(A) x16 frequency multiplier

(i) Gilbert-cell frequency doubler

The circuit schematic for Gilbert-cell based unit doubler stage is shown in Figure 6.21. This topology is chosen due to an inherent differential operation and a high conversion gain (CG) as compared to a conventional class-B bias multiplier topology [Sarm11, Hung05, Oje11]. The capacitance C_{in} couples the differential input signal from the transconductance stage (Q1, Q2) to the switching quad (Q3–Q6).

The inductors L_{b} and L_{c} are part of the input and output matching networks. These are implemented on-chip with shielded microstrip lines in the top most metal layer with lengths l_{b} and l_{c}, respectively. Shielded microstrip lines limit the electric field coupling between different parts of the circuits. The values of the matching network elements are also provided in Figure 6.21. As the input and output of each doubler stage from D1 to D4 progressively shift to higher frequency, the design of each stage is optimized along the following guidelines [Sarm14, Sarm16]:

1. The early stages D1 and D2 operate at lower frequencies and therefore the effect of parasitics is less pronounced. This allows for saving of some chip area by omitting the bias inductor L_{b} entirely in favor of a resistor R_{b}.
2. The transistor stages D1 + D2 are optimized for a high CG, which allows them to operate with a lower LO input power. This minimizes

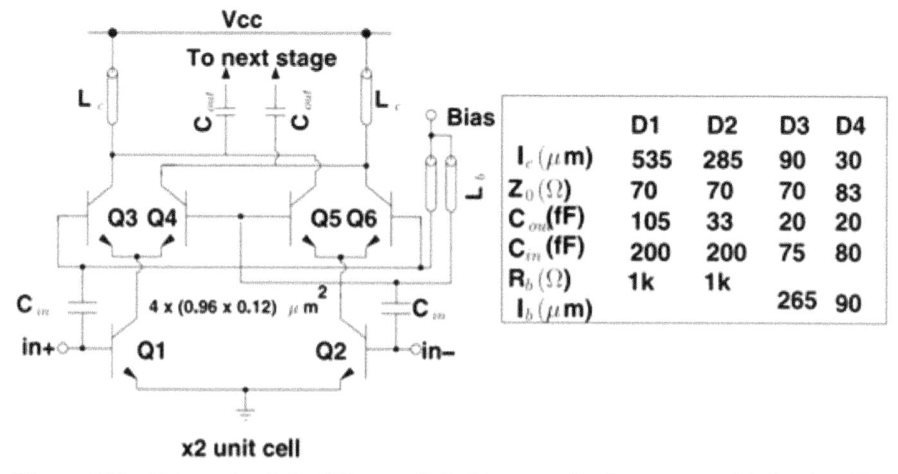

	D1	D2	D3	D4
l_c (μm)	535	285	90	30
Z_0 (Ω)	70	70	70	83
C_{out} (fF)	105	33	20	20
C_{in} (fF)	200	200	75	80
R_b (Ω)	1k	1k		
l_b (μm)			265	90

Figure 6.21 Schematic of the Gilbert-cell doubler stage for frequency multiplication. The table mentions the passive values and the lengths l_b and l_c for microstrip lines used to implement inductors L_b and L_c, respectively. For D1 and D2, the base tuning inductance, after [Sarm16].

the LO leakage associated with the inherent asymmetry of Gilbert-cell based frequency doublers, which may otherwise produce the spurious harmonics at the output of the multiplier chain.

3. The transistor sizing is determined at D4. The maximum transistor size is limited to $4 \times (0.96 \times 0.12)$ μm^2 as any further scaling will require an accurate synthesis of a very small L_C (<10 pH) which is very difficult for an on-chip BEOL environment.
4. For all the doubler stages, the transistor sizes are kept constant (same as D4) to maintain a sufficient interstage drive power. The large transistors benefit the stages D1 and D2 as they lower the inductance L_C required to tune out transistor parasitic capacitance, saving further chip area.

(ii) Interstage matching network

The design of interstage matching network among the stages D1–D4 is very crucial for achieving the desired wideband operation. The matching network must be tuned to the second harmonic of interest from the preceding stage for a doubler operation. Also, higher order even harmonics must be sufficiently attenuated; otherwise, they would exist in the pass-band of subsequent stages.

Another concern is the center frequency alignment between the stages. If the center frequencies of stages D1–D4 are perfectly aligned to consecutive second-order harmonics with similar relative bandwidth, then the overall multiplier chain frequency roll-off becomes much sharper (as in the case of higher order filters) and thus the net bandwidth is reduced. The overall bandwidth BW_{overall} of N cascaded stages is related to the bandwidth BW of single stage as $BW_{\text{overall}} = BW \times \sqrt{2^{1/N} - 1}$ [Ana04]. This implies that for a four-stage network, the overall 3 dB bandwidth corresponds to a mere 0.75 dB bandwidth of the individual stages, or the 3 dB bandwidth of the individual stages equates to the overall 12 dB bandwidth.

One trick to mitigate this limitation is to use a staggered frequency tuning, where the stages are deliberately misaligned for an overall smoother frequency roll-off [Sarm13, Sarm14]. The detailed interstage matching network used in this design is shown in Figure 6.22. The doublers D1 and D3 are tuned higher while D2 and D4 are tuned lower and this resulted in a much smoother roll-off beyond the 3 dB point. As shown in Figure 6.23, the peak CG at the output of D1, D2, D3, and D4 are at the frequencies of 35, 55, 130, and 230 GHz respectively. For D1–D2 and D2–D3, the interstage matching is such that the optimum impedance is transformed at the output of D1 and D2 at the second harmonic of interest. Additionally, it ensures that the impedance is low at the fourth and eighth harmonics for D1 (passband of D3 and D4) and at the fourth harmonic for D2 (passband of D4). For this design, the D3 output and the D4 input were matched to a 100-Ω differential impedance for the ease of breakout characterization and interfacing with other circuits. The simulated (large signal) output impedance at the output of each doubler considering the loading of the succeeding stages is shown in Figure 6.24. The low impedance at the undesired harmonics ensures sufficient harmonic rejection. The stagger

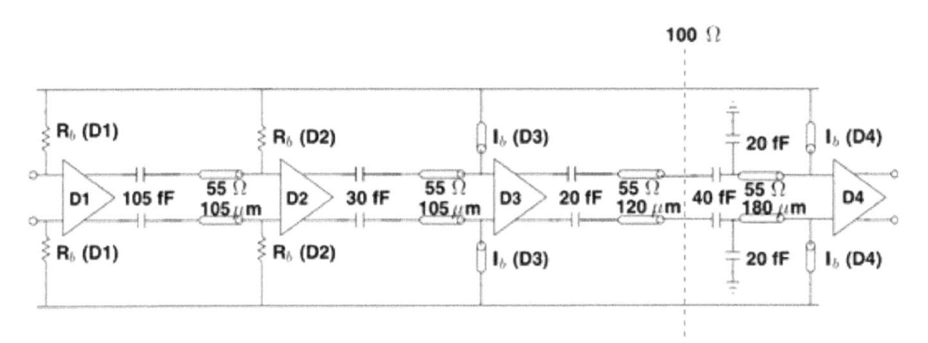

Figure 6.22 Interstage matching between the doubler stages of the multiplier chain for LO generation. The tuning inductance L_c connected to the collector output is not shown, after [Sarm16]. Other parameter values are mentioned in the table in Figure 6.21.

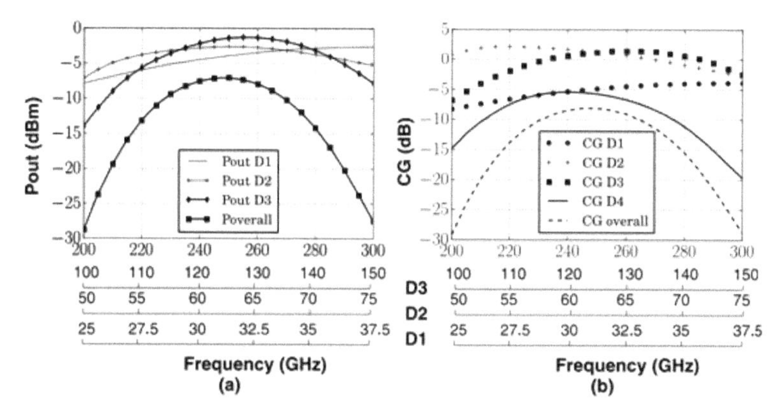

Figure 6.23 Simulated: (a) output power and (b) CG of the individual doubler stages. The doublers D1 and D3 are tuned higher, while D2 and D4 are tuned lower, and this resulted in an overall flat response, after [Sarm16].

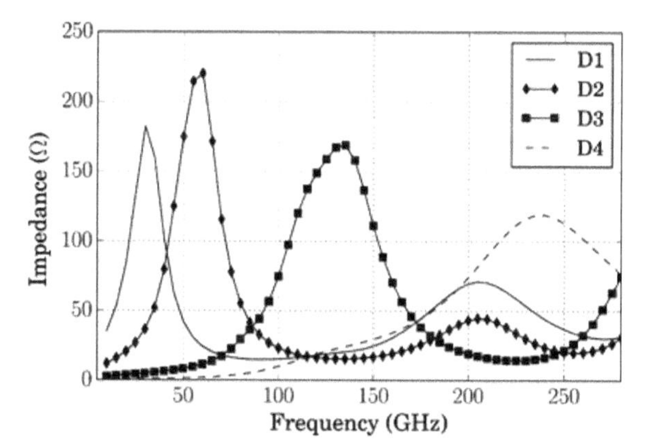

Figure 6.24 Simulated impedance at the collector outputs of D1–D4 derived from the large signal S-parameter simulations, after [Sarm16].

frequency tuning between the stages resulted in an overall simulated 3 dB bandwidth of 50 GHz (210–260 GHz) with a peak output power of –2.8 dBm at 240 GHz. The overall multiplier chain along with the active balun at the input consumes about 720 mW of power [Sarm16].

(B) Power amplifier (PA)

The $\times 16$ frequency multiplier is cascaded with a three-stage PA. A detailed schematic of the single stage of this PA is shown in Figure 6.25. The circuit architecture is based on pseudo-differential cascode amplifier. The cascode

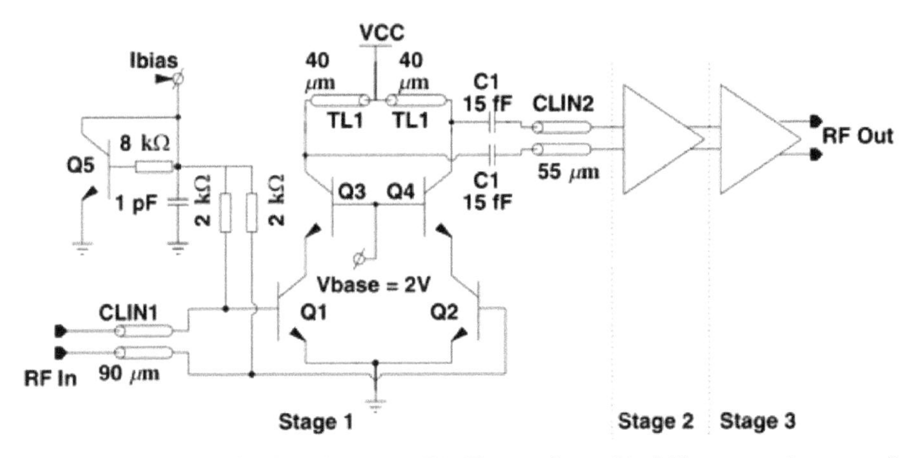

Figure 6.25 Schematic of the three-stage PA. The transistors Q1–Q4 have an emitter area of $8 \times (0.96 \times 0.12)\ \mu m^2$, after [Sarm16].

topology is a popular choice in high-frequency amplifier design. The use of common-base stage as a load to the common-emitter transistor in a cascode reduces the Miller capacitance, therefore improving the reverse isolation, stability and the ease of impedance matching. Also, the voltage swing for the PA is limited by the base–collector breakdown voltage (BV_{CBO}), which is larger than the collector–emitter breakdown voltage (BV_{CEO}) in a common-emitter configuration. A higher voltage swing allows for a higher saturated output P_{sat} from the PA [Kerh15].

A general design outline for the PA is provided in [Sarm13b]. The differential configuration leads to a virtual ground at the base of the common-base stage of the cascode amplifier, which enables efficient and compact on-chip layout by relaxing the need for extensive on-chip decoupling capacitors. However, while a differential topology is expected to have a good common mode rejection, the design of a true differential amplifier requires an active tail current source. At frequencies reaching 200 GHz, the impedance of such current source becomes very low rendering it ineffective. Inductor-based current sources are also challenging as their low self-resonance frequency (SRF) limits the maximum synthesizable inductance, and the quarter wave transmission line-based inductors are inherently narrow band. Therefore, in this design, a pseudo-differential architecture is used where the common-emitter terminal is grounded and the base biasing is provided through the current mirrors. It becomes very challenging to implement the switching PAs at frequencies extending beyond 100 GHz due to the transistor parasitics

[Song15]. In this design, a low power gain of the device at the high operating frequency (beyond $f_{\mathrm{MAX}}/2$) necessitates the use of class-A biasing for the PA.

In Figure 6.25, the emitter area of each of the transistors Q1–Q4 is $8 \times (0.96 \times 0.12)\ \mu\mathrm{m}^2$. The emitter area and subsequently the power gain is therefore limited by the device parasitics. For larger device size, the required matched tuning inductor becomes too small (less than 10 pH) for on-chip implementation. The microstrip line-based inductor TL1, capacitor C1, and the coupled microstrip line CLIN2 are part of the output match, while CLIN1 is part of the input match [Sarm16]. A decoupling capacitor of 100 fF is used at the common base of transistors Q3–Q4 (not shown here) [Sarm16c].

To maximize the power, the optimum load impedance at the collector node (R_{opt}) must ensure a simultaneous maximization of the voltage and current swing. This can be derived from the loadline analysis and depends on the breakdown voltage and the maximum allowable current density [Ref]. The output resistance of the cascode R_{o} should also be as high as possible to maximize the power delivered to the load [Ref]. However, due to the internal device parasitics, even when the reactance at the output node is tuned out, the output resistance shows a sharp reduction with frequency ($R_o \propto 1/f^2$) as shown in [Sarm13b]. In this case the loadline impedance match becomes inefficient and therefore instead a conjugate matching is used in this design.

For the multistage PA design, the device sizing is generally scaled from input to the output stages for handling progressively increasing RF power levels. However, this also requires a modified interstage matching network for each subsequent stage. In this design, the transistor sizes for all the stages are kept identical, which provides the flexibility to cascade multiple stages based on the gain requirements without a need for altering the interstage matching network. Identical stages also reduce the probability of frequency misalignment between different stages. Note that a four-stage variant of this PA is used in the Tx after the up-conversion mixer to increase the transmit power. For the interstage matching, the capacitor-coupled LC resonator technique is used [Shek06, Ana04, Nel32]. Here, the coupling capacitor C1 between the stages introduces an additional zero in the passband which improves the overall bandwidth.

The multiplier chain with the three-stage PA was characterized separately in a breakout structure using WR03 220–325 GHz ground-signal-ground (GSG) waveguide probes along with an on-chip wideband (210–280 GHz) Marchand Balun at the output. A DC-40 GHz GDG probe was used to provide a low-frequency (<20 GHz) input signal using an external frequency synthesizer. An Erickson Calorimeter equipped with a WR03 waveguide taper was used for the absolute output power measurement for two different input LO power levels (–10 dBm and 0 dBm), and the results are shown in Figure 6.26 [Sarm14, Sarm16c]. For a –10 dBm input LO power, the peak output power is 6.4 dBm at 230 GHz, and the 3 dB RF bandwidth is 50 GHz (215–265 GHz). The output power remains constant for an input LO power of 0 dBm below the input frequency of 13.75 GHz due to some spurious harmonic generation. The LO generation network also shows a phase noise degradation of around 25 dB and consumes about 0.74 W of DC power.

(C) On-chip wideband quadrature coupler

For the quadrature operation, the LO form the PA is provided to a 3 dB 90° coupler. A simplified geometry of the on-chip quadrature coupler is shown in Figure 6.27. It is implemented using three buried metal layers. The coupler exploits a combination of broadside coupling between the strip conductors located on different metallization layers and edge coupling between adjacent strip conductors located on the same layers. The design is sized to minimize the propagation loss and equalization of the propagation speed for all propagating modes to ensure maximum operation bandwidth. The coupler operates along with differential 100-Ω grounded coplanar stripline feeds implemented on a thick top metal layer. The coupler was originally designed

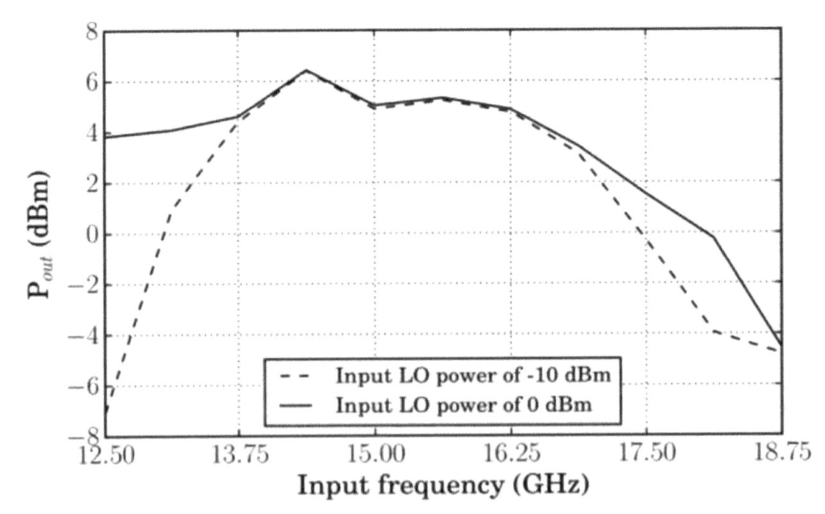

Figure 6.26 On-chip power measurements for the standalone LO generation source with an Erickson calorimeter for input LO power of –10 dBm and 0 dBm, after [Sarm16c].

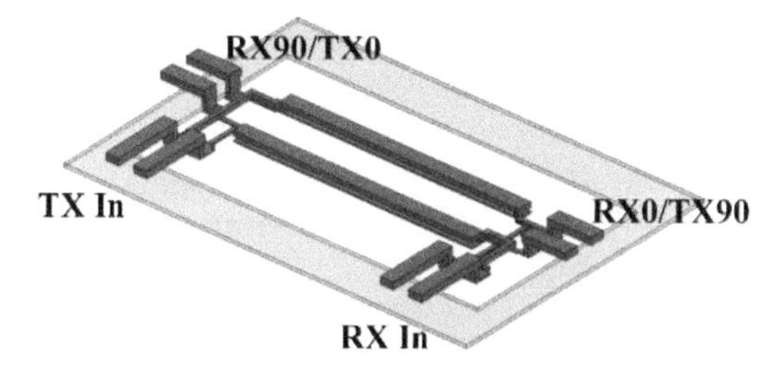

Figure 6.27 Simplified metal-level multi-layer geometry for the differential quadrature coupler, after [Grz15].

for a duplex FMCW radar chipset [Grz15], and unlike the conventional couplers, the required circuit connections here favor the placement of Through and Coupled ports on the same side. Therefore, the EM structure is carefully optimized to minimize the layout asymmetry.

When employed in the transceiver chipset, one of the input ports of the coupler is terminated with a matched 100-Ω differential load impedance. This does not result in any additional losses, as the isolation between the input ports is more than –23 dB. The input return loss is less than –26 dB for a wide 160–340 GHz bandwidth (Figure 6.28). The simulated phase and amplitude imbalance between the quadrature output ports within this frequency band are better than 3° and 1.6 dB respectively.

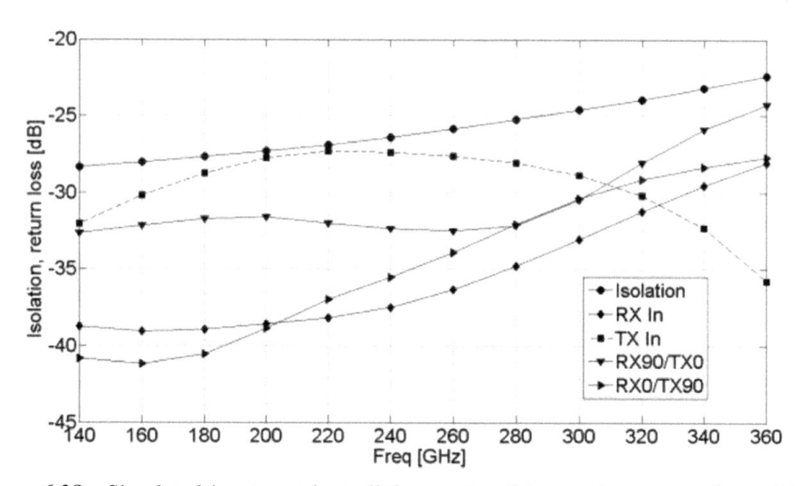

Figure 6.28 Simulated input match at all four ports of the quadrature coupler and isolation between the input ports for a differential excitation. All ports are referred to a 100-Ω differential impedance, after [Grz15].

6.2.1.2 Transmitter building blocks

Other than the common LO generation network, an up-conversion mixer at the Tx converts an external IF signal into the RF signal which is then boosted with a four-stage PA before radiating through the antenna. The relevant design criteria for the Tx thus are wideband RF and IF operation, as well as sufficient output power.

(A) Up-conversion mixer

The up-conversion mixer is based on the double-balanced Gilbert-cell topology due to its inherent differential operation, LO rejection, and high CG [Voin13]. In the circuit shown in Figure 6.29, the switching quads (Q1–Q4, Q7–Q10) are driven by the quadrature LO signal while the transconductance stages (Q5–Q6, Q11–Q12) are driven by the quadrature IF signal. The center taps for both inductors L_b and L_c are used for the DC biasing, where L_b forms the part of LO matching network. Both the inductors L_b and L_c are implemented in TM1 (second from the top) metal layer as microstrip transmission lines [Sarm16c].

Since the LO drive power is limited, minimum-sized transistors were used at the switching quad stages to allow for a stronger switching and to reduce the parasitic capacitances. Here, the transistors with emitter areas $2 \times (0.96 \times 0.12)\ \mu m^2$ and $1 \times (0.96 \times 0.12)\ \mu m^2$ were used for the

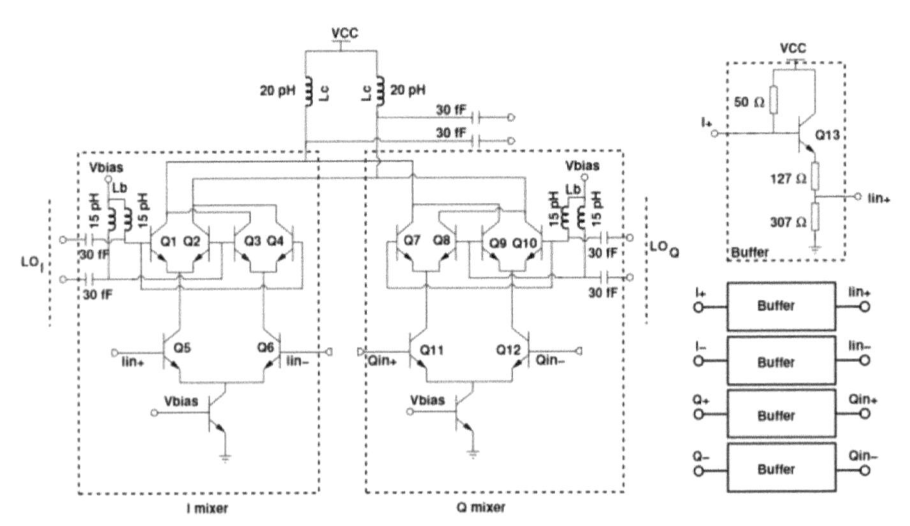

Figure 6.29 Up-conversion mixer schematic with additional buffer stages for wideband 50-Ω IF matching, after [Sarm16b].

transconductance stages and the switching quads, respectively. Also, buffer stages with a shunt resistance of 50 Ω were added for a wideband match to the external IF. For these buffer amplifiers, the transistors with an emitter area of $4 \times (0.96 \times 0.12)$ μm^2 are used for linearity reasons.

Note that the mixers are not characterized as separate breakouts, and it requires high-power LO above 200 GHz from external signal sources. To simulate this mixer, a 25 MHz, –5 dBm signal was applied at one of then IF channels. For a 240 GHz LO with 5 dBm power at the input of the 90° coupler, the simulated peak CG is 0.7 dB, P_{sat} is –5 dBm, and OP_{1dB} is –8 dBm. The simulated 3 dB IF and RF bandwidths are 38 GHz and 50 GHz, respectively [Sarm16c].

(B) Four-stage PA

The four-stage PA extends on the three-stage PA used in the LO generation network, with one identical additional stage. This PA was characterized as a separate breakout. At 230 GHz, the measured peak small-signal gain is 26 dB, 3 dB RF bandwidth is 28 GHz, and $S_{11} \leq -10$ dB between 215 and 255 GHz. For large signal measurements at 240 GHz, the PA provides a gain of 12.5 dB at compression, and the 1 dB compression points for input and output are –16.5 dBm and 3.7 dBm, respectively. Also, the measured $P_{sat} > 6$ dBm for the 220–260 GHz frequency range. Note that since the drive power of Tx PA is large, the large signal bandwidth is more applicable to communication links, and this is usually larger than the small signal bandwidth. The measured peak power-added efficiency (PAE) at 240 GHz is 1% [Sarm14].

6.2.1.3 Receiver building blocks

Noise is the primary concern for the Rx sensitivity. Therefore, along with a wideband IF and RF, a low NF at the Rx is very much desirable. At the Rx, the three-stage variant of the PA is used as a preamplifier. The center frequencies of both the three-stage and four-stage PAs are similar, and therefore the probability of frequency misalignment between the Tx and Rx becomes very low [Sarm16b]. Also, the small signal bandwidth of the preamplifier decides the Rx bandwidth when it is driven with a low power signal (as in the case of a communication system with Tx and Rx separated over a wide distance resulting in a large path loss). A three-stage PA shows a larger small signal bandwidth as compared to the four-stage PA and therefore it is preferred at the Rx, ensuring a wideband operation.

(A) Down-conversion mixer

The down-conversion mixer at the Rx end, also implemented with a double-balanced Gilbert-cell topology, is shown in Figure 6.30. The RF current at the transconductance stage (Q9–Q10) is fed by the input RF signal through the antenna and the pre-amplification PA, and this is shared between the I and Q switching quads (Q1–Q4 and Q5–Q8, respectively), which in turn are supplied with the quadrature LO signal from the on-chip wideband 90° coupler. Common-emitter buffers with a series resistance of 50 Ω are added at the IF outputs for a wideband match. The transistor sizes for the mixer and the buffers are identical to those of the up-conversion mixer at the Tx. The choice of load resistor R_C determines the trade-off between the CG and the RC time constant-limited IF bandwidth at the collector output. The load resistor of 200 Ω used in this design corresponds to a simulated 3 dB IF bandwidth of 35 GHz and a peak CG of –0.5 dB. For a 33 MHz IF and 240 GHz, 5 dBm LO at the input of the quadrature coupler, the simulation predicts a 3 dB RF bandwidth of 52 GHz and a minimum NF of 14.2 dB. Also, both the simulated minimum LO power and IP_{1dB} are around –2.5 dBm [Sarm16b, Sarm16c].

6.2.1.4 Antenna design

The linearly polarized on-chip antenna in the Tx and the Rx chipset is topologically similar to the differential wire ring topology [Grz12]. It consists of

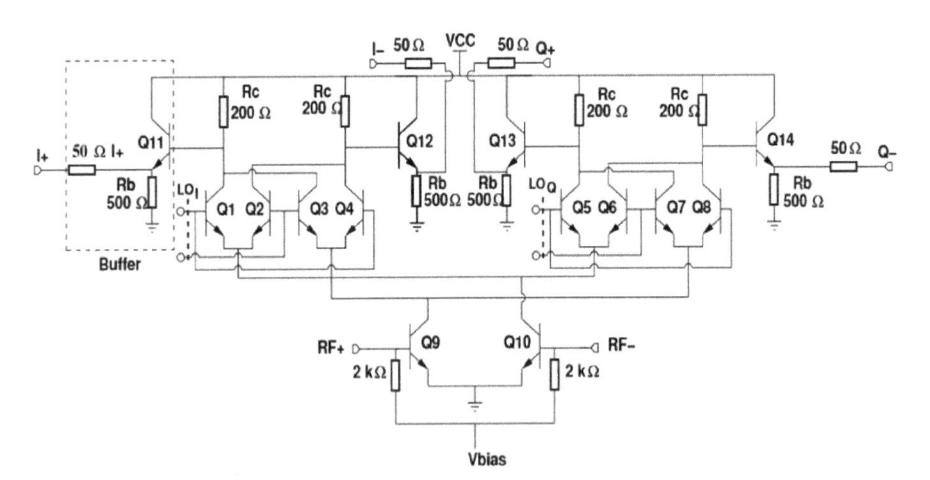

Figure 6.30 Schematic for the down-conversion mixer. Here, additional buffer stages were added to have 50-Ω input impedance required for wideband IF matching, after [Sarm16b].

two wire semi-rings connected along the center feed. For wideband operation, the feed is non-uniformly tapered using step-wise approximation [Grz17]. It is designed to illuminate a 9 mm-diameter silicon hyper-hemispherical lens through the chip backside. The lens reduces the influence of surface waves on the radiation efficiency and radiation patterns and inherently delivers a high gain to compensate for the high free-space propagation loss. The backside radiation offers significant advantages over the front-side radiation. The bandwidth is no longer limited by the distance of the ground plane (few μm) as in the case of front-side radiation. The form-factor reduction is by a factor of 3.3 ($\sqrt{11}$ for silicon), which is 39% less than in the case of front-side radiation ($\sqrt{3.9}$ for silicon-dioxide). Moreover, the ability to mount external silicon lens of different sizes gives the ability to have flexible application specific directivity. The lens extension is chosen to be close to the elliptical position with extension to radius ratio of 34.4%. The antenna provides a differential impedance of 100 Ω over a very wide bandwidth (S_{11} < –20 dB over 180–330 GHz) [Sarm16]. The simulated cross-polarization is below 20 dB for differential operation. By providing a low impedance (4–5 Ω) for the common mode, radiation from the parasitic common-mode signal is minimized. It is also optimized for the minimization of mode conversion (differential to the common mode) and the simulated mode conversion is below 40 dB. The overall directivity of the antenna with the lens is 26.4 dBi at 240 GHz.

6.2.1.5 Packaging and high-speed PCB design

The chip-on-board (COB) technology is used for both Tx and Rx packaging. The entire chip-on-lens assembly is accommodated inside a recess on a PCB, and chip pads are connected to the PCB bond pads through the wirebond process. A heat sink with direct thermal contact to the silicon lens is also added to improve the heat dissipation away from the chip (Figure 6.33).

The high-frequency IF signal is the major concern while designing the PCB. The PCB material should have a low dispersion and low dielectric loss. Therefore, materials with low relative permittivity and low loss tangent are favored. Here, the ROGERS 4350B material from Rogers Corporation is used. This material is designated for high-frequency applications and it shows a permittivity ε_R and a loss tangent $\tan\delta$ of 3.66 and 0.0037, respectively over DC-20 GHz frequency range. The choice of PCB thickness is a trade-off between the mechanical stability and the maximum allowed line width for the lowest impedance microstrip lines. Since the differential and quadrature IF routing lines must be highly symmetrical, very large trace widths cannot be tolerated within a reasonable PCB area. Here, a PCB thickness of 0.388 mm is found to be optimum, with a 50-Ω microstrip line corresponding to a 0.718 mm width.

The wirebond connecting the chip to the PCB also limits the IF bandwidth. Therefore, a phase linear wideband matching filter needs to be implemented on the PCB. While the Bessel filters show the most linear phase response, the feasible component values limit the choice to maximally flat or Butterworth filter topology which still provides a more linear phase response as compared to the Chebyshev filters. For this purpose, an 8-section lumped parameter LC filter is synthesized using Richard transformation on an iterative basis [Pozar09], which takes into account the wire bond inductance (Figure 6.31) [Sarm16b].

Figure 6.32 shows the results from the full EM simulation of the filter. For this, the PCB and the chip ground pads are included to accurately model the ground return current. This simulation predicts an insertion loss (S_{21}) of −0.5 dB in the passband with a 3 dB bandwidth of 15 GHz. The input return loss (S_{11}) is less than −10 dB for up to 14 GHz and the group delay variation is less than 10% up to 9 GHz.

6.2.1.6 Tx and Rx characterization

The chip-micrograph of the Tx and Rx chipset with the on-chip antenna is shown in Figure 6.34 [Sarm16b]. For on-wafer characterization, the Tx

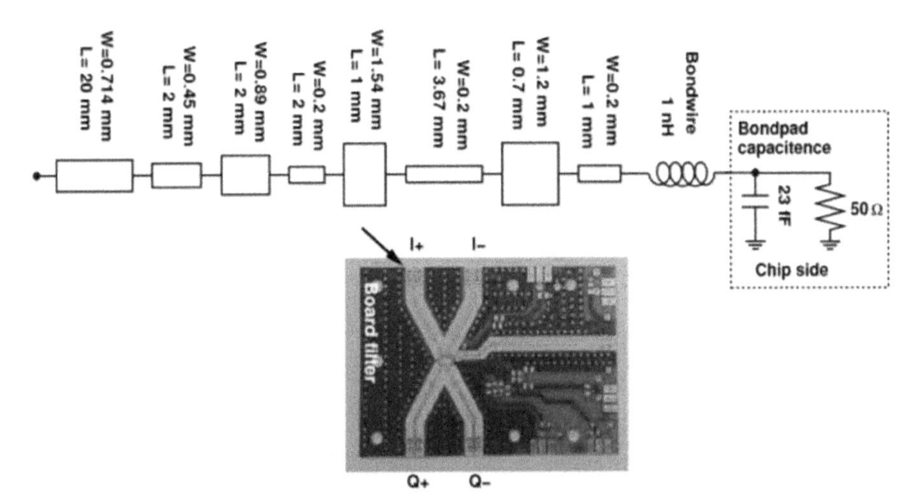

Figure 6.31 Schematic of the low-pass filter implemented on a ROGERS 4350B PCB material with a thickness of 0.338 mm. The stepped impedance filter low-pass filter is implemented with microstrip lines on the PCB, after [Sarm16b].

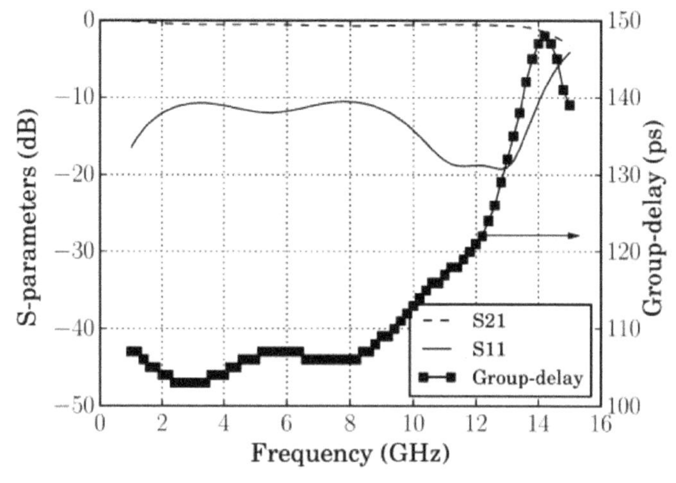

Figure 6.32 The full-EM simulation results from the filter, after [Sarm16c].

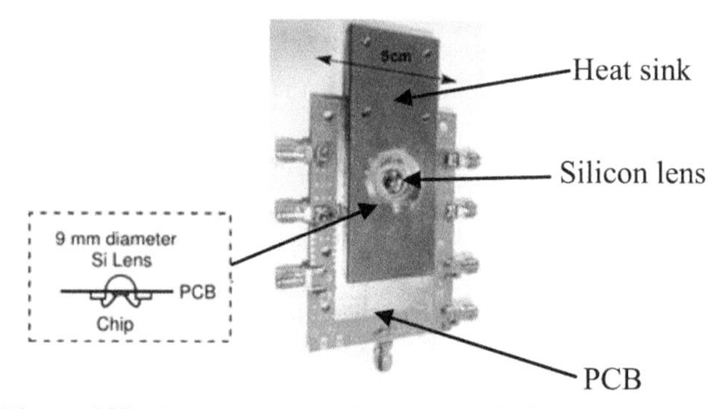

Figure 6.33 Lens mounted and packaged chip for Tx/Rx module.

and Rx have an auxiliary balun instead of an on-chip antenna, and are not shown here. A WR03 GSG waveguide probe for the 220–325 GHz band is used to measure the RF output from the Tx and to supply RF to the Rx. Using the Short-Open-Load (SOL) calibration, the loss due to the probe and the waveguide is estimated to be 7.5 dB in this band. A GSG probe (DC-40 GHz) is used to couple the low-frequency LO signal with −10 dBm output power to the chip. A 25 MHz signal from a function generator is along with external 90° and 180° hybrids are used for the differential quadrature IF signal generation at the Tx side, and the output RF power is measured using an Erickson calorimeter. For the Rx characterization, a WR03 VNA extension module is used in the transmit mode as an RF source, and its output power

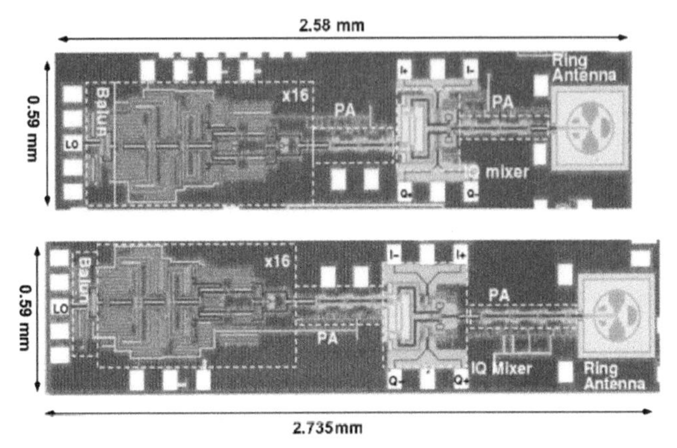

Figure 6.34 Chip-micrograph of the Tx and Rx chipset. The total chip area including the pads is (a) Tx: 1.613 mm^2 (b) Rx: 1.522 mm^2. For the on-wafer measurements, an auxiliary balun with an estimated 2.5 dB loss has been added at the output, after [Sarm16b].

is calibrated using the Erickson calorimeter. The down-converted IF power at 33 MHz is measured with the spectrum analyzer.

The on-wafer characterization results are shown in Figure 6.35 [Sarm16b]. The Tx can deliver up to 6 dBm output power at 240 GHz and the 3 dB bandwidth is 40 GHz. The peak CG for the Rx is 11 dB and the NF is 15 dB while the 3 dB RF bandwidth is 28 GHz. The NF is calculated using the direct method [Oje12] under the assumption that the input noise floor is −174 dBm/Hz (thermal noise at the room temperature).

For the IF bandwidth characterization, the packaged Tx and Rx (with on-chip antenna and the lens) are placed back to back with a distance of 90 cm. The IF inputs of the Tx and the Rx are connected to a VNA and swept IF

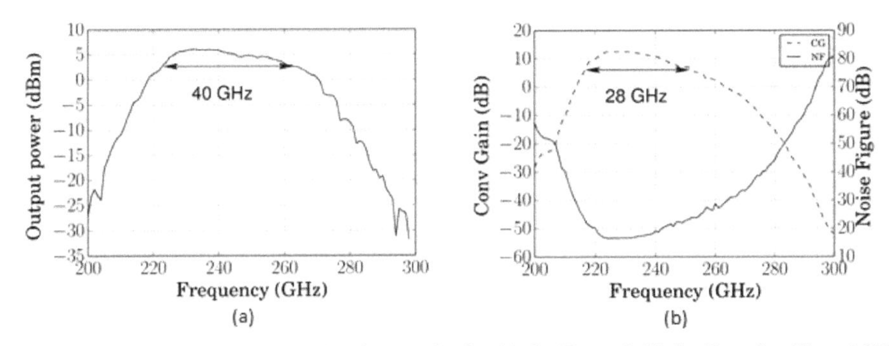

Figure 6.35 On-chip characterization results for (a) the Tx, and (b) the Rx, after [Sarm16b].

measurements are done for a fixed LO input of 240 GHz. The results indicate a 6 dB IF bandwidth of 13 GHz as shown in Figure 6.36 [Sarm16b].

6.2.1.7 Ultra-high data rate wireless communication

The measurement setup for ultra-high data rate wireless communication system is shown in Figure 6.37 [Pedro17, Grz17b]. The fully integrated and packaged Tx and Rx modules were separated by a link distance of 1 m. The differential inputs of the Tx are connected to an arbitrary waveform generator (Tektronix AWG70001A) and the IF outputs of the Rx are connected to a real-time oscilloscope (Tektronix DPO77002SX) using phase-matched cables.

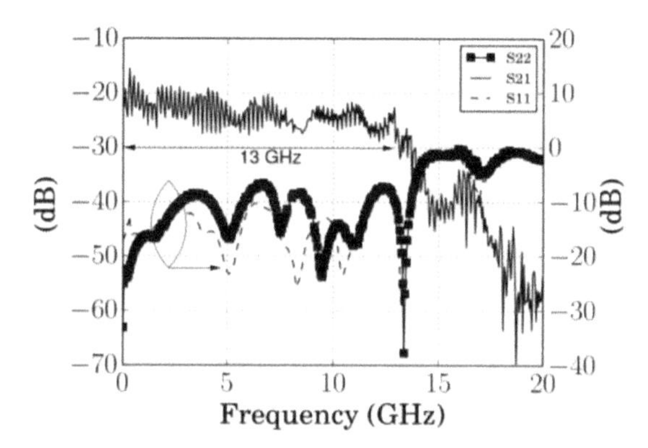

Figure 6.36 Measurement results from the IF bandwidth characterization over a link distance of 90 cm. For this measurement, the LO is fixed at 240 GHz and the measured 6 dB IF bandwidth is 13 GHz, after [Sarm16b].

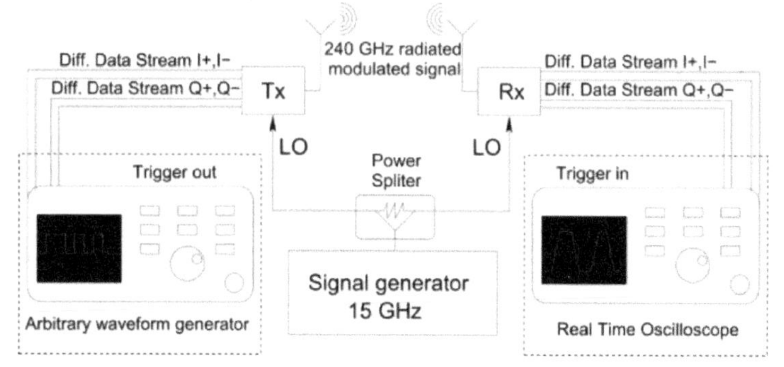

Figure 6.37 Measurement setup for the high data rate wireless communication with arbitrary waveform generator (Tektronix AWG70001A) and real-time oscilloscope (Tektronix DPO77002SX), after [Pedro17].

The external LO inputs for both Tx and Rx are driven by the same 15 GHz signal from an external synthesizer with a power splitter.

With no channel equalization applied, maximum transmission speeds of 30 Gbps with an EVM of 26% and 50 Gbps with an EVM of 29% were demonstrated for BPSK and QPSK, respectively, using PRBS9 binary sequence. To increase the reliability of the link with a smaller EVM, a second test was performed for reduced transmission speeds, resulting in an EVM of 11% for 25 Gbps and an EVM of 22% for 40 Gbps for BPSK and QPSK, respectively (Figure 6.38). The limitations in the board bandwidth as well as in the Rx RF/LO bandwidth influence the achievable EVM for the tested modulation speeds.

6.2.2 210–270 GHz Circularly Polarized Radar

Contrary to other imaging techniques, high-resolution radar-based imagers are capable of providing significant improvements in the imaging quality thanks to their range-gating capabilities [Coop08, Lian14, Dick04, Quas09, Graj15]. Similar to general-purpose transceivers operating beyond 200 GHz, they feature low integration and are thus not commonly used because they become expensive and space-inefficient [Coop08, Esse08, Bryl13]. Considering the increasing popularity of radar sensors in various high-volume consumer and industrial markets such as health care [Li13], autonomous navigation in robotic platforms [Chen08, Moal14], non-destructive testing [Karp05], and automotive systems [Maur11], the implementation costs with low weight and small form-factor are more and more relevant. By suitable

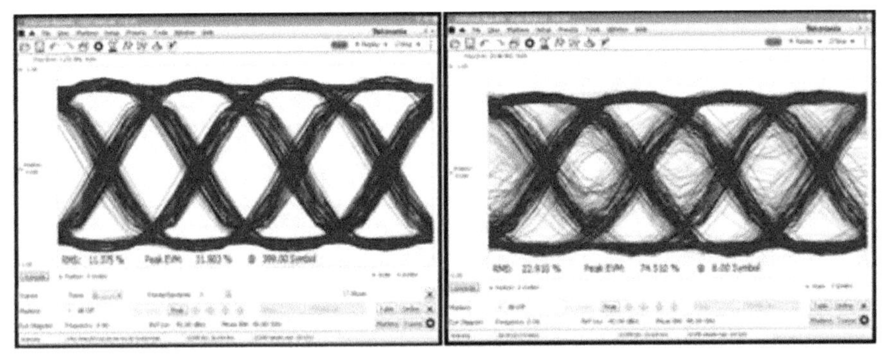

Figure 6.38 Measured eye diagrams for: 25 Gbps BPSK modulation (left); 40 Gbps QPSK modulation (right), after [Pedro17].

combination of microelectronic packaging with silicon technologies, high-integration levels of the complete radars become a reality and will develop in the future into the solution of choice for such sensors. Currently, most of Si-integrated radars are operated below 100 GHz [Shen12, Maur11] in view of the technology limitations.

Within the frame of DOTSEVEN project, a complete highly integrated FMCW homodyne monostatic radar system operating around 240 GHz with a 60 GHz bandwidth and a state-of-the-art 2.57 mm-range resolution was developed. Its RF front-end is implemented in the form of a single chip in a 0.13-μm SiGe HBT technology with f_T/f_{MAX} of 300/450 GHz from IHP. To facilitate a low-cost packaging scheme, the chip further includes a wideband lens-integrated on-chip annular-slot antenna [Grz15, Grzyb15] and is wire-bonded onto a low-cost FR4 printed-circuit board. Despite the expected lower sensitivity [Graj15] of the homodyne monostatic architecture, this radar topology was selected for implementation due to low costs of the accompanying baseband chain and the highest possible integration level on a single chip. As opposed to classical linearly polarized monostatic radar front-ends [Jahn12, Jaes13, Jaes14], this radar employs circular polarization [Kim05, Statn15] for multiple reasons. The first reason is the absence of on-chip circulators separating Tx and Rx paths. This issue is typically solved by using equivalent quasi-circulators made of on-chip directional hybrid couplers such as rat-race [Jahn12]. Such a solution suffers from an excessive 6 dB loss in SNR because of some additional power loss in the terminating loads [Jahn12, Kim05]. As shown in [Statn15], this loss can be gained back by means of a circularly polarized architecture. Circular polarization may further increase detection probability in the presence of wave depolarization [Moal14, Nash16] or reduce the influence of Rx jamming while operating multiple radar sensors simultaneously [Lian07] and of ghost targets in indoor environments [Moal14].

Figure 6.39 presents the radar chip micrograph and its block diagram. Circular polarization is provided by a broadband annular-slot antenna supporting two orthogonal polarizations [Grz15, Grzyb15] driven from a wideband quadrature coupler [Grz15]. The transceiver is implemented in a fully differential configuration. To achieve the fundamental radar operation in a wide frequency range of 210–270 GHz, the LO-generation path is realized with the ×16 multiplier-chain architecture because of the missing appropriate tuning varactors devices. Both Tx and Rx paths share the same LO-generation chain which is driven around 13.1–16.9 GHz at a power level of around 0 dBm.

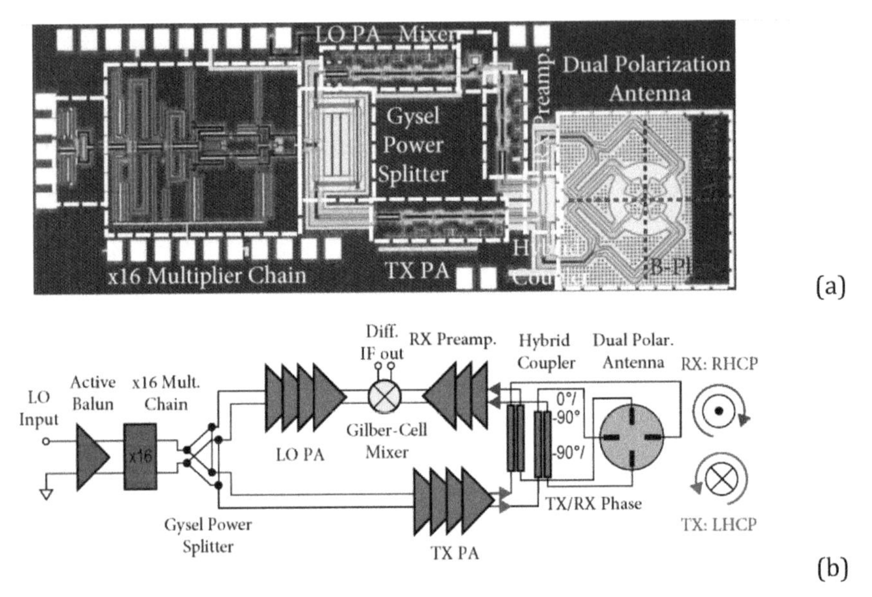

Figure 6.39 Radar transceiver chip implemented in 0.13-μm SiGe HBT technology and operating at 210–270 GHz: (a) chip micrograph, (b) block diagram; after [Stat15]. The chip size is 2.9 mm × 1.1 mm.

The LO drive is provided from the printed circuit-board level as a single-ended signal which is then converted to a differential topology by means of an active balun in front of the multiplier chain. The ×16 multiplication factor was selected in view of the limited RF performance of the regular mm-long wire-bonded interconnects. The multiplier chain comprises four cascaded Gilbert-cell frequency doublers [Sarm16] which inherently provide differential operation. The output signal from the LO-path is equally split by a novel differential Gysel power divider to drive both Tx and Rx paths. Compared to the Wilkinson divider, the chosen Gysel power-splitter is capable of providing an improved isolation between its two output ports at the operation frequency. Each of two outputs drives a four-stage power amplifier with a small-signal gain of 14 dB and a P_{sat} of around 7 dBm [Sarm16]. One of the amplifiers is connected directly to the Tx port of the circularly polarized antenna whereas the other drives the down-conversion mixer. Due to the similar impedance range at the inputs of both amplifiers, the power splitter imbalance can be minimized. Furthermore, the power amplifiers are useful in providing an improved TX-to-RX isolation from the LO-chain side. The down-converting mixer is operated fundamentally and implemented as a double-balanced Gilbert-cell topology. In order to minimize the influence of excessive mixer noise on the Rx NF, the receive signal from the antenna output port is pre-amplified with a three-stage PA. Here, the power amplifier was used instead of a regular LNA [Statn15] to maximize both the radar operation bandwidth and the linearity with similar noise performance metrics to that achievable with a silicon-integrated LNA at the operation frequency [Statn15]. Please note that the Rx linearity is crucial for the radar operation because of its monostatic architecture suffering from the TX-to-RX leakage. From previous measurements of the similar Rx paths [Sarm16], an input-referred 1 dB Rx compression point of around –9 dBm sets a reference value for finding the minimum required antenna input match to avoid Rx compression. More advanced adaptive leakage power cancellation techniques [Brook05, Beas90] can be considered in the future to improve the radar performance.

For free-space operation, the transceiver chip is mounted on the back of a high-resistivity hyper-hemispherical silicon lens with the primary on-chip feed antenna aligned with the lens optical axis. Then, the entire chip-on-lens assembly is in turn mounted in a recess of a regular FR-4 PCB surrounded by a large metal plane, as shown in Figure 6.40. The lens volume is crucial for thermal control for the chip dissipating around 1.6 W. Here, the heat is

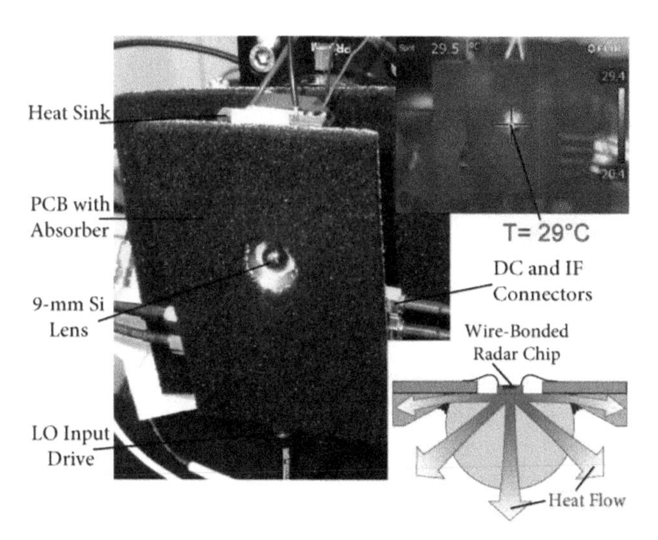

Figure 6.40 Complete radar transceiver module with a copper heat sink and a 9 mm silicon lens; after [Sta15]. The incorporated IR-image indicates that the chip-on-lens assembly is at around 29°C.

transferred to a copper heatsink through the lens attached to the PCB bottom side which stays in direct contact with the lens. The lens further allows an in-door operation of the complete radar module with no additional external optical components because of a significant increase of the antenna effective gain. Moreover, the pointing-direction errors present in a lens-integrated 2-antenna system (bistatic radar) radiating at an angular offset from the lens optical axis are eliminated in the monostatic radar architecture relying on a single on-chip antenna aligned with the lens center. The current radar implementation features a 9 mm-diameter hyper-hemispherical lens with a 1.3 mm extension to maximize the antenna directivity [Fili93]. Such a lens size provides enough volume for cooling the chip to 29°C, as shown in Figure 6.40.

The integration of a high-performance antenna on a silicon chip is one of the most challenging tasks because the typical cross section of a silicon chip comprises only few metal layers embedded in a low-refraction-index BEOL (Back-End-of-Line) dielectric stack, typically only few micrometers thick, which is located on the top of a lossy bulk silicon substrate. With such a dielectric stack, there are basically only two options for implementing on-chip antennas. The first and the most straightforward approach is to use a ground-plane support between the BEOL stack and the substrate to realize the classical microstrip-type antenna radiating to the top of a silicon chip. This solution, however, results in low radiation efficiency and narrow operation bandwidth [Jaes13, Jaes14]. With the ground-support eliminated, electromagnetic waves start penetrating the complete volume of a lossy and electrically thick substrate launching various parasitic modes such as surface waves. This leads to a very poor prediction of radiation characteristics over a large RF bandwidth, parasitic inter-element coupling, and low radiation efficiency. An alternative solution is to apply the so-called lens-integrated on-chip antennas for mm-wave and THz applications [Fili93, Jha14] which was the preferred option also for this work.

For this particular case, an isolation between the transmit path and the receive path over a wide operation frequency range is additionally required which is provided by a circularly polarized antenna. Circular polarization is achieved by suitable combination of a broadband differential quadrature coupler [Grz15] and a novel dual-polarization circular-slot antenna [Grzyb15]; as shown in Figure 6.41. Here, a left-handed polarization (LHCP) is implemented in the Tx path ('Tx out' in Figure 6.41), whereas the receive signals reflected in free-space at an odd number of times are directed to the receive port ('Rx in' in Figure 6.41) as RHCP waves.

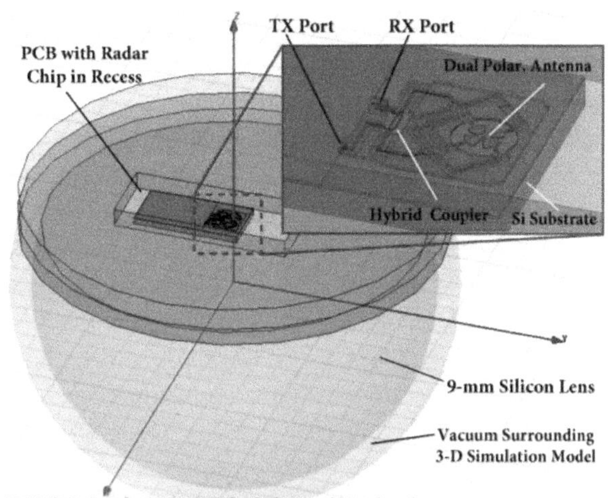

Figure 6.41 A 3-D EM simulation model of the packaged radar module with a silicon chip mounted on the back of a 9-mm lens; after [Sta15]. The chip-on-lens assembly is placed inside a rectangular recess in the PCB and surrounded by a large ground plane. The slot antenna with the differential quadrature coupler in the BEOL dielectric stack of a silicon chip is shown in the magnified view. The transmit port and the receive port are denoted as 'Tx out' and 'Rx in', respectively.

The implemented slot antenna is capable of supporting two orthogonal polarizations launched by two orthogonal pairs of patch probes located along the slot circumference (inset of Figure 6.41). The antenna is embedded in a 12-μm thick BEOL stack with seven aluminum layers on top of a 150-μm thick lossy substrate with a bulk resistivity of 50 Ωcm. Its transmit and receive ports are interconnected to the corresponding differential outputs of the quadrature coupler with two intermediate 900-μm long T-line sections implementing the mode conversion from a microstrip line configuration on the antenna side to a grounded-coplanar stripline feed on the coupler side. A 10 dB-defined input-impedance operation bandwidth of the standalone slot antenna is very broad and spans between 150 GHz and 500 GHz. The quadrature coupler driving the antenna (inset of Figure 6.41) is realized by exploiting both the side coupling between two adjacent strips and the broadside coupling between two other strips located on different metallization layers. A total coupling length is only 110 μm. In order to ensure broadband operation of the hybrid coupler [Grz15], its layout is implemented by means of three buried metal layers of the BEOL stack to equalize propagation speeds of the relevant modes. For the considered radar operation bandwidth of 210–270 GHz, the simulated phase and amplitude imbalance of the coupler are within ±0.75° and 0.2 dB, respectively, whereas the isolation between the Tx and Rx ports is superior to –23 dB within 160–340 GHz. A radiation efficiency of around 62–67% within 200–300 GHz was simulated for the complete circularly polarized antenna, comprising the slot radiator and the coupler, radiating through a 150-μm thick lossy Si-substrate into a silicon half-space [Tong94]. The parameters 3.8×10^7 and 0.02 were assumed for metal conductivity and dielectric loss tangent of the BEOL stack, respectively. The complete packaged chip-on-lens assembly, as shown in Figure 6.41, was further EM-simulated in the transmit mode to study the leakage between the Tx and Rx ports through the antenna path in the presence of reflections at the lens aperture. This leakage may potentially cause nonlinear effects in the Rx, an increase in the noise level, or even lead to the Rx saturation [Brook05, Stov92, Ondr81, Pipe95]. From Figure 6.42, the simulated TX-to-RX leakage and return loss at the TX and RX ports are better than –21 dB and –23 dB, respectively. Considering the previously estimated Rx P_{1dB} compression point and the Tx output power, such antenna isolation is not expected to result in the Rx compression. Moreover, the simulated radiation efficiency of the entire packaged lens-integrated antenna is only a few percent lower than that for the corresponding silicon half-space because of the transmit mode of operation.

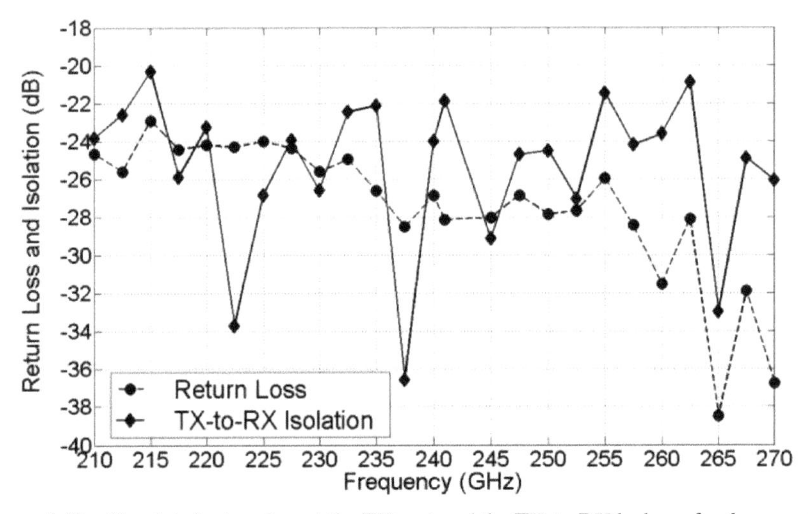

Figure 6.42 Simulated return loss at the TX port and the TX-to-RX leakage for the complete chip-on-lens packaged assembly from Figure 6.41; data from [Sta15].

The complete radar module was characterized in free-space exploiting the Friis-transmission equation in the antenna far-field zone. The following key parameters were measured: radiation patterns with antenna directivity and axial ratio, transmitted power, and Rx CG and NF. The measurements were conducted for both operation modes: transmit and receive. The antenna directivity in the TX and the RX operation mode was measured to be 25.8–27.8 dBi and 25.9–27 dBi, respectively, for the radar operation frequency range of 210–270 GHz. An exemplary chosen radiation pattern at 270 GHz for the radar module in the transmit mode is shown in Figure 6.43. The pattern shows good beam rotational symmetry with a side-lobe level of around –17 dB. An axial ratio of 1–1.45 and 1–1.35 for the transmit and the receive mode of operation, respectively, was further measured at boresight for 210–290 GHz. The frequency-dependent radiated output power is presented in Figure 6.44, where a different power level for two module orientations (see Figure 6.39 for orientation definition) can be recognized. This difference is predominantly related to the non-ideal antenna axial ratio. A peak radiated power is around 5 dBm and a –10 dB-defined RF bandwidth is 46 GHz (217–263 GHz) [Sta15].

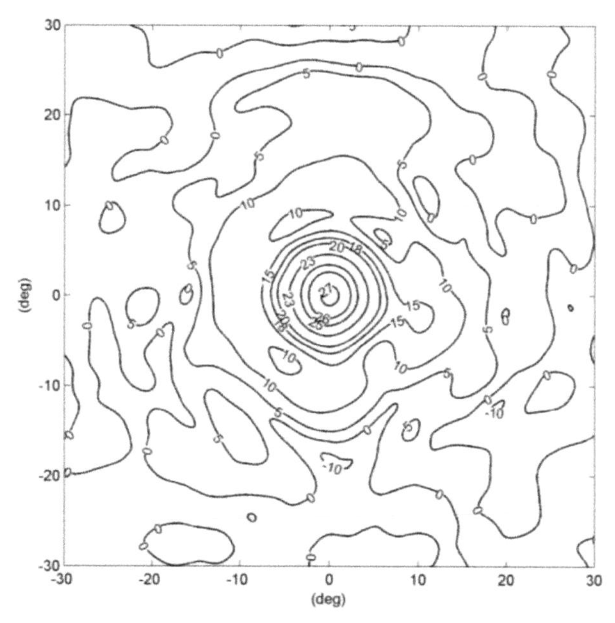

Figure 6.43 Azimuthal view of the antenna co-polar radiation pattern at 270 GHz for the radar module operating in the transmit mode.

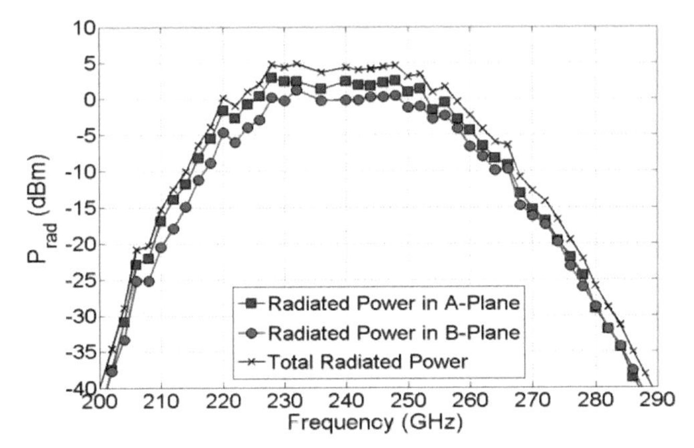

Figure 6.44 Frequency-dependent radiated power; data from [Sta15]. Both the total power and the power levels for two orthogonal antenna orientations ('A-plane' and 'B-plane') are plotted. For plane orientation, please, refer to Figure 6.39.

The CG and the NF of the radar module operating in the receive mode were measured with the pre-calibrated reference power source from OML. It should be noted that this characterization was conducted in the presence of leakage from the Tx chain operating simultaneously, resulting in the noise floor increase of the receive path. The NF was calculated indirectly from the measured CG and the noise power spectral density at the radar baseband outputs because of missing noise standards in the lab equipment at the operation frequency. The frequency dependence of the measured CG and NF is plotted in Figure 6.45. A peak CG of 12.1 dB with the corresponding minimum NF

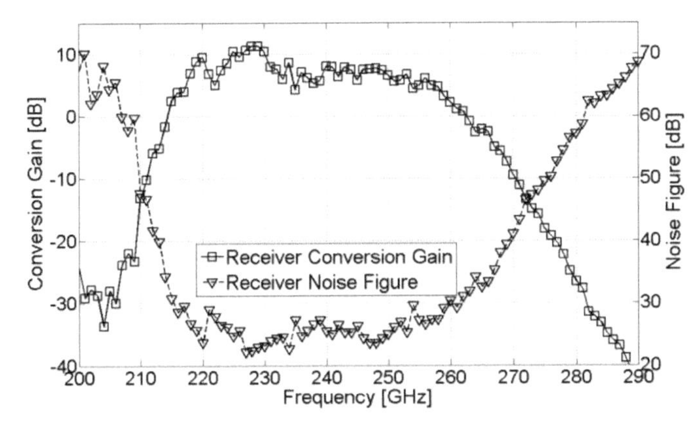

Figure 6.45 Noise figure and conversion gain of the radar module for an IF frequency of 33 MHz; data from [Sta15].

of 21.1 dB was measured. A –10 dB-defined RF operation bandwidth for the radar receive mode is around 46.3 GHz (214.8–261.1 GHz).

Besides the 210–270 GHz transceiver module, the complete radar system includes an external in-house-developed linear-frequency chirp generator, a set of differential IF amplifiers with a data-acquisition unit, and a MATLAB code for signal post-processing. Its architecture is shown in Figure 6.46.

For maximum range resolution [Meta07], fast and wideband sawtooth up-chirps of high linearity from 13.1 GHz to 16.9 GHz driving the input port of the radar RF module are first generated. The chirps can be made as short as 100 μs but the overall system was optimized for a chirp duration of 2 ms. The chirp generator relies on a hybrid architecture consisting of a direct digital synthesizer (DDS), a phase-locked loop (PLL), and a VCO. Here, the chirp signal from the DDS serves as a reference for the integer-N PLL which then up-converts the reference DDS frequency to the required 13.1–16.9 GHz. The chosen chirper topology combines the benefits of both the PLL [Zhiy13] and the DDS in terms of fast-chirp generation and spectral purity. In particular, the spurious tones from DDS are suppressed by the PLL loop-filter. In comparison with a fractional-N PLL, the DDS-based architecture offers finer frequency resolution [Stel05]. In the current implementation, the chirper is realized as a set of off-the-shelf PCB-mounted components with the DDS circuit clocked at 1 GHz from the signal generator Agilent E8257D. In the PLL, a high tuning range (13–20 GHz) VCO from Sivers IMA is used. In order to minimize the overall PLL phase noise, a comparison frequency of the phase-frequency detector was selected to be as high as possible, resulting in the PLL loop division ratio, N, of 192 (4 × 48). A third-order active filter with a 1.8 MHz low-noise operational amplifier implements the PLL loop filter. With the aid of a behavioral simulation model, the PLL loop bandwidth

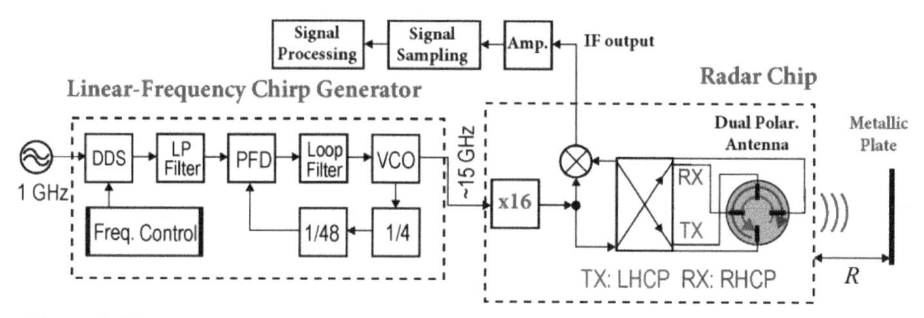

Figure 6.46 Architecture of the complete radar under test with a metallic plate located at a distance R; after [Sta15].

was set to 479 kHz with an appropriately high phase margin of the open-loop transfer function (around 58°) for minimum total integrated phase noise of the chirp generator.

In Figure 6.47, an exemplary chosen phase noise at the frequency chirper output around 15 GHz is presented. The total jitter integrated from 10 Hz to 100 MHz is 2.62° rms. The linearity of the generated frequency ramp was verified by means of Hilbert transform [Thro84, Grei93] but only indirectly at the divider output due to limitations in the lab equipment. With this approach, a complex-value analytic signal is first obtained from the real-value time-domain train acquired at the radar IF output. The instantaneous phase of such an analytic signal is then compared with the ideal phase trajectory of a linear frequency chirp and the phase deviation between the two can be computed. The corresponding frequency error results from the rate of change of this phase deviation and its root-mean square value is defined over the chirp duration time. In particular, for a frequency ramp of 210–270 GHz swept over 2 ms, the rms frequency error is below 9 kHz.

In the current radar implementation, beat signals at the IF ports are digitized by an external 16 bit sampling card from National Instruments and then post-processed using MATLAB routines. The card sampling rate is limited to 2 MHz which results in 4,000 samples for 2 ms long chirp period. For the consecutive experiments, a Hanning window will be predominantly applied

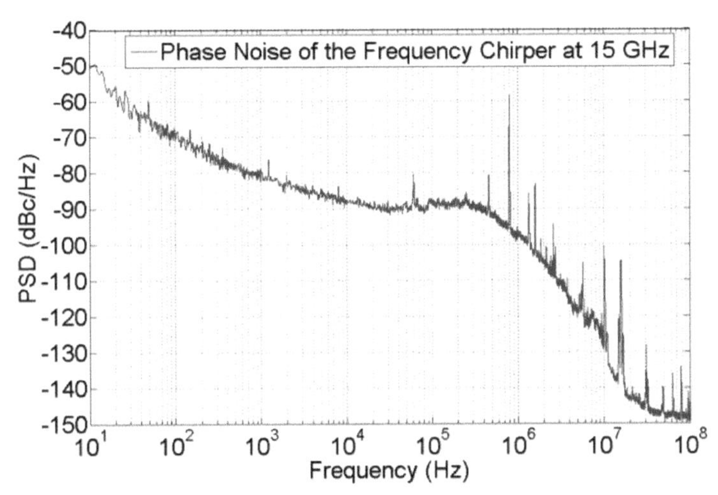

Figure 6.47 Phase noise of the frequency chirp generator at 15 GHz driven from the frequency synthesizer Agilent E8257D at 1 GHz; after [Sta15]. A frequency of 15 GHz corresponds to 240 GHz for the up-converted signal at the radar RF output.

in the FFT data post-processing as a fair compromise between selectivity and resolution [Harr78]. For a 2 ms long frequency chirp, this window corresponds to an equivalent noise bandwidth of 750 Hz.

In order to minimize the influence of the RF front-end non-idealities on the performance of the complete radar system, a 2-step calibration procedure was conducted with the aid of a metallic plate as a single-target reflector at a distance of 80 cm from the radar module, as shown in Figure 6.46. Such a target shows the $1/\lambda^2$ frequency-dependent radar cross section, where λ is the free-space wavelength. The first step of the calibration aims at removing the influence of the close-in returns resulting from the limited isolation between Tx and Rx paths which do not depend on the imaged objects and appear as low-frequency IF signals at the Rx output port. This step does not require any reference target to be placed in front of the radar antenna. However, to mimic this 'no-target' radar response in the presence of close-proximity reflections in the lab environment coming from insufficient absorber attenuation, the metal plate was tilted by 45° to the boresight of the radar antenna. In the next step, the metallic plate was set back to its perpendicular position with respect to the radar boresight and the Hilbert transform on the acquired calibration signal from the plate was applied to calibrate the influence of parasitic phase and amplitude modulations resulting from the non-ideal RF front-end characteristics. A frequency dependence of the normalized power received from the plate after the Hilbert-transformed time-domain IF calibration train is plotted in Figure 6.48. Considering the $1/\lambda^2$ frequency-dependent radar

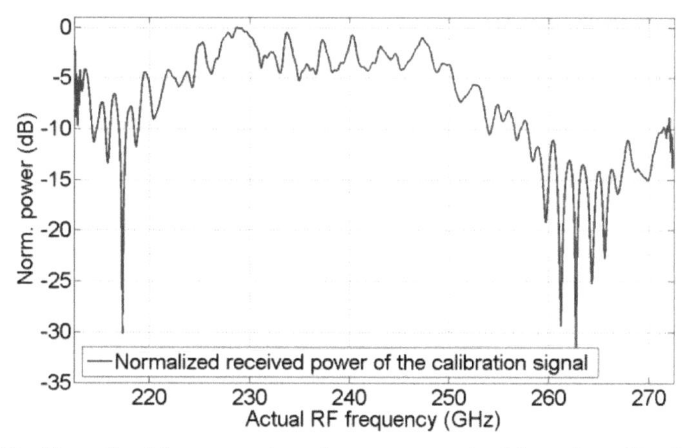

Figure 6.48 Normalized frequency-dependent power received from the calibrating metallic plate after the Hilbert-transformed time-domain IF calibration train; data from [Sta15].

cross section of the metallic-plate, a −10 dB-defined bandwidth of around 45 GHz can be estimated for the complete radar transceiver combining the characteristics of both Tx and Rx paths.

For verification purposes of the applied calibration procedure, the beat-signal time-domain trains were further acquired for different positions of the metal plate. Exemplary, the calibrated frequency-dependent radar response to the plate at a distance of 60 cm is presented in Figure 6.49.

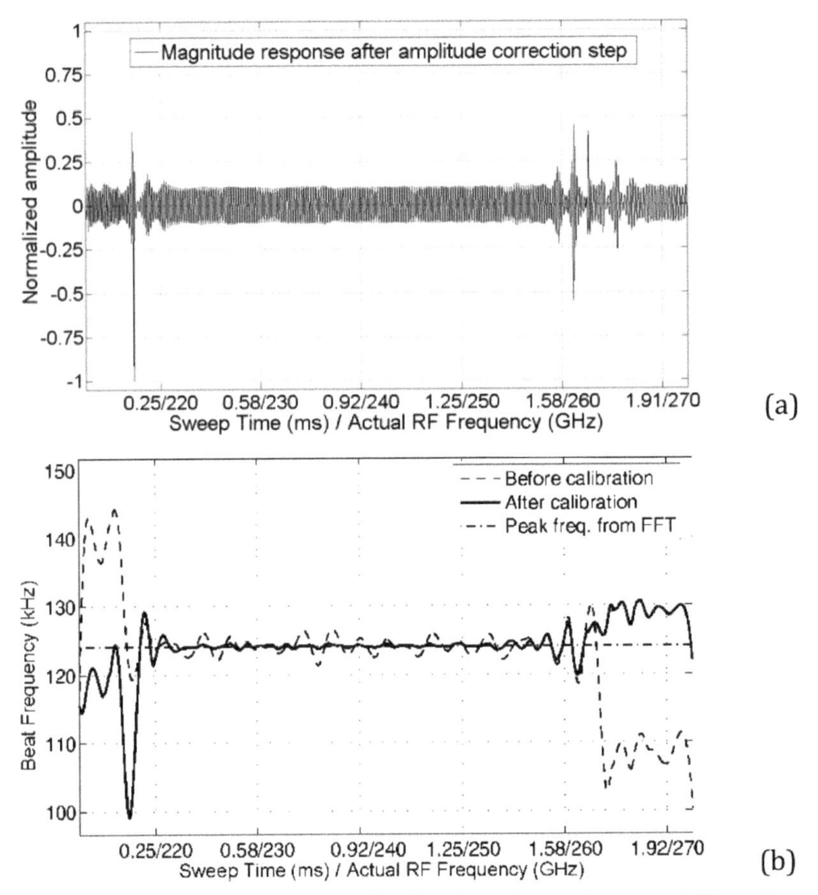

Figure 6.49 Radar response to a metallic plate located at a distance of 60 cm from the radar module; data from [Sta15]. (a) Magnitude response after amplitude calibration only, (b) instantaneous beat frequency after the amplitude and phase corrections. The beat frequency de-embedded from the peak in the IF power spectrum of the return signal is 124.1 kHz (see also Figure 6.50). Both frequency and time units are shown due to duality of the sweep time and the actual RF frequency for a linear FMCW radar.

From the amplitude-corrected magnitude response, it can be noticed that the envelope of the acquired beat signal is almost constant in the frequency range of around 60 GHz but it starts deviating below 220 GHz and beyond 260 GHz. It was verified that two parasitic harmonics with the $\times 14$ and $\times 18$ multiplication factor leaking from the multiplier chain-based LO path are mainly responsible for this behavior which could not be appropriately calibrated. Similarly, the extracted instantaneous beat frequency after amplitude and phase correction steps is influenced by the same harmonic spurs. These harmonic distortions and not the Rx noise floor are primarily limiting the achievable spurious-free dynamic range (SFDR) and the operational bandwidth of the currently implemented radar.

The corresponding FFT-computed IF power spectra of the calibrated beat signal for two RF bandwidths of 60 GHz and 45 GHz are shown in Figure 6.50. The Hanning window was applied in the computation for low spectral leakage and large dynamic range (DR) [Harr78]. Here, similar to the plots from Figure 6.49, the influence of $\times 14$ and $\times 18$ harmonic spurs at 109 kHz and 140 kHz, respectively, located around the main peak at 124.1 kHz can be recognized. Please note that for a reduced bandwidth of 45 GHz, the radar SFDR achieves around -40 dBc.

From Figure 6.50, the achievable radar range resolution can be further extracted by means of the so-called point spread function (PSF). For a 60 GHz operation bandwidth, a theoretical range resolution of $c/2B$ = 2.5 mm can be calculated for the currently implemented radar, where B is the RF bandwidth and c is the speed of light. This theoretical resolution is, however, of limited practical use. In practice, the ability of distinguishing between two close-proximity targets is more relevant. In this case, the main-lobe full-width at -6 dB of the PSF becomes the parameter of interest. For a rectangular weighting function promising the best resolution but with the highest side-lobe level of –13 dB, it results in 3 mm. With the Hanning window, commonly applied in imaging for low spectral leakage, this number becomes $2c/2B$ = 5.0 mm [Harr78]. A main-lobe full-width at -6 dB of 5.14 mm was extracted for the radar implemented here operating with the maximum considered bandwidth of 60 GHz after amplitude and phase corrections and the Hanning window applied, which is close to a theoretical limit of 2.5 mm.

A very simple 2-D scanning optical setup comprising two elliptical collimating and refocusing mirrors, as shown in Figure 6.51, was applied to demonstrate the radar 3-D imaging capabilities. Here, the scanned objects were placed in the focal point of one of the mirrors. A lateral optical resolution of around 1 mm was estimated for this setup with the aid of a

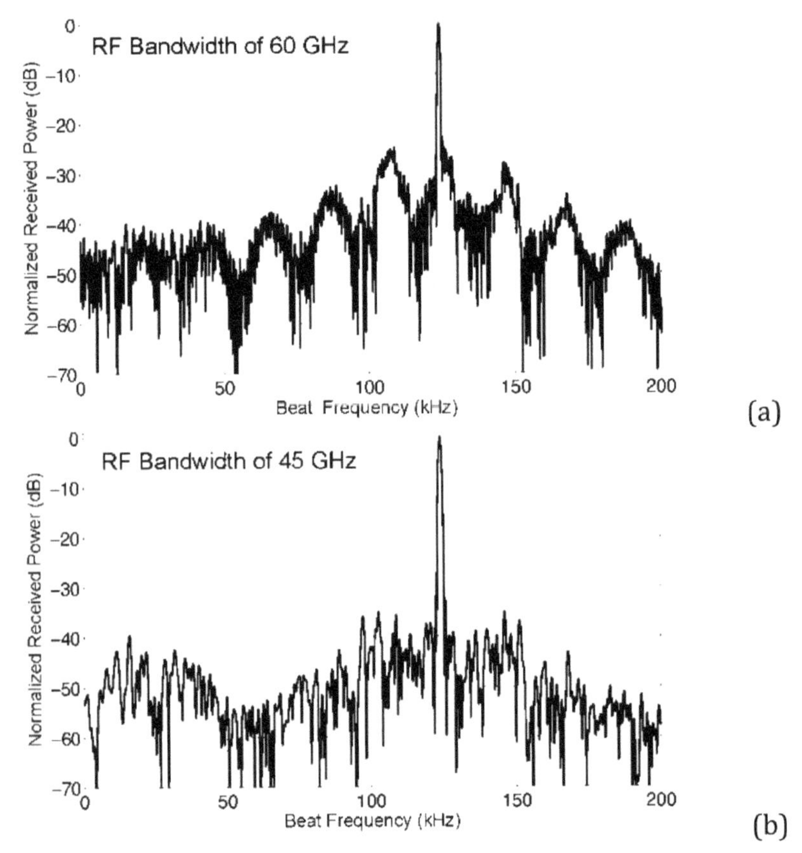

Figure 6.50 The FFT-computed IF spectrum of the calibrated beat signal corresponding to the metallic plate spaced by 60 cm from the radar module for two different operational RF bandwidths: (a) 60 GHz and (b) 45 GHz. For comparison purposes, the chirp duration was varied for both bandwidths to arrive at the same beat frequency of 124.1 kHz. For 60 GHz, it was set to 2 ms whereas for 45 GHz it was reduced accordingly. The influence of harmonic spurs can be identified around 109 kHz and 140 kHz.

simple aluminum pin-type heatsink as a resolution target. For a set of the consecutive imaging experiments, the previously mentioned Hanning weighting function was replaced by the Hamming window offering a slightly improved resolution of 4.65 mm [Harr78] for the full 60 GHz operation bandwidth.

As a scanning object, a 12 cm $\times$ 6 cm large cardboard box with a hidden blister pack of drugs and two missing tablets was chosen. The object was meander-scanned in both directions (X and Y), as sketched in Figure 6.52, and the acquired data was post-processed using 3-D data matrix routines from

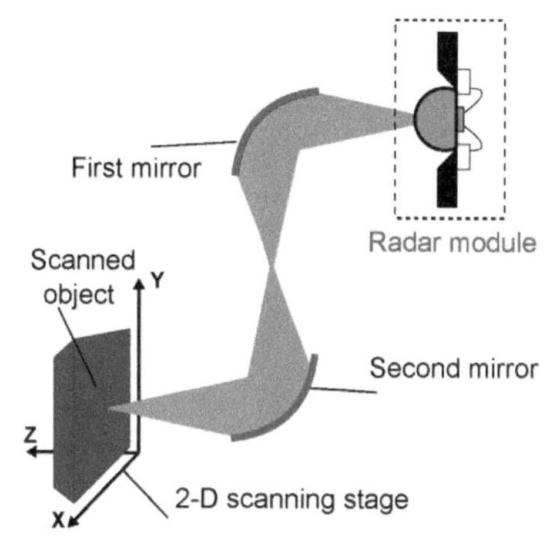

Figure 6.51 2-D optical scanning test setup for demonstration of the radar 3-D imaging capabilities. A total path length between the radar module and the object is around 780 mm.

MATLAB. Each IF signal burst took 2 ms and was sampled at two MSPS. The range profile was sampled with a resolution of 0.5 mm, whereas the pixel lateral pitch was set to 0.25 mm after interpolation, resulting in an image size of 200 $\times$ 480 $\times$ 240 voxels for a range-gated distance of $\pm$50 mm around the image center position.

Figure 6.52 presents the exemplary chosen 2-D image of the normalized power received for an object-to-radar distance of 780 mm altogether with the range profiles for two lateral positions across the cardboard which correspond to the present and the missing tablet. It can be noticed that the signal returns correlating with the positions of the lidding seal of aluminum foil, the plastic cavity, and the cardboard box can be identified due to the radar appropriate range resolution and finally the missing tablets can be detected. The corresponding 3-D surface reconstruction of the scanned object can be found in Figure 6.53. Here, the image is formed with a peak-search algorithm which identifies the positions of the highest reflected power for each lateral position and the color-scale represents the normalized received power.

6.2.3 0.5 THz Computed Tomography

Increasing the transistors f_{MAX} deep into the terahertz frequency range does not only lead to a significant performance improvement for mm-wave and sub-mm-wave circuits, it also contributes to the vision of closing the THz gap with silicon-based circuits. Traditional compound semiconductor-based

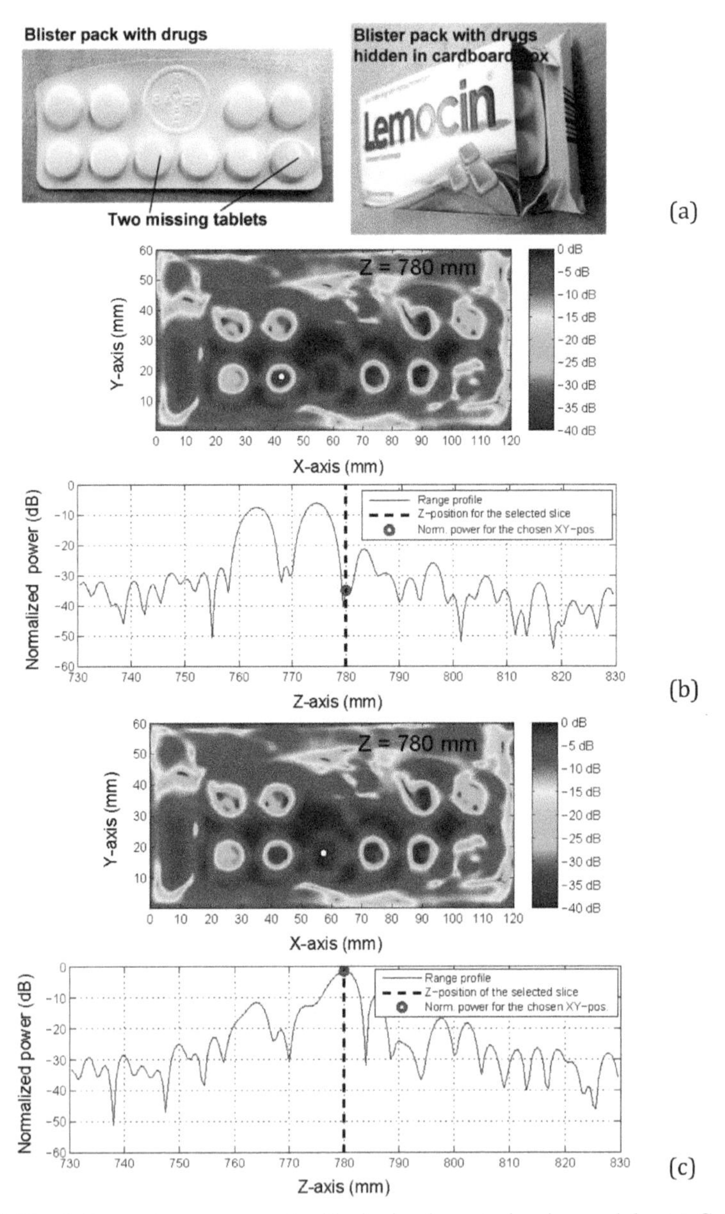

Figure 6.52 3-D imaging experiment with the implemented radar module. (a) Cardboard box with a blister pack of drugs with two missing tablets as the scanned object. (b, c) 2-D scan of the normalized power received for an object-to-radar distance of 780 mm altogether with the range profiles for two different X–Y positions across the cardboard. The positions correspond to the present and the missing tablet, respectively. Both acquired range profiles show a DR of around 50 dB.

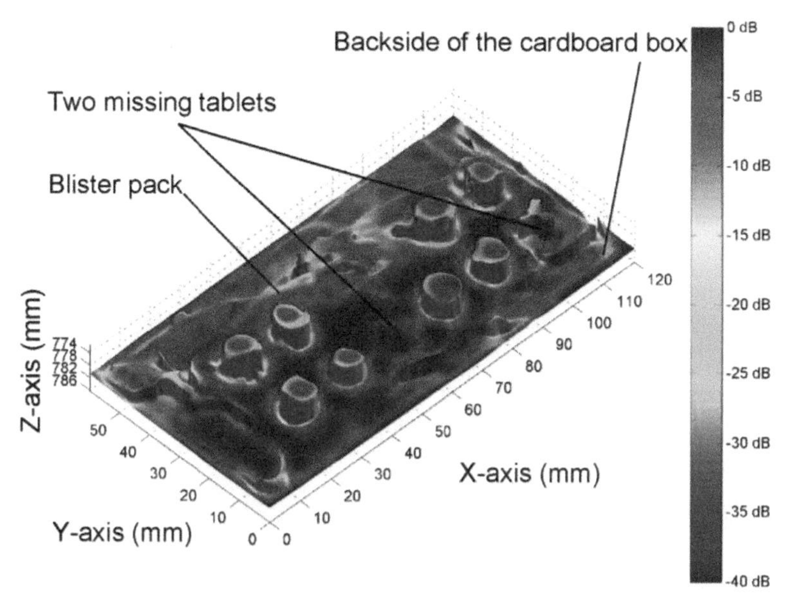

Figure 6.53 3-D surface reconstruction of the object from Figure 6.52(a) after a peak-search algorithm. The scan was appropriately range-gated (770 mm $\leq Z \leq$ 788 mm) to eliminate the influence of multi-path reflections inside the cardboard box and the plastic cavities of the blister pack as well as reflections from the front of the box.

THz imaging systems tend to be bulky and expensive and thus suffer from a poor price–performance ratio. The advances in SiGe-HBT and CMOS technology continuously increase the device power generation and detection capabilities in the THz frequency range and thus may ultimately leverage the commercial interest in THz imaging systems. Three-dimensional THz imaging based on the principle of computed tomography (THz-CT) is one of the applications that may potentially be explored in commercial environments. THz-CT offers volumetric object reconstruction with an image contrast based on the characteristic THz absorption of the illuminated material. Since THz radiation is non-ionizing and thus requires no dedicated safety measures, THz-CT represents an interesting alternative to X-ray technology for low-cost industrial quality control.

The THz-CT system implemented in this work is solely based on components built in silicon technology that was developed in the frame of DOTSEVEN. Figure 6.54 shows an illustration of the THz-CT scanner. The radiation of a SiGe-HBT source is focused in the object plane and refocused

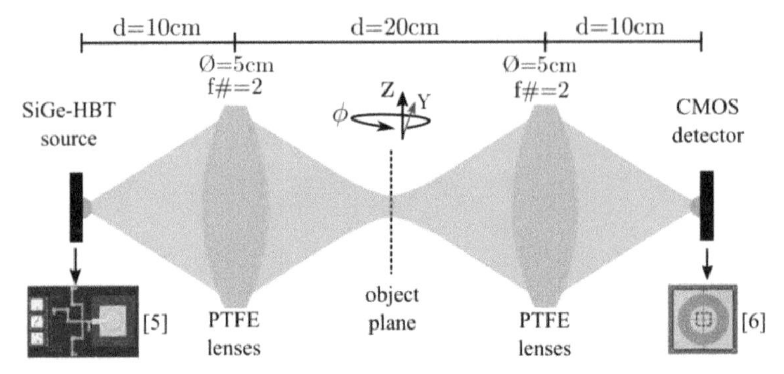

Figure 6.54 Illustration of the THz-CT scanner. The system comprises a 490 GHz SiGe-HBT source, an NMOS detector, and an optical train based on four f# = 2, 50 mm PTFE-lenses. The object is rotated (ϕ) and stepped in the 2D object plane (y,z).

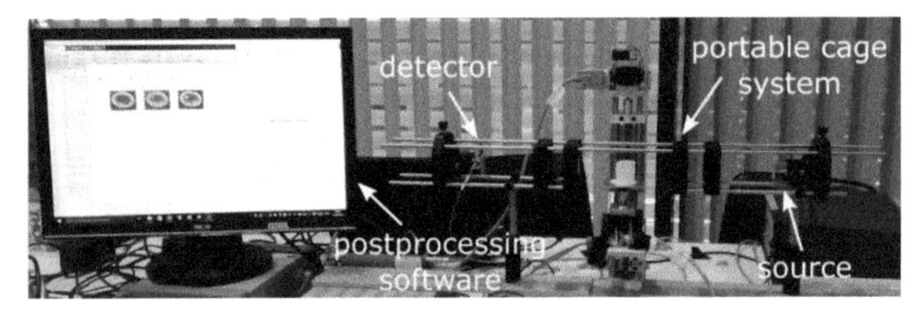

Figure 6.55 Photograph of the THz-CT scanner.

to an NMOS detector with low-cost PTFE lenses. The transmission through the object is measured along the y-axis and for different projection angles to form the sinograms of the object. This pocess is repeated along the z-axis to allow full 3D reconstruction based on a filtered back-projection algorithm. In order to facilitate measurements at multiple projection angles and positions, the location of the object at is computer-controlled by $x \times y \times \varphi$ stepper motors. For each position the detector output signal is sampled with a data-acquisition system.

6.2.3.1 Components

There are two components that define the quality of THz-CT systems. Increasing the DR, which is defined as the relation between maximum received signal without an object and the integrated noise over the readout bandwidth, relaxes the trade-off between object thickness, material composition, and scanning time. Secondly, the achievable image resolution is inversely proportional to the beam spot size in the imaging plane, which is defined by the wavelength and effective aperture and focal length of the optics. Since the output power of silicon-based radiation sources drops significantly when going beyond f_{MAX}, the trade-off between achievable DR and operational frequency becomes he bottleneck for silicon-based THz-CT systems. This fact stresses the need for a high-quality source design that maximizes the radiated power while providing sufficient directivity and Gaussicity.

A. Source design

The source in this work was implemented in Infineon DOTSEVEN 0.13 μm SiGe BiCMOS technology with an $f_{\mathrm{T}}/f_{\mathrm{MAX}}$ of 260/350 GHz [Hill15]. It comprises a broadband lens-integrated circular slot antenna coupled to a single-ended triple-push Colpitts oscillator. Figure 6.56 shows the schematic and the microgrpah of the source. An operation frequency of 490 GHz was chosen as a compromise between resolution and power generation capability of the technology. However, the output frequency is still significantly higher than the f_{MAX} of the technology, necessitating the use of harmonic generation techniques. In this design, a triple-push topology is used to extract the third harmonic at the base terminal of three Colpitts oscillators. The circular slot antenna connected to the common base node loads the circuit at the fundamental oscillation frequency and thus forces the oscillators to run 120° out of phase [Tang01]. In this mode, the third harmonic currents of all three oscillators add in phase, while the currents at the fundamental frequency superimpose destructively. Note that the symmetry of the physical design is very important since it directly impacts the extraction efficiency of the third harmonic and the rejection of the fundamental.

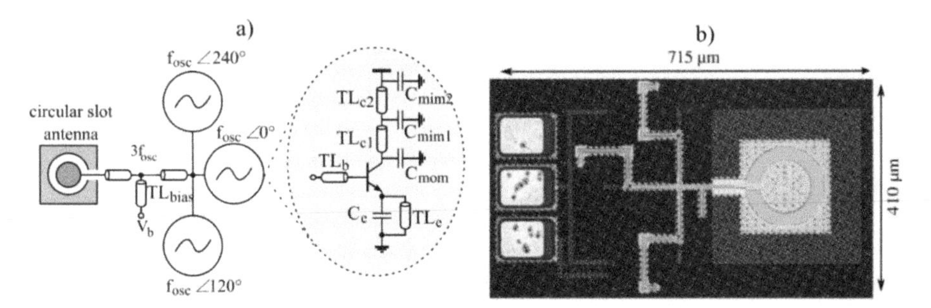

Figure 6.56 Schematic (a) and micrograph (b) of the 490 GHz SiGe-HBT radiator, after [Hill15].

The design procedure for the Colpitts oscillator can be summarized as follows. First, small-signal simulations were used to optimize the transistor size and the feedback capacitor C_e by maximizing the negative resistance at the design frequency of 163 GHz. After that, the tank inductance TL_b was sized to tune out the imaginary part of the transistor input impedance. The series–series feedback introduced by C_e increases the reverse transmission behavior of the circuit and thus the impact of the collector load impedance on the third harmonic matching at the base terminal [Pfei14]. In this design, the maximum third harmonic output power is realized by avoiding the lossy collector via stack and by using an ideal short circuit at the collector terminal. The broadband harmonic idler is realized with three capacitors (C_{mom}, C_{mim1}, C_{mim2}) that are self-resonant at the first, second, and third harmonic.

The lossy silicon die and the strickt design rules that are usually enforced upon modern silicon technologies make silicon chips a very unfavourable environment for integrated antennas. At the same time, the requirements for the antenna system in a THz-CT system with a free-running source are high. Process variations and modelling inaccuracies can lead to a shift in the oscillation frequency after manufacturing which makes a broadband antenna design inevitable if a fist-pass design is needed. Additionally, the antenna needs to have a directivity of around 20 dBi to be compliant with low-cost optical components, i.e., 5 mm-diameter PTFE-lenses. These requirements call for a broadband lens-integrated antenna system that is composed of an on-chip primary antenna and a secondary hyper-hemispherical silicon lens. In this design a multi-layer linearly polarized circular slot antenna was used to illuminate a 4 mm-diameter silicon lens. Figure 6.57 pictures the HFSS model used for 3-D EM simulation of the antenna system. The simulation results show a directivity of 22.5 dBi and 86% radiation efficiency.

The source module was fully characterized in free space. The radiation frequency of the source was measured with an 18th harmonic mixer. For the biasing conditions that optimize the radiated power for all supply voltages, the oscillation frequency is close to constant at 490 GHz. The radiated power was measured with a photo-acoustic power meter (TK). Figure 6.58 shows that the source delivers an output power of up to 38 μW with a DC-to-RF of up to 0.059%.

6.2.3.2 Detector design

The terahertz detector is a zero-bias NMOS detector fabricated in IHP's 0.13-μm SiGe-BiCMOS technology. The asymmetric NMOS device is derived from the standard 1.2 V NMOS by a modified layout of the source/drain

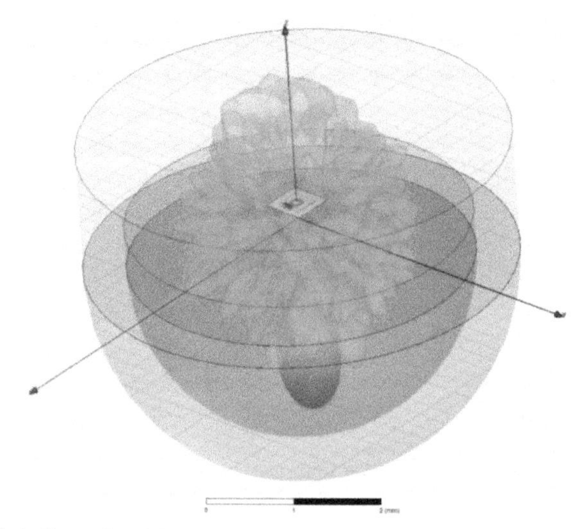

Figure 6.57 A 3-dimensional EM simulation model of the packaged source module with a silicon chip mounted on the back of a 4 mm hyper-hemispherical silicon lens.

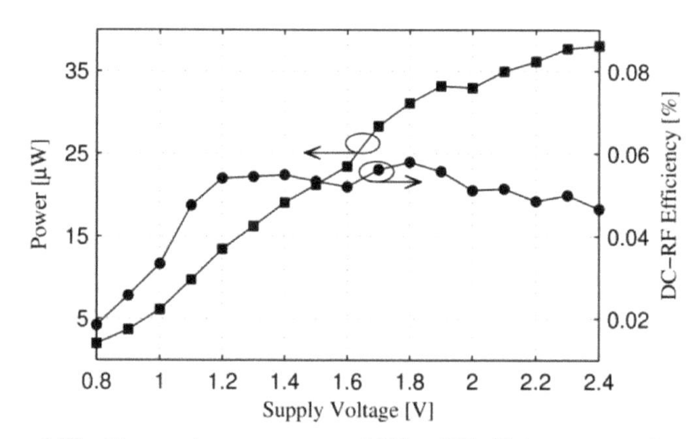

Figure 6.58 Measured output power and DC-to-RF efficiency versus frequency.

extension masks [Jain16]. The drain side comprises the normal high-dose n+ extension (HDD) and halo implants while the source side is implanted with the low-dose n-extension (LDD) from the 3.3 V I/O devices. Due to the absence of halo implants at the source side, a reduced threshold voltage of the transistor shifts the optimal gate bias point for highest sensitivity to zero volts. Similar to the source design, the detector antenna is comprised of a primary on-chip antenna and a secondary silicon-lens. Figure 6.59 shows the

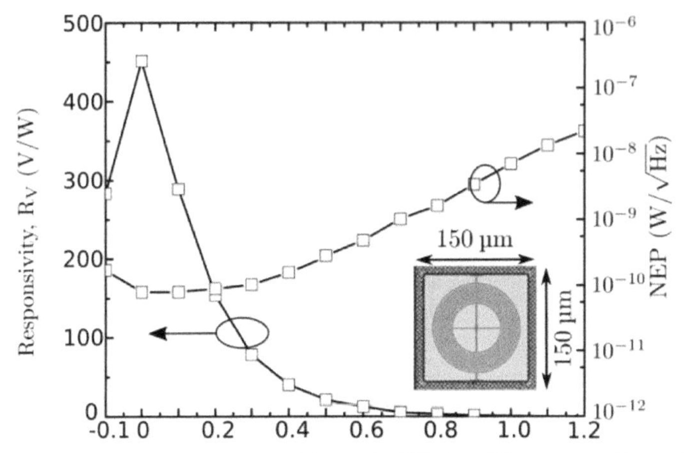

Figure 6.59 Measured voltage responsivity and NEP for different gate bias voltages and micrograph of the detector. The modification of the source/drain extension masks shifts the bias point for optimum sensitivity to zero volts, after [Jain16].

measured voltage responsivity and noise-equivalent power (NEP) of the zero-bias NMOS detector for different gate bias voltages. The responsivity peaks at zero bias with 450 V/W and the detector shows an NEP of 80 pW/$\sqrt{Hz}$.

6.2.3.3 THz-CT results

The THz-CT system can be operated in two modes. A rapid acquisition mode with CW illumination with continous object rotation and a rotary encoder-based angle allocation can be used for rapid data acquisition. Furthermore, a acquisition mode with a chopped source and stepped object rotation can be used when high accuracy and DR are needed. Figure 6.60 shows the measured DR for different source chopping frequencies for 1 ms lock-in time constant. The system offers 38 dB DR in CW mode and around 60 dB for chopping frequencies higher than 1 kHz. The spot size and related image resolution was measured using the knife edge method [Gonz13]. A knife was mounted to a high-precision translation stage and was moved into to spot to block off a 2D plane from further propagation through the optical train. Figure 6.61 shows the normalized measured power received by the detector for y- and z-axis knife translation in the focus point and the corresponding result of the Gaussian fit obtained with

$$P = \frac{P_{max}}{2}[1 + \text{erf}(\sqrt{2}\frac{(x - x_0)}{w})],$$

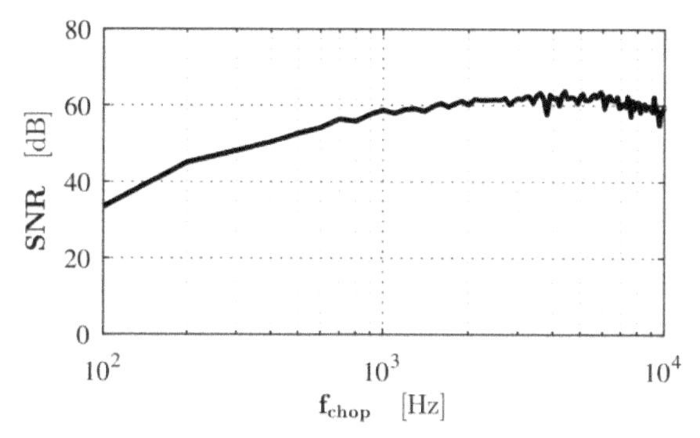

Figure 6.60 Dynamic range of the THz-CT system for different chopping frequencies measured with a lock-in amplifier with 1 ms time constant.

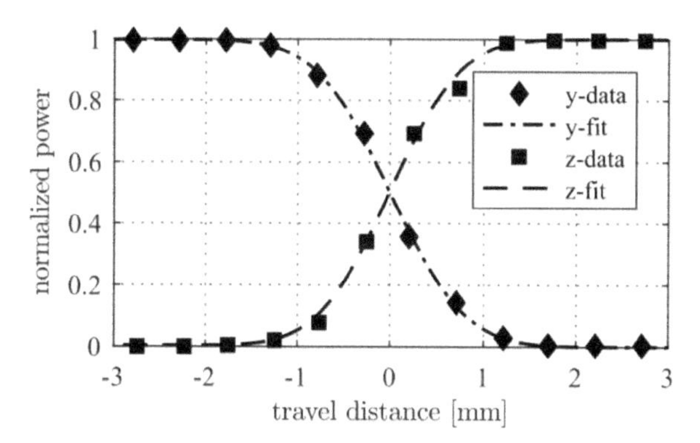

Figure 6.61 Measured and fitted normalized power at the detector for a knife translation in y- and z-directions. The Gaussian beam waists are 2.54 mm in y-direction and 2.40 mm in z-direction.

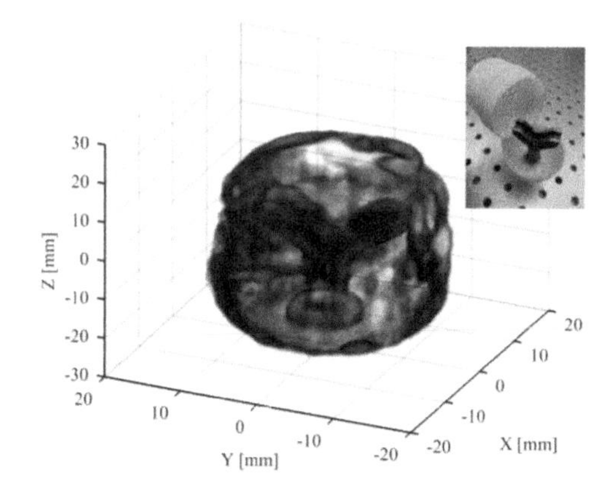

Figure 6.62 Tomographic reconstruction of a Y-shaped hook driver inside a polyethylene container. The image was recorded with 1 mm spatial and 9° angular resolution within a 250 min acquisition time.

where P_{max} is the maximum received power, $x - x_0$ is the relative position from the beam center, and w is the $1/e^2$ beam radius [Gonz13]. The resulting Gaussian beam waists in y- and z-directions are 2.54 mm and 2.40 mm. The values closely reassemble the theoretically estimated value of 2.26 mm for an effective lens aperture of 30 mm. Figure 6.62 shows the result of the tomographic reconstruction of a Y-shaped hook driver that is hidden inside a polyethylene container with a size of 26 mm $\times$ 48 mm. The image was recorded in stepped acquisition mode with 1 kHz chopping and a spatial and angular resolution of 1 mm and 9°, respectively. The total scanning time in this scanning time is still quite high with 250 min. However, with a continuous object rotation and CW detection, the overall acquisition time can be reduced to as low as 20 min.

References

[Abba11] Abbasi, M., Member, S., Gunnarsson, S. E., Wadefalk, N., Kozhuharov, R., et al. (2011). Single-Chip 220 GHz active heterodyne receiver and transmitter MMICs with on-chip integrated antenna. *IEEE Trans. Microw. Theory Tech.* 59, 466–478.

[Afsh06] Afshar, B., and Niknejad, A. M. (2006). "X/Ku band CMOS LNA design techniques," in *IEEE Custom Integr. Circuits Conf.* pp. 389–392, 2006.

[Agar14] Agarwal, P., and Sah, S. P., and Heo, D. (2014). "A 4.8 m W, 4.4 dB NF, Wideband LNA Using Positively Coupled Transformer for V-band Applications" *IEEE MTT-S Int. Microw. Symp. Dig.*, pp. 1–3, 2014.

[Ana04] Analui, B., and Hajimiri, A. (2004). Bandwidth enhancement for transimpedance amplifiers. *IEEE J. Solid State Circuits* 39, 1263–1270.

[Armi89] Armijo, C. T., and Meyer, R. G. (1989). A new wide-band Darlington amplifier. *IEEE J. Solid State Circuits* 24, 1105–1109.

[Beas90] Beasley, P., Stove, A., Reits, B., and As, B. (1990). "Solving the problems of a single antenna frequency modulated CW radar," in *Proceedings of the IEEE International Conference on Radar*, Arlington, VA, 391–395.

[Brook05] Brooker, G. M. (2005). "Understanding millimetre wave FMCW radars," in *Proceedings of the 1st International Conference on Sensing*, Palmerston North, 152–157.

[Bryl13] Bryllert, T., Drakinskiy, V., Cooper, K. B., and Stake, J. (2013). Integrated 200-240 GHz FMCW radar transceiver module. *IEEE Trans. Microw. Theory Tech.* 61, 3808–3815.

[Ceti12] Cetinoneri, B., and Atesal, Y. A., Fung, A., and Rebeiz, G. M. (2012). W-Band Amplifiers With 6 dB Noise Figure and Milliwatt-Level 170–200 GHz Doublers in 45 nm CMOS. *IEEE Trans. Microw. Theory Tech.* 60, 692–701.

[Chat04] Chattopadhyay, G., Schlecht, E., Ward, J., Gill, J., Javadi, H., Maiwald, F., et al. (2004). An all-solid-state broad-band frequency multiplier chain at 1500 GHz. *IEEE Trans. Microw. Theory Tech.* 52, 1538–1547.

[Chen08] Chen, V. (2008). "Detection and analysis of human motion by radar," in *Proceedings of the 2008 IEEE Radar Conference*, Rome, 1–4.

[Chen10] Chen, Z., and Heydari, P. (2010). "An 85–95.2 GHz transformer-based injection-locked frequency tripler in 65 nm CMOS," in *Proceedings of the IEEE MTT-S International Microwave Symposium Digest*, Rome, 776–779.

[Chen12] Chen, Z., Wang, C.-C., Yao, H.-C., and Heydari, P. (2012). A BiCMOS W-band 2 X 2 focal-plane array with on-chip antenna. *IEEE J. Solid State Circuits* 47, 2355–2371.

[Chi13] Chiang, P. Y., Momeni, O., and Heydari, P. (2013). A 200 GHz Inductively Tuned VCO With –7 dBm Output Power in 130 nm SiGe BiCMOS. *IEEE Trans. Microw. Theory Tech.* 61, 3666–3673.

[Coop08] Cooper, K., Dengler, R., Llombart, N., Bryllert, T., Chattopadhyay, G., Schlecht, E., et al. (2008). Penetrating 3-D imaging at 4- and 25 m range using a submillimeter-wave radar. *IEEE Trans. Microw. Theory Tech.* 56, 2771–2778.

[Dick04] Dickinson, J. C., Goyette, T. M., and Waldman, J. (2004). "High resolution imaging using 325 GHz and 1.5 THz transceivers," in *Proceedings of the 15th International Symposium on Space Terahert: Technology*, Northampton, MA, 373–380.

[Esse08] Essen, H., Stanko, S., Sommer, R., Wahlen, A., Brauns, R., Wilcke, J., et al. (2008). "A high performance 220 GHz broadband experimental radar," in *Proceedings of the 33rd International Conference on Infrared, Millimeter and Terahertz Waves (IRMMW-THz)*, Mainz, 1.

[Fili93] Filipovic, D., Gearhart, S., and Rebeiz, G. (1993). Double-slot antennas on extended hemispherical and elliptical silicon dielectric lenses. *IEEE Trans. Microw. Theory Tech.* 41, 1738–1749.

[Gilr11] Gilreath, L., Jain, V., Yao, H.-C., Zheng, L., and Heydari, P. (2011). Design and analysis of a W-band SiGe direct-detection-based passive imaging receiver. *IEEE J. Solid State Circuits* 46, 2240–2252.

[Gonz13] González-Cardel, M., Arguijo, P., and Díaz-Uribe, R. (2013). Gaussian beam radius measurement with a knife-edge: a polynomial approximation to the inverse error function. *Appl. Opt.* 52, 3849–3855.

[Graj15] Grajal, J., Badolato, A., Rubio-Cidre, G., Ubeda-Medina, L., Mencia-Oliva, B., Garcia-Pino, A., et al. (2015). 3-D high-resolution imaging radar at 300 GHz with enhanced FoV. *IEEE Trans. Microw. Theory Tech.* 63, 1097–1107.

[Gray08] Gray, P. R., and Meyer, R. G. (1964). *Analysis and Design of Analog Integrated Circuits*, 2nd edn. New York, NY: Wiley.

[Grei93] Grein, N., and Winner, H. (1993). FMCW radar system with linear frequency modulation. Patent US 5252981.

[Grz12] Grzyb, J., Sherry, H., Cathelin, V, Kaiser, A., and Pfeiffer, U. R. (2012). "On the co-design between on-chip antennas and THz MOSFET direct detectors in CMOS technology," in *Proceedings of the 2012 37th International Conference on Infrared, Millimeter, and Terahertz Waves*, Wollongong, NSW, 1–3.

[Grz15] Grzyb, J., Statnikov, K., Sarmah, N., and Pfeiffer, U. R. (2015). "A wideband 240 GHz lens-integrated circularly polarized on-chip annular slot antenna for a FMCW radar transceiver module in SiGe technology," in *Proceedings of the 2015 SBMO/IEEE MTT-S International Microwave and Optoelectronics Conference (IMOC)* (Brazil: Porto de Galinhas), 1–4.

[Grz16] Grzyb, J., Statnikov, K., Sarmah, N., Heinemann, B., and Pfeiffer, U. R. (2016). A 210–270 GHz Circularly Polarized FMCW Radar with a Single-Lens-Coupled SiGe HBT Chip. *IEEE Trans. Terahertz Sci. Technol.* 6, 771–783.

[Grz17] Grzyb, J., Vazquez, P. R., Sarmah, N., Förster, W., Heinemann, B., and Pfeiffer, U. (2017). "High data-rate communication link at 240 GHz with on-chip antenna-integrated transmitter and receiver modules in SiGe HBT technology," in *Proceedings of the 11th European Conference on Antennas and Propagation (EUCAP) 2017*, Paris, 1369–1373.

[Grz17b] Grzyb, J., Vazquez, P. R., Sarmah, N., Heinemann, B., and Pfeiffer, U. R. (2017). "A 240 GHz high-speed transmission link with highly-integrated transmitter and receiver modules in SiGe HBT technology," in *Proceedings of the 42nd International Conference on Infrared, Millimeter and Terahertz Waves (IRMMW-THz)*, Cancun.

[Grzy12] Grzyb, J., Sherry, H., Zhao, Y., Al Hadi, R., Cathelin, A., Kaiser, A., et al. (2012). "Real-time video rate imaging with a 1k-pixel THz

CMOS focal plane array," in *Proceedings of the SPIE 8362, Passive and Active Millimeter-Wave Imaging*, Baltimore, MD, 919218.

[Grzy13] Grzyb, J., Hadi, R. A., Zhao, Y., and Pfeiffer, U. (2013). "Towards room-temperature all-silicon integrated THz active imaging," in *Proceedings of the IEEE European Conference on Antennas and Propagation (Eucap)*, Paris, 1740–1744.

[Grzy15] Grzyb, J., and Pfeiffer, U. (2015). THz direct detector and heterodyne receiver arrays in silicon nanoscale technologies. *J. Infrared Millim. Terahertz Waves* 36, 998–1032.

[Grzyb15] Grzyb, J., Statnikov, K., Sarmah, N., and Pfeiffer, U. R. (2015). "A broadband 240 GHz lens-integrated polarization-diversity on-chip circular slot antenna for a power source module in SiGe technology," in *Proceedings of the 45th European Microwave Conference (EuMC)*, Paris, 570–573.

[Harr78] Harris, F. J. (1978). On the use of windows for harmonic analysis with the discrete Fourier transform. *Proc. IEEE* 66, 51–83.

[Hill15] Hillger, P., Grzyb, J., Lachner, R., and Pfeiffer, U. (2015). "An antenna-coupled 0.49 THz SiGe HBT source for active illumination in terahertz imaging applications," in *Proceedings of the 2015 10th European Microwave Integrated Circuits Conference (EuMIC)*, Paris, 180–183.

[Hung05] Hung, J.-J., Hancock, T. M., and Rebeiz, G. M. (2005). High-power high-efficiency SiGe Ku- and Ka-band balanced frequency doublers. *IEEE Trans. Microw. Theory Tech.* 53, 754–761.

[Hung10] Hung, C.-C., Chiong, C.-C., Chen, P., Tsai, Y.-C., Tsai, Z.-M., and Wang, H. (2010). "A 72–114 GHz fully integrated frequency multiplier chain for astronomical applications in 0.15-μm mHEMT process," in *Proceedings of the European Microwave Conference (EuMC),* Berlin, 81–84.

[IHP03] IHP SG13G2 Process Specification Rev. 0.3 (2013). *IHP Micro-Electronics, Im Technologiepark, 25, 15236 Frankfurt (Order), Germany*. Available at: https://www.ihp-microelectronics.com/en/start.html

[Ina14] Inanlou, F., Khan, W., Song, P., Zeinolabedinzadeh, S., Schmid, R. L., Chi, T., et al. (2014). "Compact, low-power, single-ended and differential SiGe W-band LNAs," in *Proceedings of the European Microwave Integrated Circuit Conference (EuMIC)*, Rome, 452–455.

[Inan14] Inanlou, F., Coen, C. T., and Cressler, J. D. (2014). A 1.0 V, 10–22 GHz, 4 mW LNA utilizing weakly saturated SiGe HBTs for singlechip, low power, remote sensing applications. *IEEE Microw. Wirel. Comp. Lett.* 24, 890–892.

[Jaes13] Jaeschke, T., Bredendiek, C., and Pohl, N. (2013). "A 240 GHz ultra-wideband FMCW radar system with on-chip antennas for high resolution radar imaging," in *Proceedings of the IEEE Int. Microwave Symposium Digest (IMS)*, Seattle, WA, 1–4.

[Jaes14] Jaeschke, T., Bredendiek, C., and Pohl, N. (2014). "3D FMCW SAR imaging based on a 240 GHz SiGe transceiver chip with integrated antennas," in *Proceedings of the 2014 German Microwae Conference (GeMiC)*, Aachen, 1–4.

[Jahn12] Jahn, M., and Stelzer, A. (2012). A 120 GHz FMCW radar frontend demonstrator based on a SiGe chipset. *Int. J. Microw. Wirel. Technol.* 4, 309–315.

[Jain16] Jain, R., Rücker, H., and Pfeiffer, U. R. (2016). "Zero gate-bias terahertz detection with an asymmetric NMOS transistor," in *Proceedings of the 2016 41st International Conference on Infrared, Millimeter, and Terahertz waves (IRMMW-THz)*, Copenhagen, 1–2.

[Jha14] Jha, K. R., and Singh, G. (2014). *Terahertz Planar Antennas for Next Generation Communication*. Berlin: Springer.

[Kall10] Kallfass, I., Tessmann, A., Massler, H., Pahl, P., and Leuther, A. (2010). "An all-active MMIC-based chip set for a wideband 260–304 GHz receiver," in *Proceedings of the 5th European Microwave Integrated Circuits Conference (EuMIC)*, Paris, 53–56.

[Kall11] Kallfass, I., Antes, J., Schneider, T., Kurz, F., Lopez-Diaz, D., Diebold, S., et al. (2011). All active MMIC-based wireless communication at 220 GHz. *IEEE Trans. THz Sci. Technol.* 1, 477–487.

[Kang14] Kang, S., Thyagarajan, S. V., and Niknejad, A. M. (2014). "A 240GHz wideband QPSK transmitter in 65 nm CMOS," in *Proceedings of the 2014 IEEE Radio Frequency Integrated Circuits Symposium*, Tampa, FL, 353–356.

[Karp05] Karpowicz, N., Zhong, H., Zhang, C., Lin, K.-I., Hwang, J.-S., Xu, J., et al. (2005). Compact continuous-wave subterahertz system for inspection applications. *Appl. Phys. Lett.* 86, 054105.

[Kerh15] Kerherve, E., and Belot, D. (2015). *Linearization and Efficiency Enhancement Techniques for Silicon Power Amplifiers: From RF to mmW*. Amsterdam: Elsevier.

[Kim05] Kim, J.-G., Sim, S.-H., Cheon, S., and Hong, S. (2005). "24 GHz circularly polarized Doppler radar with a single antenna," in *Proceedings of the 35th European Microwave Conference (EuMC)*, Paris.

[Kiss10] D. Kissinger, K. Aufinger, T. F. Meister, L. Maurer, and R.Weigel, “A high-linearity broadband 55–77 GHz differential low-noise amplifier with 20 dB gain in SiGe technology,” in *Proc. Asia–Pacific Microwave Conference.*, pp. 1501–1504. Dec. 2010,

[Ko96] Ko, B. K., and Lee, K. (1996). A comparative study on the various monolithic low noise amplifier circuit topologies for RF and microwave applications. *IEEE J. Solid State Circuits* 31, 1220–1225.

[Koch10] Koch, S., Guthoerl, M., Kallfass, I., Leuther, A., and Saito, S. (2010). A 120–145 GHz heterodyne receiver chipset utilizing the 140 GHz atmospheric window for passive millimeter-wave imaging applications. *IEEE J. Solid State Circuits* 45, 1961–1967.

[Li13] Li, C., Lubecke, V. M., Boric-Lubecke, O., and Lin, J. (2013). A review on recent advances in doppler radar sensors for noncontact healthcare monitoring. *IEEE Trans. Microw. Theory Tech.* 61, 2046–2060.

[Lian07] Liang, J. L. J., Liang, Q. L. Q., and Zhou, Z. Z. Z. (2007). “Radar sensor network design and optimization for blind speed alleviation,” in *Proceedings of the 2007 IEEE Wireless Communications and Networking Conference*, Hong Kong, 2643–2647.

[Lian14] Liang, M. Y., Zhang, C. L., Zhao, R., and Zhao, Y. J. (2014). Experimental 0.22 THz stepped frequency radar system for ISAR imaging. *J. Infrared Millim. Terahertz Waves* 35, 780–789.

[Lia17] Liang, W., Pawlak, A., Sakalas, P., and Schröter, M. (2017). ‘96 GHz 4.7 mW low-power frequency tripler with 0.5 V supply voltage”, acc. for public. *Electron. Lett.* 53, 1308.

[Lin07] Lin, Y.-T., Chen, H.-C., Wang, T., Lin, Y.-S., and Lu, S. S. (2007). 3–10 GHz ultra-wideband low-noise amplifier utilizing miller effect and inductive shunt-shunt feedback technique. *IEEE Trans. Microw. Theory Tech.* 55.

[Liu13] Liu, G., and Schumacher, H. (2013). Broadband millimeter-wave LNAs (47–77 GHz and 70–140 GHz) using a T-type matching topology. *IEEE J. Solid State Circuits* 48, 2022–2029.

[Maes10] Maestrini, A., et al. (2010). A frequency-multiplied source with more than 1mW of power across the 840-900 GHz band. *IEEE Trans. Microw. Theory Tech.* 58, 1925–1932.

[Maur11] Maurer, L., Haider, G., and Knapp, H. (2011). “77 GHz SiGe based bipolar transceivers for automotive radar applications – An industrial perspective,” in *Proceedings of the 2011 IEEE 9th International New Circuits and Systems Conference (NEWCAS)*, Bordeaux, 257–260.

[May10] May, J. W., and Rebeiz, G. M. (2010). Design and characterization of W-Band SiGe RFICs for passive millimeter-wave imaging. *IEEE Trans. Microw. Theory Tech.* 58, 1420–1430.

[May10] May, J. W., and Rebeiz, G. M. (2010). Design and characterization of W-band SiGe RFICs for passive millimeter-wave imaging. *IEEE Trans. Microw. Theory Tech.* 58, 1420–1430.

[Mea91] Mead, J., and McIntosh, R. (1991). Polarimetric backscatter measurements of deciduous and coniferous trees at 225 GHz. *IEEE Trans. Geosci. Remote Sens.* 29, 21–28.

[Mead91] Mead, J., Langlois, P., Chang, P., and McIntosh, R. (1991). "Polarimetric scattering from natural surfaces at 225 GHz," in *Proceedings of the International Symposium on Antennas and Propagation (AP-S)*, London, ON, 756–759.

[Meta07] Meta, A., Hoogeboom, P., and Ligthart, L. P. (2007). Signal processing for FMCW SAR. *IEEE Trans. Geosci. Remote Sens.* 45, 3519–3532.

[Moal14] Moallem, M., and Sarabandi, K. (2014). Polarimetric study of MMW imaging radars for indoor navigation and mapping. *IEEE Trans. Antennas Propag.* 62, 500–504.

[Mukh16] Mukherjee, A., Lin, C., Schröter, M. (2016). "The broadband Darlington amplifier as a simple benchmark circuit for compact model verification at mm-wave frequency," in *Proceedings of the BCTM*, Miami, FL, 102–105.

[Nash16] Nashashibi, A. Y., Ibrahim, A. A., Cook, S., and Sarabandi, K. (2016). Experimental characterization of polarimetric radar backscatter response of distributed targets at high millimeter-wave frequencies. *IEEE Trans. Geosci. Remote Sens.* 54, 1013–1024.

[Nel32] Nelson, J. R. (1932). A theoretical comparison of coupled amplifiers with staggered circuits. *Proc. Institute Radio Eng.* 20, 1203–1220.

[Nema88] Nemarich, J., Wellman, R., and Lacombe, J. (1988). Backscatter and attenuation by falling snow and rain at 96, 140, and 225 GHz. *IEEE Trans. Geosci. Remote Sens.* 26, 319–329.

[Oje11] Ojefors, E., Heinemann, B., and Pfeiffer, U. R. (2011). Active 220- and 325 GHz frequency multiplier chains in an sige hbt technology. *IEEE Trans. Microw. Theory Tech.* 59, 1311–1318.

[Oje12] Ojefors, E., Heinemann, B., and Pfeiffer, U. R. (2012). Subharmonic 220- and 320 GHz SiGe HBT receiver front-ends. *IEEE Trans. Microw. Theory Tech.* 60, 1397–1404.

[Ondr81] Ondria, J., and Cardiasmenos, A. G. (1981). "Desensitization of spread spectrum radar systems by far off the carrier noise generation in millimeter sources," in *Proceedings of the 2nd Conference on Military microwaves '80*, London, 292–299.

[Pedro17] Vazquez, P. R., Grzyb, J., Sarmah, N., Pfeiffer, U. R., and Heinemann, B. (2017). "Towards THz high data-rate communication: a 50 Gbps all-electronic wireless link at 240 GHz," in *Proceedings of the 4th ACM International Conference on Nanoscale Computing and Communication*, Washington, DC, 27–29.

[Pfei14] Pfeiffer, U. R., *et al.*, (2014). A 0.53 THz reconfigurable source module with up to 1 mW radiated power for diffuse illumination in terahertz imaging applications. *IEEE J. Solid State Circuits* 49, 2938–2950.

[Pipe95] Piper, S. (1995). "Homodyne FMCW radar range resolution effects with sinusoidal nonlinearities in the frequency sweep," in *Proceedings of the International Radar Conference*, Alexandria, VA, 563–567.

[Pozar09] Pozar, D. M. (2009). *Microwave Engineering*. Hoboken, NJ: John Wiley & Sons.

[Quas09] Quast, H., and Loffler, T. (2009). "3D-terahertz-tomography for material inspection and security," in *Proceedings of the 34th International Conference on Infrared, Millimeter, and Terahertz Waves (IRMMW-THz)*, Busan, 1–2.

[Sah14] Sah, S. P., and Heo, D. (2014). "An 8th sub-harmonic injection locked V-band VCO for low power LO routing in mm-Wave Beamformers," in *Proceedings of the IEEE Radio Frequency Integrated Circuits Symposium*, San Francisco, CA, 177–180.

[Sarm11] Sarmah, N., Schmalz, K., Winkler, W., Scheytt, C. J., and Glisic, S. (2011). "122 GHz transmitter using frequency doublers," in *Proceedings of the 2011 IEEE 11th Topical Meeting on Silicon Monolithic Integrated Circuits in RF Systems*, Phoenix, AZ, 157–160.

[Sarm13] Sarmah, N., Heinemann, B., and U. R. Pfeiffer, U. R. (2013). "A 135–170 GHz power amplifier in an advanced sige HBT technology," in *Proceedings of the 2013 IEEE Radio Frequency Integrated Circuits Symposium (RFIC),* Seattle, WA, 287–290.

[Sarm13b] N. Sarmah, P. Chevalier and U. R. Pfeiffer, "160 GHz Power Amplifier Design in Advanced SiGe HBT Technologies with Psat in Excess of 10 dBm," in IEEE Transactions on Microwave Theory and Techniques, vol. 61, no. 2, pp. 939–947, Feb. 2013.

[Sarm14] Sarmah, N., Heinemann, B., and Pfeiffer, U. R. (2014). "235–275 GHz (x16) frequency multiplier chains with up to 0 dBm peak output power and low DC power consumption," in *Proceedings of the 2014 IEEE Radio Frequency Integrated Circuits Symposium*, Tampa, FL, 181–184.

[Sarm16] Sarmah, N., et al. (2016) A fully integrated 240 GHz direct-conversion quadrature transmitter and receiver chipset in SiGe technology. *IEEE Trans. Microw. Theory Tech.* 64, 562–574.

[Sarm16b] Sarmah, N., Vazquez, P. R., Grzyb, J., Foerster, W., Heinemann, B., and Pfeiffer, U. R. (2016). "A wideband fully integrated SiGe chipset for high data rate communication at 240 GHz," in *Proceedings of the 2016 11th European Microwave Integrated Circuits Conference (EuMIC)*, London, 181–184.

[Sarm16c] Sarmah, N. (2015). *Wideband Circuit Design Techniques for Ultra-High Data-Rate Wireless Communication in Silicon Technologies*, Ph.D thesis, University of Wuppertal, Wuppertal.

[Sato10] Sato, M., Takahashi, T., and Hirose, T. (2010). 68-110 GHz-band low-noise amplifier using current reuse topology. *IEEE Trans. Microw. Theory Tech.* 58, 1910–1916.

[Schm13] Schmalz, K., Wang, R., Borngraber, J., Debski, W., Winkler, W., and Meliani, C. (2013). "245 GHz SiGe transmitter with integrated antenna and external PLL," in *Proceedings of the IEEE MTT-S Int. Microwave Symposium Digest (IMS)*, Seattle, WA, 1–3.

[Schr05] Schröter, M., Wittkopf, H., and Kraus, W. (2005). "Statistical modeling of bipolar transistors," in *Proceedings of the BCTM*, Santa Barbara, CA, 54–61.

[Schr99] Schröter, M., Rein, H.-M., Rabe, W., Reimann, R., Wassener, H.-J., and Koldehoff, A. (1999). Physics- and process-based bipolar transistor modeling for integrated circuit design. *IEEE J. Solid State Circuits* 34, 1136–1149.

[Seth11] Seth, S., Poh, C. H. J., Thrivikraman, T., Arora, R., and Cressler, J. D. (2011). "Using saturated SiGe HBTs to realize ultra-low voltage/power X-band low noise amplifiers," in *Proceedings of the IEEE Bipolar/BiCMOS Circuits and Technology Meeting*, Minneapolis, MN, 103–106.

[Sev10] Severino, R. R., Taris, T., Deval, Y., Belot, D., and Begueret, J. B. (2010). "A SiGe:C BiCMOS LNA for 94 GHz band applications," in *Proceedings of the IEEE Bipolar/BiCMOS Circuits and Technology Meeting (BCTM)*, Atlanta, GA.

[Shek06] Shekhar, S., Walling, J. S., and Allstot, D. J. (2006). Bandwidth extension techniques for CMOS amplifiers. *IEEE J. Solid State Circuits* 41, 2424–2439.

[Shen12] Shen, T.-M., Kao, T.-Y. J., Huang, T.-Y., Tu, J., Lin, J., and Wu, R.-B. (2012). Antenna design of 60 GHz Micro-radar system-in-package for noncontact vital sign detection. *IEEE Antennas Wirel. Propag. Lett.* 11, 1702–1705.

[Song15] Songm P., et al. (2015). A Class-E tuned W-Band SiGe power amplifier with 40.4% power-added efficiency at 93 GHz. *IEEE Microw. Wirel. Componen. Lett.* 25, 663–665.

[Sta15] Statnikov, K., Grzyb, J., Sarmah, N., Heinemann, B., and Pfeiffer, U. R. (2015). "A lens-coupled 210–270 GHz circularly polarized FMCW radar transceiver module in SiGe technology," in *Proceedings of the 45th European Microwave Conference (EuMC)*, Paris, 550–553.

[Stat15] Statnikov, K., Grzyb, J., Heinemann, B., and Pfeiffer, U. R. (2015). 160 GHz to 1 THz multi-color active imaging with a lens-coupled SiGe HBT chip-set. *IEEE Trans. Microw. Theory Tech.* 63, 520–532.

[Statn15] Statnikov, K., Grzyb, J., Sarmah, N., Malz, S., Heinemann, B., and Pfeiffer, U. R. (2015). A 240 GHz circularly polarized FMCW radar based on a SiGe transceiver with a lens-coupled on-chip antenna. *Int. J. Microw. Wirel. Technol.* 1–9.

[Stel05] Stelzer, A., Kolmhofer, E., and Scheiblhofer, S. (2005). "Fast 77 GHz chirps with direct digital synthesis and phase locked loop," in *Proceedings of the 2005 Asia-Pacific Microwave Conference (APMC)*, Suzhou, 2165–4727.

[Stov92] Stove, A. (1992). "Linear FMCW radar techniques," Radar and Signal Processing. *IEEE Proc. F* 139, 343–350.

[Tang01] Tang, Y.-L., and Wang, H. (2001). Triple-push oscillator approach: theory and experiments. *IEEE J. SolidState Circuits* 36, 1472–1479.

[Thro84] Throne, N. (1984) The Hilbert transform. *Tech. Rev.* 3, 3–15.

[Thya14] Thyagarajan, S., Kang, S., and Niknejad, A. (2014). "A 240 GHz wideband QPSK receiver in 65 nm CMOS," in *Proceedings of the IEEE Radio Frequency Integrated Circuits Symposium*, Tampa, FL, 357–360.

[Tong94] Tong, C., and Blundell, R. (1994). An annular slot antenna on a dielectric half-space. *IEEE Trans. Antennas Propag.* 42, 967–974.

[Vass13] Vassilev, V., Zirath, H., Furtula, V., Karandikar, Y., and Eriksson, K. (2013). "140–220 GHz imaging front-end based on 250 nm InP/InGaAs/InP DHBT process," in *Proceedings of the SPIE 8715, Passive and Active Millimeter-Wave Imaging*, Baltimore, MD, 871502.

[Vera13] Vera, L., and Long, J. R. (2013). "Benchmarking circuits for model verification," in *Proceedings of the 13th HICUM Workshop*, Delft.

[Vera14] Vera, L., Long, J. R., Gross, B. J. (2014). "A low-power SiGe feedback amplifier with over 110 GHz bandwidth," in *Proceedings of the BCTM*, Miami, FL.

[Vigi16] Vigilante, M., and Reynaert, P. (2016). "20.10 A 68.1-to-96.4GHz variable-gain low-noise amplifier in 28 nm CMOS," in *Proceedings of the IEEE International Solid-State Circuits Conference (ISSCC)*, San Francisco, CA, 360–362.

[Vish12] Vishnipolsky, A., and Socher, E. (2012). "F-Band injection locked tripler based on Colpitts oscillator," in *Proceedings of the 2012 IEEE 12th Topical Meeting on Silicon Monolithic Integrated Circuits in RF Systems (SiRF), San Juan, PR,* 13–16.

[Voin13] Voinigescu, S. (2013). *High-Frequency Integrated Circuits*. Cambridge: Cambridge University Press.

[Wang12] Wang, C.-C., Chen, Z., and Heydari, P. (2012). W-band silicon-based frequency synthesizers using injection-locked and harmonic triplers. *IEEE Trans. Microw. Theory Tech.* 60, 1307–1320.

[Yang13] Yang, Y., Cacina, S., and Rebeiz, G. M. (2013). "A SiGe BiC-MOS W-Band LNA with 5.1 dB NF at 90 GHz," in *Proceedings of the IEEE Compound Semiconductor Integrated Circuit Symposium (CSICS)*, Austin, TX, 1–4.

[Yeh13] Yeh, Y.-L., and Chang, H.-Y. (2013). A W-Band wide locking range and low DC power injection-locked frequency tripler using transformer coupled technique. *IEEE Trans. Microw. Theory Tech.* 61, 860–870.

[Zein14] Zeinolabedinzadeh, S., Kaynak, M., Khan, W., Kamarei, M., Tillack, B., Papapolymerou, J., et al. (2014). "A 314 GHz, fully-integrated SiGe transmitter and receiver with integrated antenna," in *Proceedings of the 2014 IEEE Radio Frequency Integrated Circuits Symposium (RFIC)*, Tampa, FL, 361–364.

[Zhan12] Zhang, B., Xiong, Y.-Z., Wang, L., Hu, S., and Li, L.-W. (2012). Gain-enhanced 132–160 GHz low-noise amplifier using 0.13μ SiGe BiCMOS. *Electron. Lett.* 48, 257–259.

[Zhiy13] Zhiyong, Z., Xiangyang, L., Wenge, C., and Gaowei, J. (2013). "A novel DDS-PLL hybrid structure to generate the LFM signal," in *Proceedings of the 5th International Conference on Advances in Satellite and Space Communications (SPACOMM)*, Venice, 58–62.

7

Future of SiGe HBT Technology and Its Applications

M. Schröter[1,2], U. Pfeiffer[3] and R. Jain[3]

[1]Chair for Electron Devices and Integrated Circuits, Technische Universität Dresden, Germany
[2]Department of Electrical and Computer Engineering, University of California at San Diego, USA
[3]Institute for High-Frequency and Communication Technology (IHCT), University of Wuppertal, Germany

7.1 Introduction

The results of DOTSEVEN described in the previous chapters of this book mark a milestone in the development of SiGe HBT technology. This chapter reflects on how this milestone fits into the overall picture of semiconductor technologies with potential for high-speed/high-frequency applications. Furthermore, as any milestone is a temporary state, the possible future prospects of SiGe HBT device performance will be presented in terms of a roadmap. Finally, obstacles on the path toward the perceived performance limits of this technology are discussed.

7.2 Technology Comparison

Circuits operating at mm-wave frequencies have traditionally been implemented in III–V semiconductors due to the higher mobility in these materials. However, the high mobility occurs only at relatively low electric fields, which are easily exceeded under circuit-relevant bias conditions, especially when trying to generate high output power. Nevertheless, the fastest HBTs today have been fabricated in InP technology with the respective prototyping processes reaching (f_T, f_{max}) values around (0.5, 1.1) THz for emitter widths of 130 nm [Urte11] and 200 nm [Bol16]. Compared to these

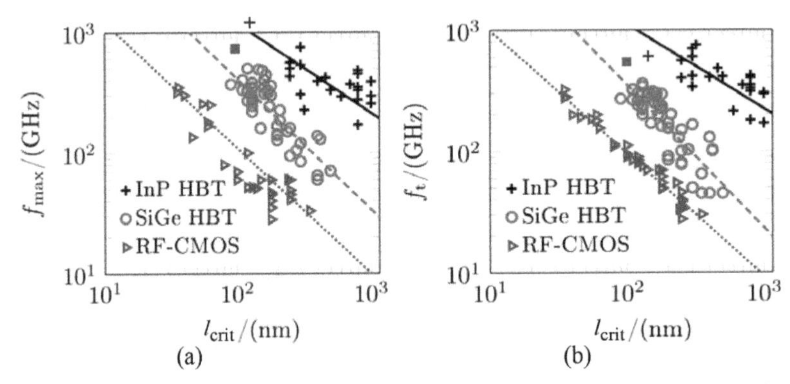

Figure 7.1 Operating speed comparison between SiGeC HBTs, InP HBTs, and MOSFETs vs. critical lithography dimensions (i.e., emitter width or channel length): (a) maximum oscillation frequency and (b) transit frequency. The lines represent LSQ fits of the data, the red filled squares DOTSEVEN results, and the larger crosses the best InP HBT data.

devices, DOTSEVEN SiGe HBTs with $(f_T, f_{max}) = (0.5, 0.72)$ THz are about one generation behind in terms of power gain cutoff frequency$_X$. This performance difference is displayed in Figure 7.1, which includes f_T and f_{max} data of the three mainstream technology contenders gathered from many publications [Ros16]. The DOTSEVEN results have been marked by the red squares.

Although InP HBTs do have certain advantages over SiGeC HBTs such as higher breakdown voltage (at the same speed) and the potential of combined optical/electronic operation, it was shown in [Voi04] that "at comparable f_T and f_{max}, there is very little difference in their performance in narrow-band mm-wave and in broadband and high-speed digital circuits"; i.e., for circuit applications, the advantage of the higher breakdown voltage in InP HBTs appears to be relatively small.

A main disadvantage of III–V technologies is their fairly low integration level and yield as well as the difficulty of structural downscaling due to strong surface recombination and the resulting low current gain. Hence, it will be difficult to leverage these technologies to their full extent to enable both more complex mm-wave systems and, in particular, mass-market wafer volume in the future. As a consequence, III–V material-based technologies do not achieve functionality and energy efficiency increases comparable to those of silicon technology, which in turn makes it difficult for them to compete also on cost. In addition, the relatively large process variations and the lack of accurate modeling tools have notoriously hampered (cost-) efficient III–V circuit design. Therefore, the III–V semiconductor industry and the related investment have been focused strongly on high-frequency/high-speed low-volume niche applications. The market success of more recent approaches toward a heterogeneous integration with completed CMOS wafers (e.g., [Ram12]) remains unclear yet. Among the issues are certainly the very different process qualification criteria in the silicon (especially the digital CMOS) world and in the III–V world.

According to Figure 7.1 the cutoff frequencies of the best RF-CMOS processes come close to those of SiGe HBTs, but at the expense of a significantly more advanced lithography, typically at least three lithography nodes. Achieving the DOTSEVEN results though would require a CMOS process with about a 14 nm channel length (which does not correspond to the 14 nm node!), assuming that progress in device speed continues for CMOS as sketched by the corresponding dashed line in Figure 7.1. As will be discussed later below, this assumption is unlikely to be the case.

The considerations so far have centered around *device*-related operating frequencies. However, in a circuit the transistors are connected to other devices, which along with the connection represent more or less large capacitive, inductive, and resistive loads. For instance, high-quality passive devices for RF applications have to be placed in the uppermost metalization levels. Figure 7.2(a) displays a 3D view of a typical connection between a transistor and the upper metal layer. The impact of this connection on transistors having the same speed after deembedding (about 300 GHz) is shown in Figure 7.2(b): the MOSFET's f_T decreases by about a factor of two, while the HBT looses

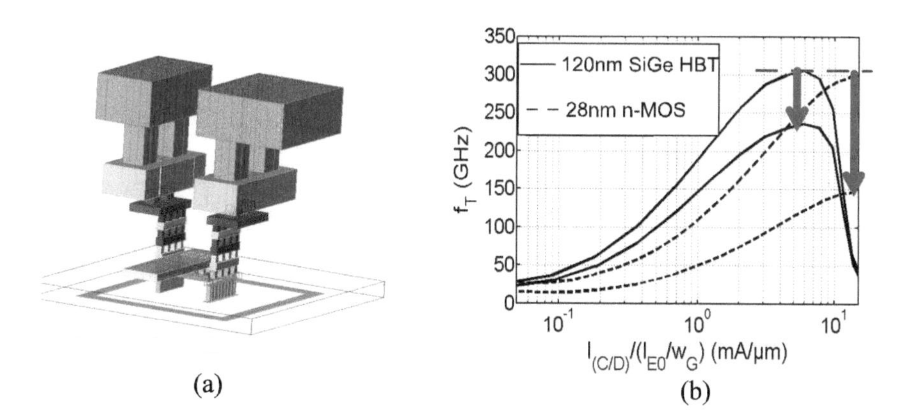

(a) (b)

Figure 7.2 Impact of device connections to other circuit elements (a) on the transit frequency of a SiGeC HBT with 120 nm emitter window width and a MOSFET of the 28 nm node (b). The upper lines in (b) represent the pad and pad-device connection line deembedded data, while the lower lines represent the un-deembedded data.

about 20% of its speed. Similar observations have been noted in [Ina11], where the peak operating frequencies (f_T, f_{max}) of 45 nm MOSFETs drop by a factor of two, once the metallization necessary for building circuits is included. The reason why HBTs do not show this severe deterioration is their much higher transconductance g_m and corresponding drive capability. In other words, devices with the same ratio of g_m and input capacitance have the same f_T, but devices with a higher g_m will fundamentally fare better in circuits since the device capacitance there will only be a more or less small fraction of the total capacitance.

Based on the observations described above, it is therefore instructive to look at the values of g_m when comparing process technology performance, since g_m represents a better indication of achievable circuit speed. Figure 7.3 shows the measured and predicted values of g_m for a variety of incumbent and emerging process technologies. For comparison, g_m has been normalized to emitter length (for HBTs), gate width (for planar MOSFETs), or channel perimeter length (for nano-wire and -tube FETs[1]). It is clearly visible that HBTs have a much higher transconductance per finger length than FETs. Furthermore, FETs appear to be unable to ever catch up to HBTs in terms of transconductance, even when assuming ideal performance scaling according to the approximation of existing data (dashed lines).

Figure 7.3 also includes the transconductance values for future technology nodes, which have been predicted based on detailed TCAD simulations and compact models for complete 3D transistors with all relevant parasitics included [Sch17, Voi17] and have become the basis for the 2014 and 2015 ITRS tables. For MOSFETs, the peaks of g_m (around 4 mS/µm) and f_T (around 600 GHz) are predicted for a channel length of 10 nm, while beyond that length the strong impact of surface scattering in the extremely thin channel layer will lead to a drastic decrease (to, e.g., $g_m \approx 1.5$ mS/µm, $f_T \approx 200$ GHz). This is in stark contrast to HBT scaling, which significantly benefits from smaller dimensions in all aspects of electrical performance until a critical base width is reached and the small emitter area leads to a high emitter contact resistance.[2] The ultimate performance of SiGe HBTs was investigated in [Sch11a, Sch11b] using detailed TCAD simulation and

[1]Normalizing to the perimeter takes into account (roughly) that screening effects would lead to a lower g_m value if the wires or tubes were placed next to each other, which would correspond to a normalization to the diameter (or minimum footprint for a gate-all-around channel).

[2]Note though that the impact of contact resistances increases with smaller device dimensions in all technologies and becomes visible first in highly scaled FETs.

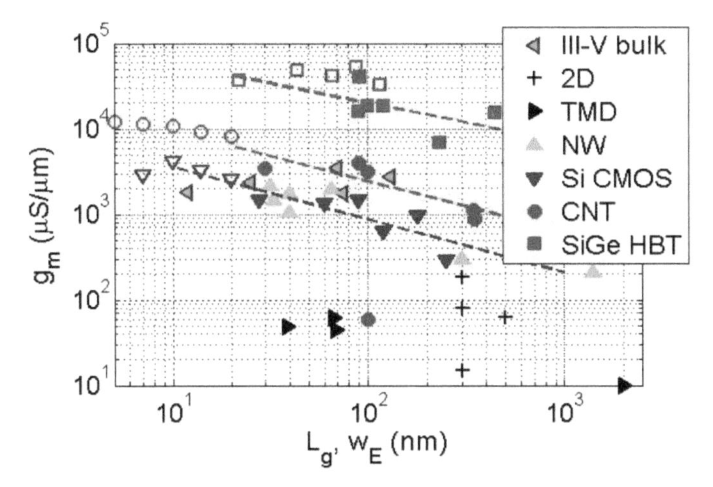

Figure 7.3 Comparison of the terminal (or extrinsic) transconductance of transistors from a large variety of process technologies. Filled symbols represent measured data and open symbols represent predicted (roadmap) data. For FETs, the legend designations correspond to the channel material and structure: planar III–V such as InGaAs (III–V bulk); planar single- or few atomic layers such as black phosphorus or germanane (2D); planar transition metal dichalcogenide such as MoS_2 (TMD); planar or FinFET silicon (Si-CMOS); nano-wire silicon or III–V (NW); carbon nano-tube (CNT).

accurate compact models with all known physical effects and device-related parasitics included. The predictions resulted in a transit frequency between 0.8 and 1 THz and maximum oscillation frequency around 2 THz, depending on the assumed contact resistivities especially for the emitter. For the ITRS tables (see also [Sch17] for more details), quite conservative values have been assumed, which lead to somewhat reduced device operating frequencies and transconductances compared to [Sch11b].

The doping profile of the technology node N3, the performance of which was predicted in 2013, served as the guideline for the DOTSEVEN process development. During process development, accurate physics-based model parameters were determined based on which the evaluation of the impact of the various physical and parasitic effects on device performance [Kor15, Paw17] is possible. This strategy provided valuable insight for prioritizing the process development tasks. It is interesting to note that the electrical parameters, such as base sheet resistance and capacitance per area, of the final process version of DOTSEVEN meet those of the roadmap node N3 predicted earlier quite well. This validates the accuracy of the predictions and the employed approach.

The predicted progress in device speed is closely linked to the increase in current densities required for achieving peak operating frequencies. The corresponding transconductance at the device terminals is more or less strongly reduced by the emitter series (contact) resistance. With conservative assumptions about contact resistivities, transconductance values of at least 40 mS/μm can be expected according to Figure 7.3 at the present end of the roadmap. This value remains an order of magnitude higher than that predicted for the best MOSFETs and also still higher than that predicted for the best CNTFETs by a factor of four. Notice that due to the lack of sufficient hardware, the predictions of the latter are associated with much higher uncertainty than those for MOSFETs.

An important aspect for mm-wave and THz applications is the achievable output power at a given frequency. For advanced MOSFET technologies, the latter is impacted significantly by the decrease of the voltage gain, which is caused by the increased output leakage and associated output conductance. HBTs do not show this decrease unless they are scaled (vertically) beyond the last node presently shown in the ITRS tables. Output power can also be increased with device size. While this increases the device-related capacitances in both HBTs and MOSFETs, an increase in emitter length alone reduces all series resistances in HBTs but a corresponding gate width increase in FETs leads to a larger gate resistance. Moreover, the maximum allowed drain–source voltage in advanced RF-MOSFETs is lower than the maximum collector–emitter voltage in advanced high-speed HBTs. Increasing the number of parallel devices in MOSFET-based power amplifiers leads to larger interconnect parasitics per unit drain width, while stacked power amplifier architectures require a larger number of devices and thus again more passives and parasitics per unit drain width. Another important building block in (sub-)mm-wave systems is the VCO. Its phase noise strongly depends on the flicker ($1/f$) noise of the transistors. Here, HBTs have much lower corner frequencies than MOSFETs, in which $1/f$ noise keeps increasing for more advanced nodes. Overall, HBTs appear to have a distinctive advantage in the area of mm-wave and THz amplifiers from a technical circuit design perspective.

Finally, a few words on cost. With the development and added masks for the required RF-passives, an RF-CMOS process becomes significantly more costly than the corresponding digital process. Thus, a depreciated CMOS process with high-speed SiGe HBTs integrated (i.e., a BiCMOS process) can often be cheaper and thus very cost competitive to advanced RF-CMOS. Such a BiCMOS process not only provides better RF front-end performance (in terms of analog HF features and energy efficiency) but is also earlier available on the market. In view of these aspects, RF-CMOS based on an advanced node makes sense only for applications that require (i) higher digital functionality and density than an already available BiCMOS process with comparable front-end (i.e., HBT) performance and (ii) very high product volume.

Due to the already existing large investment and associated infrastructure in 200 and 300 mm silicon wafer fabs around the world, SiGe:C HBTs with operating speed in the THz range are desirable. Implemented into depreciated (and hence low-cost) digital CMOS platforms, the associated SiGe:C BiCMOS single-chip technologies are capable of covering (at reasonable wafer cost) medium- and large-volume applications with mm-wave and THz analog front-ends, which would be either far too expensive when implemented in most advanced CMOS technology or would not even be possible to realize there due to the inferior analog characteristics of future advanced MOSFETs as discussed above. Therefore, a relatively small investment in SiGe:C HBT process development will yield large gains in terms of (ultimately commodity) market coverage.

7.3 Future Millimeter-wave and THz Applications

The terahertz or sub-millimeter frequency range, roughly defined as extending from 300 GHz to 3 THz, has so far resisted attempts to broadly harness its potential for everyday applications. This led to the expression THz gap, loosely describing the lack of adequate technologies to effectively bridge this transition region between microwaves and optics – both readily accessible via well-developed electronic and laser-based approaches – by, e.g., integrated and cost-efficient electronics. THz technology is an emerging field which has demonstrated a wide-ranging potential. Extensive research in the last years has identified many attractive application areas and paved the technological path toward broadly usable THz systems. THz technology is currently in a pivotal phase and will soon be able to radically expand our analytic capabilities via its intrinsic benefits.

Applications of mm-wave and THz frequencies led by the silicon integrated technologies can be subdivided into communication, radar, imaging, and sensing areas, as illustrated in Figure 7.4. In the following sections, we describe some of these applications and the recent, state-of-the-art hardware developments in the corresponding areas as it relates to DOTSEVEN.

Wireless:
- Personal and Local Area Networks (PANs/LANs)
- Consumer electronic devices
- High bandwidth Wireless backhaul with 100Gb/s data transmission
- Inter-building communication, E-band (71-76 GHz, 81-86 GHz)
- Large bandwidth data kiosks
- Board-to-board wireless links for High Density Computing (HDC)
- Secure links and surveillance
- Space and inter-satellite communication

Digital:
- High-speed interconnects
- Data switches (Mux/DeMux)
- Broadband ADCs with 50-100 GS/s and > 25GHz bandwidth at 5-6 bit resolution

Radar Applications

Automotive:
- Long Range Radar (LRR)
  - for collision avoidance, automated cruise control (77GHz)
- Short Range Radar (SRR)
  - Advanced Driver Assistant Systems (ADAS)
  - Road condition detection
- Precision Radars (120GHz, 240GHz)
  - Industrial sensors and automation

Consumer:
- Wireless gesture recognition for Consumer applications

Space:
- Aviation safety in extremely poor visibility (94GHz)
- Airport ground control (94GHz)

Industrial:
- Distance measurement and 3-D scanning
- Alarm systems and motion detection

mmWave, THz Imaging and Sensing

Security:
- Non-invasive imaging
- Drug and explosive detection
- Material spectroscopy and characterization

Industrial
- Gas sensing
- Non-destructive testing (NDT)
- Low cost computed-tomography (CT) imaging (Computational Imaging)

Sensing:
- Earth sensing and climate control
- Industrial process control
- Astronomy
- Microwave background

Biotechnology:
- Medical imaging, tumor recognition (super-resolution imaging)
- Genetic screening

Figure 7.4 Potential applications for silicon integrated mm-wave and THz circuits.

7.3.1 Communication

To improve the data capacity beyond 10 Gbps for wireless transmission, improving the spectral efficiency with advanced modulation schemes is no longer sufficient and higher bandwidth becomes an absolutely necessity. The mm-wave and THz bands, due to their large relative bandwidth, show great potential for future wireless communication [Son11]. Such large frequencies, however, suffer from greater atmospheric attenuation as compared to the traditional radio waves. Still, the atmospheric windows near 90 GHz, 140 GHz, and 240 GHz are being considered for future communication and radar applications [Fed10].

Due to a larger atmospheric attenuation, the THz waves are expected to be first used for indoor wireless personal-area (WPAN) and local-area (WLAN) networks, where the range is limited to a few tens of meters at the most. Along with the high-capacity, the high directivity of THz waves is also considered as an advantage to ensure secure (requiring line-of-sight) or high-density, non-interfering data networks. Some of the possible applications of THz communication are:

1. Wireless distribution of HDTV content in an in-home network. While the current 60 fps, progressive full HD (1080p60) content requires data transmission rates near 3 Gbps, future ultra HD 8K content would require wireless data transmission rates in the ballpark of 24 Gbps [Son11].
2. High-bandwidth wireless backhauls in high-density mesh multipoint to point/multipoint (first-mile and last-mile) networks may require data rates reaching 100 Gbps.
3. Rapid uploading and downloading of large files from a server which can serve as a public data kiosk.
4. Inter- and intra-building communication networks for manufacturing floor automation, etc.

Silicon–germanium HBT technology is driving the front-end development for mm-wave and THz bands, and several key components as well as fully integrated transceiver systems have been demonstrated. Figure 7.5 shows the output power versus frequency for some recently published SiGe mm-wave/THz sources. A 0.53 THz, 0 dBm free-space radiating source composed of 16 on-chip, non-locked radiating pixels, each with a differential triple-push oscillator (TPO) and on-chip ring antenna, was demonstrated in [Pfe14]. A similar, single-ended TPO design for a single 0.49 THz radiator was also shown in [Hil15]. [Hil17] demonstrated another approach of using a fundamental differential Colpitts oscillator with second harmonic extraction at 215 GHz and feeding it to a frequency doubler for 430 GHz signal radiation with –6.3 dBm of output power.

Cascaded multiplier chains with larger multiplication factors have also been demonstrated. In this approach, an external, phase-stable signal is fed to the multiplier chain to generate the THz signal. In [Eri11], the $\times 18$ multiplier chain consists of two cascaded tripler stages followed by a balanced doubler and shows a peak output power of –3 dBm at 325 GHz.

Further signal amplification is done by on-chip integrated power amplifiers (PAs). In [Nee13], a 160 GHz PA with 20 dB gain and 10 dBm

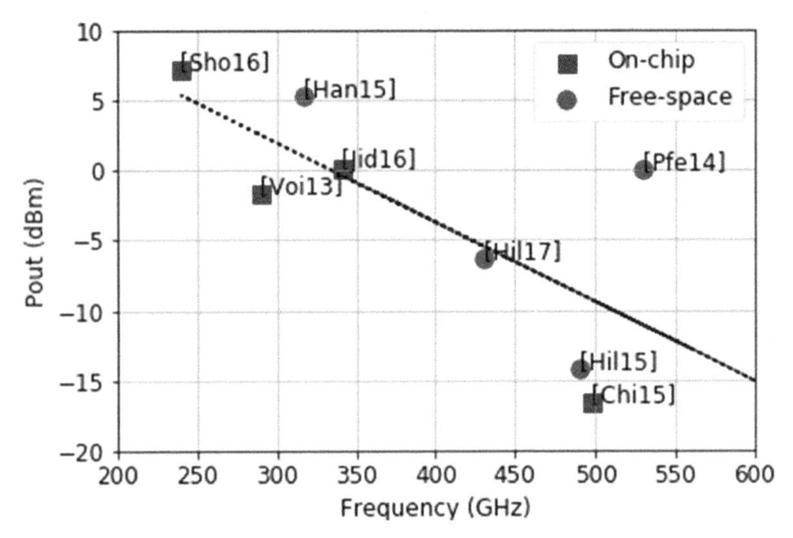

Figure 7.5 Frequency versus output power of some recently published SiGe integrated mm-wave/THz sources.

saturated output power (P_{sat}) was demonstrated. A parallel power-combination approach with four PA cores was used in [Nee16] to show 25 dB gain and 9 dBm P_{sat} for 200–225 GHz.

In the recent years, fully integrated silicon RF front-ends operating above 200 GHz have become feasible due to the continuous technology improvement. Both CMOS [Kan15, Thy15, Par12, Tak17] and SiGe [Fri17, Rod17, Rod18] front-ends have been reported in the literature. Circuits using sub-harmonic techniques with doubler or tripler as last stage before the antenna like a 240 GHz transceiver chipset utilizing QPSK in 65 nm CMOS can be found in the combination of [Kan15] and [Thy15]. A 260 GHz OOK transceiver in 65 nm bulk CMOS was reported in [Par12]. [Tak17] presents a CMOS transmitter working at 300 GHz capable of communicating over 5 cm at 56 Gbps. A complete SiGe chipset for 50 Gbps at mm distance is presented by [Fri17]. In [Rod17], a SiGe integrated 240 GHz transceiver chipset in the DOTSEVEN technology (as detailed in Chapter 6) is reported with a data-rate of 50 Gbps communicating over 100 cm distance. The data rate was further improved to 65 Gbps by post-process IF filtering at the receiver-end in [Rod18]. A comparison of these front-ends is presented in Table 7.1. Extensive research in device technology and circuits continues to improve the performance further, indicating that SiGe HBTs are becoming a formidable

Table 7.1 Si-integrated wireless communication links above 200 GHz

Reference	Frequency [GHz]	Tx/Rx	Data [Gbps]	Error (Modulation)	Distance
[Kan15]	240	Tx	16	–	–
[Thy15]	240	Rx	10/16	BER $< 10^{-6}$ $/10^{-4}$ (QPSK)	–
[Par12]	260	Tx, Rx	14	– (OOK)	4 cm
[Fri17]	190	Tx, Rx	50	BER $< 10^{-3}$ (BPSK)	0.6 cm
[Rod17]	240	Tx, Rx	50	EVM 29% (QPSK)	100 cm
[Tak17]	300	Tx	56	EVM 13.4% (16 QAM)	5 cm
[Rod18]	240	Tx, Rx	65	BER $<10^{-4}$ (QPSK)	100 cm

alternative to III–V technologies for mm-wave and THz communication, and SiGe HBT technology is quickly extending toward 100 Gbps data-rates.

7.3.2 Radar

Radar systems are used for distance and velocity sensing, and they also benefit by moving to higher frequencies. A larger available absolute bandwidth at higher frequencies improves the overall range resolution. Also, higher frequencies allow for compact radar apertures.

One extremely popular commercial usage for high-frequency radars is for automotive applications with 76–77 GHz and 79–81 GHz allocated frequency bands. These automotive radars are being considered for a wide range of Advance Driver Assist Systems (ADAS), including (i) long-range (high-directivity, narrow forward looking beam) systems for applications such as adaptive cruise control, (ii) medium-range (medium directivity, beam-width) systems for applications such as cross-traffic alert, and (iii) short-range (direct proximity) systems for applications such as obstacle avoidance and parking assist. While such systems are already deployed at present [Has12], continuous improvement in technology would allow for a better performance (power consumption/noise figure), ultimately leading to a cost reduction and improved reliability. Similarly, 94 GHz constitutes another frequency band which is popular for aerospace and aviation radars, for application such as for displaying runway image in poor weather conditions [Gos09], for airport ground control systems [Mar16], and weather-cloud investigations [Mar08].

Even higher frequencies allow for millimeter-range resolution. Such precision radars can be used for industrial imaging, inspection, and automation. In addition, novel consumer applications such as gesture recognition [Arb13], [GSoli] have also started to emerge.

Due to a similar coherent nature, the hardware advancements discussed for the communication chipsets also benefit the radar systems similarly. Within the frame of the DOTSEVEN project, a complete highly integrated FMCW homodyne monostatic radar system operating around 240 GHz with a 60 GHz bandwidth and a state-of-the-art 2.57 mm range resolution was developed [Statn15]. The radar module was used for 3-D imaging of cardboard boxes with a dynamic range of around 50 dB in the acquired range profiles. Key to the success of silicon technologies is a low-cost packaging scheme which needs to be developed to support low cost on the sub-component level. Hence, the DOTSEVEN radar chips were further packaged together with wideband lens-integrated on-chip annular-slot antenna [Grzyb15] and wire bonded onto a low-cost FR4 printed-circuit board.

7.3.3 Imaging and Sensing

Three-dimensional THz imaging based on the principle of computed tomography (THz-CT) is one of the emerging applications that may be explored in commercial imaging and sensing applications. THz-CT offers volumetric object reconstruction with an image contrast based on the characteristic THz absorption of the illuminated material. Since THz radiation is non-ionizing and thus requires no dedicated safety measures, THz-CT represents an interesting alternative to X-ray technology for low-cost industrial quality control. A THz-CT system solely based on components built in DOTSEVEN technology was built and evaluated. A SiGe-HBT source [Hill15] was focused in the object plane and refocused to an NMOS [Jain16] detector with low-cost PTFE lenses. The 3D image was reconstructed from measurements at multiple projection angles and positions based on a filtered back-projection algorithm. The system offers a dynamic range of up to 60 dB at 490 GHz. The Gaussian beam waist sizes are 2.54 mm in the y-direction and 2.40 mm in the z-direction. These results show that Terahertz 3D CT imagers can be entirely implemented in a silicon process technology.

References

[Arb13] Arbabian, A., Callender, S., Kang, S., Rangwala, M., and Niknejad, A. M. (2013). A 94 GHz mm-wave-to-baseband pulsed-radar transceiver with applications in imaging and gesture recognition. *IEEE J. Solid State Circ.* 48, 1055–1071.

[Bol16] Bolognesi, C. R., Flückinger, R., Alexandrova, M., Quan, W., Lövblom, R., and Ostinelli, O. (2016). InP/GaAsSb DHBTs for THz applications and improved extraction of their cutoff frequencies. *IEDM Tech. Digest* 6, 723–726.

[Eri11] Ojefors, E., Heinemann, B., and Pfeiffer, U. R. (2011). Active 220- and 325 GHz frequency multiplier chains in an SiGe HBT technology. *IEEE Trans. Microw. Theory Tech.* 59, 1311–1318.

[Fed10] Federici, J., Moeller, L. (2010). Review of terahertz and subterahertz wireless communications. *J. Appl. Phys.* 107, 111101.

[Fri17] Fritsche, D., Stärke, P., Carta, C., and Ellinger, F. (2017). A low-power SiGe BiCMOS 190 GHz transceiver chipset with demonstrated data rates up to 50 Gbit/s using on-chip antennas. *IEEE Trans. Microw. Theory Tech.* 65, 3312–3323.

[Gos09] Goshi, D. S., Liu, Y., Mai, K., Bui, L., and Shih, Y. (2009). "Recent advances in 94 GHz FMCW imaging radar development," in *Proceedings of the 2009 IEEE MTT-S International Microwave Symposium Digest*, Boston, MA, 77–80.

[Grzyb15] Grzyb, J., Statnikov, K., Sarmah, N., and Pfeiffer, U. R. (2015). "A broadband 240 GHz lens-integrated polarization-diversity on-chip circular slot antenna for a power source module in SiGe technology," in *Proceedings of the 45th European Microwave Conference (EuMC)*, Paris, 570–573.

[GSoli] Project Soli (2017). Available at: https://atap.google.com/soli/ accessed November 1, 2017.

[Has12] Hasch, J., Topak, E., Schnabel, R., Zwick, T., Weigel, R., and Waldschmidt, C. (2012). Millimeter-wave technology for automotive radar sensors in the 77 GHz frequency band. *IEEE Trans. Microw. Theory Tech.* 60, 845–860.

[Hil15] Hillger, P., Grzyb, J., Lachner, R., and Pfeiffer, U. (2015). "An antenna-coupled 0.49 THz SiGe HBT source for active illumination in terahertz imaging applications," in *Proceedings of the 2015 10th European Microwave Integrated Circuits Conference (EuMIC)*, Paris, 180–183.

[Hill15] Hillger, P., Grzyb, J., Lachner, R., and Pfeiffer, U. (2015). "An antenna-coupled 0.49 THz SiGe HBT source for active illumination in terahertz imaging applications," in *Proceedings of the 2015 10th European Microwave Integrated Circuits Conference (EuMIC)*, Paris, 180–183.

[Ina11] Inac, O., Cetinoneri, B., Uzunkol, M., Atesal, Y., and Rebeiz, G. M. (2011). "Millimeter-wave and THz circuits in 45 nm SOI CMOS," in *Proceedings of the Compound Semiconductor Integrated Circuit Symposium [CSICS]* (Rome: IEEE), 1–4.

[Jain16] Jain, R., Rücker, H., and Pfeiffer, R. (2016). "Zero gate-bias terahertz detection with an asymmetric NMOS transistor," in *Proceedings of the 2016 41st International Conference on Infrared, Millimeter, and Terahertz Waves (IRMMW-THz)*, Copenhagen, 1–2.

[Kan15] Kang, S., Thyagarajan, S. V., and Niknejad, A. M. (2015). A 240 GHz fully integrated wideband QPSK transmitter in 65 nm CMOS. *IEEE J. Solid State Circ.* 50, 2256–2267.

[Kor15] Korn, J., Ruecker, H., Heinemann, B., Pawlak, A., Wedel, G., and Schroter, M. (2015). "Experimental and theoretical study of fT for SiGe HBTs with a scaled vertical doping profile," in *Proceedings of the IEEE BCTM*, Boston, 117–120.

[Mar08] Marchand, R., Mace, G. G., Ackerman, T., and Stephens, G. (2008) "Hydrometeor detection using cloudsat—an earth-orbiting 94 GHz cloud radar. *J. Atmos. Oceanic Technol.* 25, 519–533.

[Mar16] Martinez, A., Lort, M., Aguasca, A., and Broquetas, A. (2015). "Submillimetric motion detection with a 94 GHZ ground based synthetic aperture radar," in *Proceedings of the IET International Radar Conference 2015*, Hangzhou, 1–5.

[Nee13] Sarmah, N., Chevalier, P., and Pfeiffer, U. R. (2013). 160 GHz power amplifier design in advanced SiGe HBT technologies with Psat in excess of 10 dBm. *IEEE Trans. Microw. Theory Tech.* 61, 939–947.

[Nee16] Sarmah, N., Aufinger, K., Lachner, R., and Pfeiffer, U. R. (2016). "A 200–225 GHz SiGe power amplifier with peak Psat of 9.6 dBm using wideband power combination," in *Proceedings of the ESSCIRC Conference 2016: 42nd European Solid-State Circuits Conference*, Lausanne, 193–196.

[Par12] Park, J. D., Kang, S., Thyagarajan, S. V., Alon, E., and Niknejad, A. M. (2012). "A 260 GHz fully integrated CMOS transceiver for wireless chip-to-chip communication," in *Proceedings of the 2012 Symposium on VLSI Circuits (VLSIC)*, Honolulu, HI, 48–49.

[Paw17] Pawlak, A., Heinemann, B., Schröter, M. (2017). "Physics-based modeling of SiGe HBTs with fT of 450 GHz with HICUM Level 2," in *Proceedings of the IEEE BCTM*, Miami, FL, 134–137.

[Pfe14] Pfeiffer, U. R. et al. (2014). A 0.53 THz reconfigurable source module with up to 1 mW radiated power for diffuse illumination in terahertz imaging applications. *IEEE J. Solid State Circ.* 49, 2938–2950.

[Ram12] Raman, S., Dohrmann, C., and Chang, T.-H. (2012). "Heterogeneous BiCMOS technologies and circuits and the DARPA DAHI program," in *Proceedings of the IEEE BCTM*, Portland, OR, 127–132.

[Rod17] Vazquez, P. R., Grzyb, J., Sarmah, N., Pfeiffer, U. R., and Heinemann, B. (2017). "Towards THz high data-rate communication: a 50 Gbps all-electronic wireless link at 240 GHz," in *Proceedings of the 4th ACM International Conference on Nanoscale Computing and Communication*, Washington DC.

[Rod18] Rodriguez Vazquez, P., Grzyb, J., Sarmah, N., Heinemann, B., and Pfeiffer, U. R. (2018). *A 65 Gbps QPSK One Meter Wireless Link Operating at a 225–255 GHz Tunable Carrier in a SiGe HBT Technology.* Anaheim, CA: RWW.

[Ros16] Rosenbaum, T. (2017). *Performance Prediction of a Future SiGe HBT Technology using a Heterogeneous Set of Simulation Tools and Approaches.* Ph.D. dissertation, TU Dresden, Dresden.

[Sch11a] Schroter, M., Wedel, G., Heinemann, B., Jungemann, C., Krause, J., Chevalier, P., and Chantre, A. (2011). Physical and electrical performance limits of high-speed SiGeC HBTs – Part I: vertical scaling. *IEEE Trans. Electron Dev.* 58, 3687–3696.

[Sch11b] Schroter, M., Krause, J., Rinaldi, N., Wedel, G., Heinemann, B., Chevalier, P., et al. (2011). Physical and electrical performance limits of high-speed SiGeC HBTs – Part II: lateral scaling. *IEEE Trans. Electron Dev.* 58, 3696–3706.

[Sch17] Schröter, M., Rosenbaum, T., Chevalier, P., Heinemann, B., Voinigescu, S., Preisler, E., et al. (2017). SiGe HBT technology: future trends and TCAD based roadmap. *Proc. IEEE* 105, 1068–1086.

[Son11] Song, H. J., and Nagatsuma, T. (2011). Present and future of terahertz communications. *IEEE Trans. Terahertz Sci. Technol.* 1, 256–263.

[Statn15] Statnikov, K., Grzyb, J., Sarmah, N., Malz, S., Heinemann, B., and Pfeiffer, U. R. (2015). A 240 GHz circularly polarized FMCW radar based on a SiGe transceiver with a lens-coupled on-chip antenna. *Int. J. Microw. Wireless Technol.* 8, 1–9.

[Tak17] Takano, K., Katayama, K., Amakawa, S., Yoshida, T., and Fujishima, M. (2017). "56 Gbit/s 16 QAM wireless link with 300 GHz-band CMOS transmitter," in *Proceedings of the 2017 IEEE MTT-S International Microwave Symposium (IMS)*, Honolulu, HI, 793–796.

[Thy15] Thyagarajan, S. V., Kang, S., and Niknejad, A. M. (2015). A 240 GHz fully integrated wideband QPSK receiver in 65 nm CMOS. *IEEE J. Solid State Circ.* 50, 2268–2280.

[Urt11] Urteaga, M., Pierson, R., Rowell, P., Jain, V., Lobisser, E., and Rodwell, M. J. W. (2011). "130nm InP DHBTs with *ft* >0.52THz and *fmax* >1.1THz," in *Proceedings of the 69th Development Research Conference*, Washington, DC, 281–282.

[Voi04] Voinigescu, S. P., Dickson, T. O., Beerkens, R., Khalid, I., and Westergaard, P. (2004). "A comparison of si CMOS, SiGe BiCMOS, and In P HBT technologies for high-speed and millimeter-wave ICs," in *Proceedings of the Topical Meeting Silicon Monolithic Integrate Circuits in RF Systems*, Austin, TX, 111–114.

[Voi17] Voinigescu, S. P., Shopov, S., Bateman, J., Farooq, H., Hoffman, J., Vasilakopoulos, K. (2017). Silicon millimeter-wave, terahertz, and high-speed fiber-optic device and benchmark circuit scaling through the 2030 ITRS horizon. *Proc. IEEE* 105, 1087–1104.

Index

About the Editors

Niccolò Rinaldi graduated (cum laude) from the University of Naples "Federico II," Italy, in 1990, and received the Ph.D. degree in 1994. In February 1994, he became a Research Assistant at the University of Naples "Federico II." From July 1996 to December 1996, he was Research Fellow at the University of Delft, The Netherlands, working on the modeling of high-speed bipolar devices. In November 1998 he was appointed Associate Professor at the University of Naples "Federico II." Since November 2002 he has been Full Professor at the University of Naples. He was a member of the Executive Committee of the IEEE Bipolar/BICMOS Circuits and Technology Meeting, and vice-chairman of the IEEE Electron Device Chapter (Central & South Italy Section). From 2009 to 2013 he was Coordinator of the Ph.D. program in Electronics and Telecommunications Engineering at the Faculty of Engineering. From 2010 to 2012 he was the Coordinator of the Italian Ph.D. School in Electronics Engineering. In 2012 he was awarded the Medal of Meritorious from the Technical University of Lodz (Poland). He was a workpackage leader in the European Projects DOTFIVE and DOTSEVEN. His present research interests include the modeling of bipolar and power MOS transistors, self-heating effects in solid-state circuits and devices, electro-thermal simulation, and design of RF and microwave circuits and devices. He has authored or co-authored more than 100 publications in international journals and conferences.

Michael Schröter received his Dr.-Ing. degree (scl) in electrical engineering and the "venia legendi" on semiconductor devices in 1988 and 1994, respectively, from the Ruhr-University Bochum, Germany. He was with Nortel and Bell Northern Research, Ottawa, Canada, as a Team Leader and Advisor until 1996 when he joined Rockwell (later Conexant), Newport Beach (CA), where he managed the RF Device Modeling Group. Dr. Schröter has been a Full Professor at the University of Technology at Dresden, Germany, since 1999, and has an adjunct affiliation with UC San Diego, USA. He is the author of the bipolar transistor compact model HICUM, a worldwide standard since 2003, and has co-authored a textbook entitled "Compact hierarchical modeling of bipolar transistors with HICUM" as well as over 220 peer reviewed publications and four invited book chapters.

Dr. Schröter was a co-founder of XMOD Technologies in Bordeaux, France. During a two-year Leave of Absence from TUD (2009–2011) he was the Vice President of RF Engineering at RFNano, where he was responsible for the device design of the first 4" wafer-scale carbon nanotube FET process technology. He was the Technical Project Manager for DOTFIVE (2008–2011) and DOTSEVEN (2012–2016), which were EU funded research projects for advancing high-speed SiGe HBT technology towards THz applications, and has been leading the Carbon Path project within the German Excellence Cluster CfAED. Since 2013, he has been a member of the ITRS/IRDS RF-AMS subcommittee.